Environmental temperature varies in time and space on time-scales ranging from a few hours to long-term climate change. Organisms are therefore continually challenged to regulate and maintain functional capacities as their thermal environment changes. Species have been found to exhibit considerable plasticity in their responses to changing temperature, depending on its severity and duration. This plasticity is evident both phenotypically for individual organisms and genotypically for populations and species during evolutionary adaptation to diverse thermal environments. The acute temperature dependence of biological rate processes of an individual organism may be modified and sometimes ameliorated by acclimation or acclimatisation over a period of weeks or months at different temperatures. Additionally, the physiological phenotype may be permanently affected by temperature at some critical ontogenic phase. Over longer (evolutionary) time periods, populations may undergo genetic adaptation as their thermal environment changes due to migration or climatic change.

This volume brings together many of the leading workers in thermal biology, with backgrounds spanning the disciplines of molecular biology, cell biology, physiology, zoology, ecology and evolutionary biology, to discuss the responses of a wide range of species to temperature change at all scales of organisation, ranging through the molecular, cellular, organismal, population and ecosystem levels. The volume provides an important and comprehensive novel contribution to the study of temperature adaptation which, given the current concern about global climate change, will provide much to interest a wide range of biologists.

SOCIETY FOR EXPERIMENTAL BIOLOGY
SEMINAR SERIES: 59

ANIMALS AND TEMPERATURE
PHENOTYPIC AND EVOLUTIONARY ADAPTATION

SOCIETY FOR EXPERIMENTAL BIOLOGY SEMINAR SERIES

A series of multi-author volumes developed from seminars held by the Society for Experimental Biology. Each volume serves not only as an introductory review of a specific topic, but also introduces the reader to experimental evidence to support the theories and principles discussed, and points the way to new research.

6. Neurones without impulses: their significance for vertebrate and invertebrate systems. *Edited by A. Roberts and B.M.H. Bush*
8. Stomatal physiology. *Edited by P.C.G. Jarvis and T.A. Mansfield*
16. Gills. *Edited by D.F. Houlihan, J.C. Rankin and T.J. Shuttleworth*
31. Plant canopies: their growth, form and function. *Edited by G. Russell, B. Marshall and P.G. Jarvis*
33. Neurohormones in invertebrates. *Edited by M. Thorndyke and G. Goldsworthy*
34. Acid toxicity and aquatic animals. *Edited by R. Morris, E.W. Taylor, D.J.A. Brown and J.A. Brown*
35. Division and segregation of organelles. *Edited by S.A. Boffey and D. Lloyd*
36. Biomechanics in evolution. *Edited by J.M.V. Rayner and R.J. Wootton*
37. Techniques in comparative respiratory physiology: An experimental approach. *Edited by C.R. Bridges and P.J. Butler*
38. Herbicides and plant metabolism. *Edited by A.D. Dodge*
39. Plants under stress. *Edited by H.C. Jones, T.J. Flowers and M.B. Jones*
40. *In situ* hybridisation: application to developmental biology and medicine. *Edited by N. Harris and D.G. Wilkinson*
41. Physiological strategies for gas exchange and metabolism. *Edited by A.J. Woakes, M.K. Grieshaber and C.R. Bridges*
42. Compartmentation of plant metabolism in non-photosynthesis tissues. *Edited by M.J. Emes*
43. Plant growth: interactions with nutrition and environment. *Edited by J.R. Porter and D.W. Lawlor*
44. Feeding and texture of foods. *Edited by J.F.V. Vincent and P.J. Lillford*
45. Endocytosis, exocytosis and vesicle traffic in plants. *Edited by G.R. Hawes, J.O.D. Coleman and D.E. Evans*
46. Calcium, oxygen radicals and cellular damage. *Edited by C.J. Duncan*
47. Fruit and seed production: aspects of development, environmental physiology and ecology. *Edited by C. Marshall and J. Grace*
48. Perspectives in plant cell recognition. *Edited by J.A. Callow and J.R. Green*
49. Inducible plant proteins: their biochemistry and molecular biology. *Edited by J.L. Wray*
50. Plant organelles: compartmentation of metabolism in photosynthetic cells. *Edited by A.K. Tobin*
51. Oxygen transport in biological systems: modelling of pathways from environment to cell. *Edited by S. Eggington and H.F. Ross*
52. New insights in vertebrate kidney function. *Edited by J.A. Brown, R.J. Balment and J.C. Rankin*
53. Post-translational modifications in plants. *Edited by N.H. Battey, H.G. Dickinson and A.M. Hetherington*
54. Biomechanics and cells. *Edited by F.J. Lyall and A.J. El Haj*
55. Molecular and cellular aspects of plant reproduction. *Edited by R. Scott and A.D. Stead*
56. Amino acids and their derivatives in plants. *Edited by R.M. Wallsgrove*
57. Toxicology of aquatic pollution: physiological, cellular and molecular approaches. *Edited by E.W. Taylor*
58. Gene expression and adaptation in aquatic organisms. *Edited by S. Ennion and G. Goldspink.*

ANIMALS AND TEMPERATURE
Phenotypic and Evolutionary Adaptation

Edited by

Ian A. Johnston
Professor of Comparative Physiology, Director of the Gatty Marine Laboratory, University of St Andrews

Albert F. Bennett
Professor of Ecology and Evolutionary Biology, University of California, Irvine

CAMBRIDGE UNIVERSITY PRESS
Cambridge, New York, Melbourne, Madrid, Cape Town, Singapore, São Paulo

Cambridge University Press
The Edinburgh Building, Cambridge CB2 8RU, UK

Published in the United States of America by Cambridge University Press, New York

www.cambridge.org
Information on this title: www.cambridge.org/9780521496582

First published 1996
This digitally printed version 2008

A catalogue record for this publication is available from the British Library

Library of Congress Cataloguing in Publication data

Animals and temperature : phenotypic and evolutionary adaptation /
edited by Ian A. Johnston and Albert F. Bennett.
p. cm. – (Society for Experimental Biology seminar series;
59)
Includes index.
ISBN 0 521 49658 6
1. Body temperature – Regulation. 2. Evolution (Biology)
I. Johnston. Ian A. II. Bennett, Albert F. III. Series: Seminar
series (Society for Experimental Biology (Great Britain)); 59.
QP135.A54 1996
591.54′2 – dc20 96-13366 CIP

ISBN 978-0-521-49658-2 hardback
ISBN 978-0-521-05061-6 paperback

Contents

Contributors

ATKINSON, D.
Department of Environmental & Evolutionary Biology, University of Liverpool, PO Box 147, Liverpool L69 3BX, UK
BENNET, A.F.
Department of Ecology & Evolutionary Biology, School of Biological Sciences, University of California, Irvine, CA 92717, USA
BERRIGAN, D.
Department of Zoology, University of Washington, Box 351800, Seattle, WA 98195-1800, USA
BICUDO, J.E.P.W.
Departamento de Fisiologica, Instituto de Biociências, Universidade de São Paulo, 05508-900 São Paulo, SP Brazil
CLARKE, A.
British Antarctic Survey, High Cross, Madingley Road, Cambridge CB3 0ET, UK
COSSINS, A.R.
Department of Zoology, University of Liverpool, Liverpool L69 3BX, UK
DAHLHOFF, E.
Department of Biology, Sonoma State University, Rohnert Park, CA 95401, USA
DI PRISCO, G.
Instituto di Biochemica delle Proteine ed Enzimologia, National Research Council, Via Marconi 10, 80125 Napoli, Italy
EPPLEY, Z.A.
Department of Zoology, 5751 Murray Hall, University of Maine, Orono, ME 04469-5751, USA
FEDER, M.E.
Department of Anatomy, The University of Chicago, 1025 East 57th Street, Chicago, IL 60637, USA
FRENCH, V.
Institute of Cell, Animal and Population Biology, University of Edinburgh, Kings Buildings, West Mains Road, Edinburgh EH9 3JT, UK

GIARDINA, B.
Institute of Chemistry, Faculty of Medicine, Catholic University, Largo F. Vito 1, I-00168 Rome, Italy

GRACEY, A.Y.
Department of Zoology, University of Liverpool, Liverpool L69 3BX, UK

GUDERLEY, H.E.
Départment de Biologie, Université Laval, Québec, G1K 7P4, Canada

HAWKINS, A.J.S.
Plymouth Marine Laboratory, Natural Environment Research Council, West Hoe, Plymouth PL1 3DH, UK

HILL, J.
Gatty Marine Laboratory, School of Biological & Medical Sciences, University of St Andrews, St Andrews, Fife KY16 8LB, UK

HUEY, R.B.
Department of Zoology, University of Washington, Box 351800, Seattle, WA 98195-1800, USA

JOHNSTON, I.A.
Gatty Marine Laboratory, School of Biological & Medical Sciences, University of St Andrews, St Andrews, Fife KY16 8LB, UK

LENSKI, R.E.
Center for Microbial Ecology, Plant and Soil Sciences Building, Michigan State University, East Lansing, MI 48824-1325, USA

LIN, J.J.
Hopkins Marine Station, Stanford University, Oceanview Blvd, Pacific Grove, CA 93950-3094, USA

LOGUE, J.
Department of Zoology, University of Liverpool, Liverpool L69 3BX, UK

MONGOLD, J.A.
Center for Microbial Ecology, Plant and Soil Sciences Building, Michigan State University, East Lansing, MI 48824-1325, USA

PARTRIDGE, L.
Department of Biology, University College London, Wolfson House, 4 Stephenson Way, London NW1 2HE, UK

RUBEN, J.
Department of Zoology, Oregon State University, Corvallis, OR 97331-2917, USA

SOMERO, G.N.
Hopkins Marine Station, Stanford University, Oceanview Blvd, Pacific Grove, CA 93950-3094, USA

ST PIERRE, J.
Départment de Biologie, Université Laval, Québec, G1K 7P4, Canada

TIKU, P.E.
Department of Zoology, University of Liverpool, Liverpool L69 3BX, UK

VIEIRA, V.L.A.
Departmento de Pesca, Universidade Federal Rural de Pernambuco, Av. D Manoel de Medeiros, Dois Irmãos, Recife PE, CEP 52 171-900, PE. Brazil

Preface

Studies of thermoregulation and thermal effects have been one of the most prolific and enduring topics in functional biology for the past century (for reviews, see Hochachka & Somero, 1984; Prosser, 1986; Cossins & Bowler, 1987). The popularity of thermal biology stems from its great significance for biological systems. Temperature has pervasive effects on biological rate processes: rapid changes in body temperature alter nearly all physiological functions by approximately 6–10% per degree Celsius over a broad thermal range. Organismal and population-level traits that depend on those processes, such as energy utilisation, growth and reproduction, are also therefore greatly affected by temperature change. Thus virtually everything that an organism does is influenced by and dependent on its thermal condition.

Despite the major impact of temperature on functional capacities, biological systems have evolved in and adapted to almost every thermal environment on earth. Adult metazoans have successfully colonised environments ranging from −70 °C to 50 °C, while the thermal tolerance of dormant stages and prokaryotes is even greater. Individual organisms and their offspring usually encounter a range of temperatures during their life-cycle, varying from daily cycles to seasonal or longer-term climate change. Responses to a given temperature or temperature change can vary markedly between species and between different life-history stages of the same species. Much of the literature of thermal biology has been concerned with determining thermal tolerances and describing patterns of function at different temperatures. The latter may be considered reaction norms, the term given to the range of phenotypes produced when organisms (or particular stages of organisms) are exposed to different environments, such as a temperature gradient. The differences in phenotype produced in distinct environments are termed phenotypic plasticity. Physiologists and cell biologists have been particularly interested and successful in investigating mechanisms that organisms use to cope with extreme thermal environments (e.g. freezing avoidance in polar fishes, hibernation in mammals) or mechanisms associated with the functioning of tissues with a specialised thermoregulatory function (e.g. brown adipose tissue, counter-current heat exchangers, billfish brain heaters) (for other examples, see Schmidt-Neilsen, 1983). For a few species, comparative physiologists have had some success in understanding some of the cellular and molecular mechanisms underlying phenotypic plasticity to temperature change. For example, changes in the maximum swimming performance of fish following several weeks acclimation to a new temperature regimen have been explained in terms of changes in muscle structure and function: integrating studies at the organismal, physiological and biochemical levels (Johnson & Bennett, 1995; Johnston *et al.*, 1995). Much less tractable has been an understanding of the evolutionary or adaptive significance of phenotypic plasticity. Very few studies have rigorously attempted to test hypotheses about the impact of temperature acclimation on fitness (see Huey and Berrigan, pp. 205–238.

Over the past decade, there has been great interest in studies that explicitly attempt to analyse the evolution of functional characters, an interest signalled by the emergence of the term evolutionary physiology. There are many possible approaches to studying the evolutionary basis of patterns of temperature tolerance and thermal physiology. Most of our knowledge on this subject is derived from comparative studies of different natural taxa, generally populations or species. Recent emphasis has been placed on phylogenetically-based comparative studies that enable adaptive hypotheses to be tested, putative ancestral conditions to be assigned and rates of evolution to be calculated (Feder, 1987; Huey, 1987). These comparisons rely on an independently-derived set of taxonomic relationships and depend on assumptions about parsimony which may not hold true in particular cases.

Another approach to studying evolution is provided by quantitative genetics. In studies relating to thermal responses, genetically-related individuals (isogenic lines, clones, half- or full sib families) are typically exposed to different thermal environments and the response of particular traits is plotted as a reaction norm. Statistical techniques are then used to partition phenotypic variation into effects related to genotype, environment and genotype–environmental interactions and to estimate genetic correlations and heritabilities (de Jong, 1995). Multivariate extensions of quantitative genetic equations that describe responses to selection can yield predictions about the rates and direction of evolution, given certain simplifying assumptions (Turelli, 1988).

Direct experimental studies of evolution are also possible, in which replicated experimental populations are exposed to different environments, in this case temperatures, for many generations, while maintaining control populations for comparison. This approach enables a statistical analysis of genetically-based phenotypic changes attributable to evolution in the new thermal environment. The power of such an approach is that the experimenter can directly measure characters in their initial, immediate and derived conditions. Models used for experimental evolution are necessarily restricted to organisms with short generation times which can be readily maintained in long-term culture, such as bacteria and fruit flies. A short-coming of 'natural selection' in the laboratory is that environmental conditions are often overly simplistic if the intention is to model evolution in natural environments, which are highly variable with many covarying biotic and abiotic factors.

Molecular genetics offers a variety of mechanistic approaches to the study of the plasticity of genes (Pigliucci, 1996). Recently, a number of studies have begun to map genes coding for quantitative trait loci, thereby going beyond the purely statistical analysis of populations (Mitchell-Odds, 1995). Mutants can be produced that delete or interfere with a well-characterised pattern of phenotypic plasticity. Alternatively, animals or cells can be exposed to different temperatures, and specific changes in RNA or protein patterns can be identified using differential PCR, subtraction *in situ* hybridisation or 2-D gel electrophoresis. A growing number of genes have been identified that respond to temperature change with far reaching consequences for the organism; examples include genes responsible for sex determination in reptiles and the heat shock gene family which confer thermotolerance (for a review, see Pigliucci, 1996).

Recent discoveries in developmental biology are also emerging as candidates to explain both microevolutionary and macroevolutionary events. It has been suggested that the sudden emergence of the major metazoan body plans during the late Neoproterozoic and early Cambrian was related to innovations in developmental control mechanisms that included the *Hox* gene cluster (Valentine *et al.*, 1996). In addition, homologous developmental pathways have been documented in numerous embryonic processes which may be organised in discrete regions or morphogenetic fields. It has been proposed that changes in the properties of these fields can lead to evolutionary novelty as can changes in the amount, type, or duration of gene products or mutations affecting DNA-protein binding (for a recent review on the synthesis of evolutionary and developmental biology, see Gilbert *et al.*, 1996). Environmental factors, including temperature, also have the potential to alter the properties of morphogenic fields, i.e. the range of diffusion of molecules or the activity of certain components (cells and proteins) leading to changes in the resulting phenotypes. While not necessarily adaptive in itself the phenotypic variation generated during development may be subject to strong selection.

Recent concerns about the impact of human industrialisation on the rate of climate change have placed studies of thermal biology centre stage in the latter part of this century. It is clear that future progress in understanding phenotypic and evolutionary adaptations to temperature change will require a synthesis from hitherto more or less separate fields of evolutionary biology, molecular genetics, developmental biology, ecology, comparative physiology and organismal biology; no single approach will be sufficient. As a contribution to this synthesis, we invited some of the leading workers in thermal biology from across the spectrum of these disciplines to a meeting of the Society of Experimental Biology held in St Andrews on the occasion of the Centenary of the Gatty Marine Laboratory. This volume is the written record of that symposium. Although the emphasis of the volume is on animals, studies on bacteria and plants have been included in several of the chapters, where they help to illustrate general principles. We are grateful for the financial support of the Company of Biologists and the National Science Foundation which enabled us to assemble such high quality speakers.

Ian A. Johnston (St Andrews) Albert F. Bennett (Irvine)

References

Cossins, A.R. & Bowler, K. (1987). *Temperature Biology of Animals*, p. 339. London & New York: Chapman and Hall.

De Jong, G. (1995). Phenotypic plasticity as a product of selection in a variable environment. *American Naturalist*, **145**, 493–512.

Feder, M.E. (1987). The analysis of physiological diversity: the prospects for pattern documentation and general questions in ecological physiology. In *New Directions in Ecological Physiology* (ed. M.E. Feder, A.F. Bennett, W.W. Burggren & R.B. Huey), pp. 38–75. Cambridge: Cambridge University Press.

Gilbert, S.F., Opitz, J.M., Raff, R.A. (1996). Resynthesizing evolutionary and developmental biology. *Developmental Biology* **173**, 357–72.

Hochachka, P.W. & Somero, G.N. (1984). *Biochemical Adaptation*. Philadelphia: Saunders.

Huey, R.B. (1987). Phylogeny, history and the comparative method. In *New Directions in Ecological Physiology* (ed. M.E. Feder, A.F.

Bennett, W.W. Burggren & R.B. Huey), pp. 76–98. Cambridge: Cambridge University Press.

Johnson, T.P. & Bennett, A.F. (1995). The thermal acclimation of burst escape performance in fish: an integrated study of molecular and cellular physiology and organismal performance. *Journal of Experimental Biology* **198**, 2165–75.

Johnston, I.A., van Leeuwen, J.L., Davies, M.L. & Beddow, T. (1995). How fish power predation fast-starts. *Journal of Experimental Biology* **198**, 1851–61.

Mitchell-Olds, T. (1995). The molecular basis of quantitative genetic variation in natural populations. *Trends in Evolution and Ecology* **10**, 324–8.

Pigliucci, M. (1996). How organisms respond to environmental changes: from phenotypes to molecules (and vice versa). *Trends in Evolution and Ecology* **11**, 168–73.

Prosser, C.L. (1986). *Environmental and Metabolic Animal Physiology. Comparative Animal Physiology*, 4th edn, p. 578. New York: Wiley-Liss.

Schmidt-Neilsen, K. (1983). *Animal Physiology: Adaptation and Environment*, p. 619. Cambridge: Cambridge University Press.

Turelli, M. (1988). Phenotypic evolution, constant covariances and the maintenance of additive variance. *Evolution* **42**, 1342–7.

Valentine, J.W., Erwin, D.H. & Jablonski, D. (1996). Developmental evolution of metazoan body plans: the fossil evidence. *Developmental Biology* **173**, 373–81.

A.Y. GRACEY, J. LOGUE, P.E. TIKU
and A.R. COSSINS

Adaptation of biological membranes to temperature: biophysical perspectives and molecular mechanisms

Introduction

Of all the cellular responses that occur during thermal acclimation of fish, the most widespread and consistent is a change in the composition of cellular lipids. Since the early part of this century it has been known that cold acclimation leads to increased levels of unsaturated fatty acids, both in depot and cellular lipids. Since that time there have been a very large number of published studies demonstrating this effect in different animal groups, as well as in plants and in microorganisms. It is a matter of common observation that unsaturated fats are more fluid than saturated fats and the idea that the cold-induced increase in unsaturation compensated for the cold-induced rigidification of cellular lipids was implicit even in the earliest studies. This concept, however, was specifically applied to phospholipid cellular membranes and to the thermal acclimation of poikilothermic animals by Johnston & Roots (1961) who ascribed the adaptive significance to the preservation of some unidentified physical condition.

Spectroscopic and calorimetric techniques to quantify the physical properties of order or of molecular motion in phospholipid bilayers became available in the early 1970s and it was not long until they were used to compare the membranes isolated from bacteria grown at different temperatures. Sinensky (1974) showed that the membranes of *E. coli* grown at low temperatures were more 'fluid', as judged by electron spin resonance spectroscopy, than those of bacteria grown at higher temperatures. He termed this response 'homeoviscous adaptation' (HA) to emphasise the compensatory and adaptive significance of the response. Shortly after this similar responses were observed in the protozoan *Tetrahymena* (Nozawa *et al.*, 1974) and in brain membranes of the common goldfish, *Carassius auratus* (Cossins, 1977), both using fluorescence polarisation spectroscopy. More recently, a number of studies in fish tissues have demonstrated that although HA was widespread it was not always developed to the same extent. Thus, the degree of compensation, termed 'homeoviscous efficacy' (HE) varied from 75% in the basolateral membranes of carp intestinal mucosa to 25% in brain myelin, while in enterocyte brush border and in the sarcoplasmic reticulum no responses were observed (i.e. HE = 0%).

Over the past 20 years there has been a progressive improvement in the conceptual understanding of the physical structure of biological membranes as newer and more powerful methods used for its measurement have been developed. Thus the early descriptive terms of 'microviscosity' and 'fluidity' have been largely superseded by motional rate parameters and measures of static molecular ordering. Similarly, early ideas on a more-or-less continuous bulk membrane lipid phase have been superseded by a recognition of microdomains, of laterally differentiated physical states, of the importance of membrane thickness and the considerable compositional asymmetry between the inner and outer monolayers (Gennis, 1989; Hazel, 1995). These emerging concepts are obviously dependent on the development of new spectroscopic techniques used to define them (McElhaney, 1994). The existence of a conserved physical condition in biological membranes, which lies at the heart of the concept of homeoviscous adaptation, has so far not been challenged. What has changed, and will continue to evolve, is our appreciation of the details of the response and how they compensate for specific aspects of membrane function. To some extent this was recognised some years ago when the compensation of phase transition temperatures as opposed to 'fluidity' was observed in the mycoplasma *Acholeplasma laidlawii*, the response being termed 'homeophasic adaptation' (McElhaney, 1974, 1994; Williams, 1990; Hazel, 1995).

Insights from interspecies comparisons of 'fluidity' and lipid composition

Perhaps the most impressive evidence of thermal adaptation of biomembranes comes from studies of species that have evolved in particular thermal habitats or maintain elevated and constant body temperatures, as in birds and mammals (Cossins & Prosser, 1977; Behan-Martin *et al.*, 1993). Fig. 1 compares membranes isolated from a variety of teleost fish species and from a representative mammal and bird. In this study the measure of membrane physical condition was the fluorescence anisotropy of the fluorescence probes, 1,6-diphenyl-1,3,5-hexatriene (DPH) and *trans*-parinaric acid, a technique which provides information principally on the degree of hindrance to the free rotational motion of the rod-shaped fluorescent probe (Van der Meer, 1984). This is

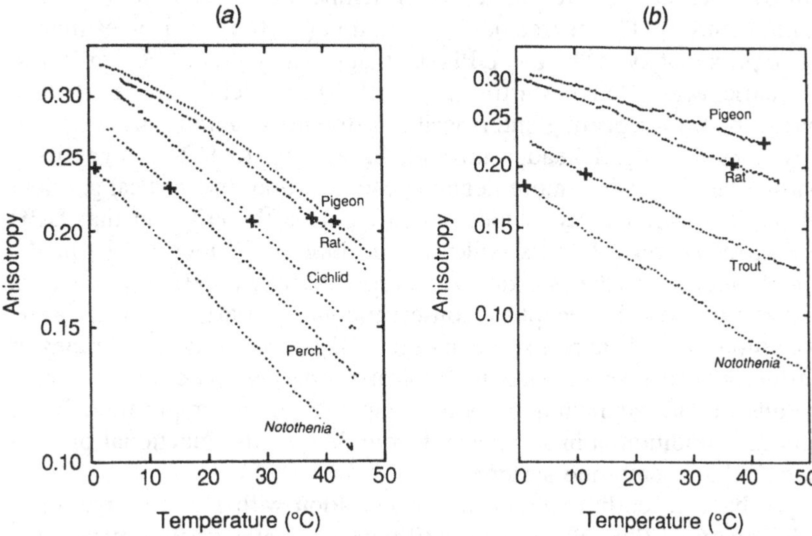

Fig. 1. A comparison of membrane physical structure of various fish, avian and mammalian species for a brain synaptic membrane fraction. Membrane physical structure was measured by the fluorescence anisotropy of 1,3-diphenyl-1,3,6-hexatriene (*a*) and *trans*-parinaric acid (*b*) during warming at approximately 0.75 °C min^{-1}. Each data point is a separate measurement. The crosses represent the anisotropy for the respective body or habitat temperature, as appropriate, for each species. (After Behan-Martin *et al.*, 1993.)

indicative of the degree of hydrocarbon ordering within the bilayer rather than a measure of the rate of wobbling or of segmental motion of the hydrocarbon chain. Thus, strictly speaking it is a static order parameter rather than a measure of motional rates (i.e., 'fluidity').

Figure 1 illustrates the temperature profiles for fluorescence anisotropy using the two probes (Behan-Martin *et al.*, 1993). In both cases there is a clear relationship between the position of the curve and the temperature to which the species has been adapted. Thus membranes from the Antarctic *Notothenia* were less ordered than equivalent membranes from the perch, *Perca fluviatilis*, and so on. Interestingly, the mammalian and avian species possess relatively ordered membranes which matches their high body temperatures of 37 °C and 41 °C, respectively. The crosses in Fig. 1 indicate the value of anisotropy at the respective body temperature for each species. Despite the obvious temperature dependence of anisotropy for each species it is clear that anisotropies at their respective body temperatures were very similar, if not identical. The degree of compensation (%HE) has been estimated at approximately 70% for DPH and approximately 100% for *trans*-parinaric acid (Behan-Martin *et al.*, 1993). Evidently there is some probe bias in reporting interspecific differences. *Trans*-parinaric acid possesses a charged headgroup and, in contrast to DPH, aligns itself within the bilayer in a more defined position within the vertical gradient of order. Consequently it does not suffer from the criticism that DPH has attracted regarding its undefined positional distribution within the membrane. Nevertheless, the comparison between species using each probe supports the adaptive conservation of a particular membrane condition. Thus it appears that nature designs the brain membranes of vertebrates to give equivalent physical properties under *in vivo* conditions of environmental or body temperature, the implication being that the condition is in some way favourable for the functional properties of that membrane system.

We have recently explored in collaboration with Dr A.L. De Vries (University of Illinois) the compositional correlates of the interspecies differences of membrane physical order to define more clearly the broad trends which may underlie genotypic membrane adaptation. Fig. 2*a* shows preliminary results for phosphatidylcholine and how the different groups of unsaturated fatty acid (i.e. saturated, monoene, polyene) vary with changing levels of saturation in brain membranes from different fish species, rat and turkey. Increasing membrane disorder in cold-adapted temperate and Antarctic species is clearly associated with increasing levels of polyunsaturates, while the monounsaturates remain comparatively unaltered. By contrast, the equivalent graph for phosphatidylethanolamine (not shown) shows that increasing membrane disorder is linked with increasing levels of monounsaturation while polyunsaturation remains roughly constant. It appears, therefore, that the two major phosphoglycerides contribute to increasing membrane disorder during evolutionary cold adaptation by distinctive compositional adjustments. Interestingly, a similar difference between the fatty acid composition of phosphatidylcholine and phosphatidylethanolamine is evident in the liver microsomes of thermally-acclimated carp (Gracey, unpublished results), suggesting that the interspecific observations have some more general significance during both phenotypic and genotypic adaptation.

Figure 2*b* shows for phosphatidylcholine the relationship between the unsaturation index (UI, a measure of the number of unsaturation bonds in 100 weight %) and the proportion of fatty acids that were unsaturated. Narrow construction lines have been drawn indicating the

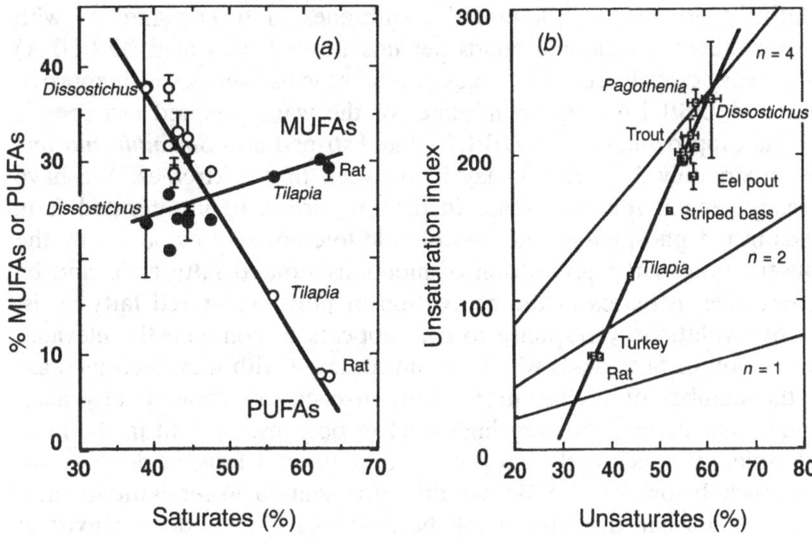

Fig. 2. Interspecies comparisons of membrane acyl saturation for phosphatidylcholine isolated from a brain synaptic membrane fraction. (*a*) Shows the variation between the proportion of fatty acids which are monounsaturated (MUFAs) or polyunsaturated (PUFAs) as a function of saturation. Interspecific variations in the proportion of saturated fatty acids were linked predominantly with changes in the proportion of polyunsaturated fatty acids while monounsaturated fatty acids were largely unaltered. This analysis excluded plasmalogens. (*b*) Shows how the unsaturation index (UI) varies with the proportion of unsaturated fatty acids. UI was calculated as the sum of the % weight of each fatty acid multiplied by the number of double bonds. Thin construction lines indicate the expected relationship if each unsaturated fatty acid had one double bond (n = 1) or two double bonds (n = 2), and so on. Increasing unsaturated fatty acid content in cold adapted fish membranes is associated with an increase in the UI above that expected from a constant number of unsaturation bonds, meaning that the average number of double bonds per unsaturated fatty acid increases.

expected relationship if all of the unsaturated fatty acids were monoenoic (n = 1), dienoic (n = 2) and tetraenoic (n = 4). The thick line is drawn by inspection connecting data points for cold-adapted fish species, two warm-adapted fish species, rat and turkey. The fact that all species fall on a single straight line suggests there is a general relationship between UI and levels of unsaturation extending from the relatively ordered and saturated membranes in turkey and rat with approximately 1.6 double bonds per unsaturated fatty acid (DB/UFA) to the very disordered membranes of Antarctic fish species with approximately 4 DB/UFA. The membranes of the warm-adapted fish species *Tilapia* (approximately 2.6 DB/UFA) and striped bass *Saxantilis morone* (approximately 3 DB/UFA) lay in between these extremes. We have already seen that the increase in the proportion of unsaturated fatty acids in the phosphatidylcholines of cold-tolerant species occurs by the conservation of the proportion of monounsaturated fatty acids and by progressive increases in the proportion of polyunsaturated fatty acids, so the evolutionary response to cold appears to combine the elevated proportion of fatty acids which are unsaturated with a marked increase in the number of unsaturation bonds in each unsaturated fatty acid. Notice that despite the very high level of polyunsaturation in the most cold-tolerant species, the proportion of saturated fatty acids does not fall much below 40%. Whether this represents a lower limit to saturation is unclear from this work because fish species do not exist at lower temperatures.

Comparative studies such as these provide a useful insight into the compositional correlates of temperature-compensated membranes of cold-tolerant species. They also provide a very useful perspective with which to appreciate the relatively saturated composition and ordered physical structure of membranes of birds and mammals. This appreciation could not have become apparent without a comparison of this sort, and the excellent correlation between compositional features of the poikilothermic and homeothermic species suggests that they conform to a common principle of design. Unfortunately, we have not been able to compare birds and mammals with fish species which experience body temperatures close to 37–41 °C, such as the desert pupfish *Cyprinodon nevadensis* or the East African cichlid, *Oreochromis magadi*. Consequently the idea that the interclass differences discovered to date are simple temperature adaptations without some phylogenetic component has been difficult to test.

Problems of current studies of membrane adaptation

The principal developments over the past 20 years have been to establish the details of the changes in lipid composition and to link these with changes in the physical structure of the bilayer and, in fewer cases, with changes in membrane function (Hazel, 1989; Cossins, 1994). Thus, increased phospholipid unsaturation has been linked with decreased membrane order and this was linked with improved function, all of which is logical and consistent with what is known about the effects of altered unsaturation from studies of model membrane systems. This logical consistency has been invariably taken as confirming their causal link and it is important to recognise that all of these studies are correlative and correlation does not necessarily imply causation.

A further problem is that the details of the compositional adjustment are complex, with changes in levels of different polyunsaturated fatty acids as well as shifts in the balance between saturated, monounsaturated and polyunsaturated fatty acids, changes in the phospholipid head group composition and in some cases with changes in cholesterol content. It is also apparent that the changes in physical structure that accompany this compositional adjustment may also be complex, not least because the physical structure of biological membranes is complex. Are all of the detailed compositional adjustments important or are some more important than others? Are some membrane adjustments of adaptive significance specifically to temperature adaptation while others may be non-thermal adaptations? Can the adaptive significance of any one of these responses be tested in terms of improved whole organism fitness? The compositional and structural complexity makes establishing truly causal relationships extremely difficult, if not impossible, because it is not possible to separate experimentally the different compositional alterations in order to assess their respective contributions to the observed change in physical structure. As a result, questions such as those just outlined are not amenable to analysis by the conventional comparative approach.

Fortunately, modern molecular genetics offers some alternative and less equivocal approaches for establishing a degree of experimental control over membrane adaptation, simply by intervening in the expression of enzymes involved in the response. The induced expression in the absence of temperature change or the overexpression or abolition of a specific enzyme allows its role in the adaptive regulation of membrane structure to be demonstrated more confidently. Moreover, methods are available for altering the genotype and for examining the extent to which any particular alteration affects the phenotype either at the subcellular and cellular level or in whole organisms. At present this control is most conveniently implemented in cultured cells by transient gene transfection, but techniques for transgenic manipulation of whole animals are rapidly improving and they will certainly have a major impact in temperature biology as their use becomes routine.

In examining the mechanisms underlying homeoviscous response it is necessary to distinguish the proximal mechanism which produces the change in enzyme activity (e.g. translational or transcriptional upregulation) from the ultimate controlling process that regulates this mechanism. It is likely that this latter process incorporates a feedback element and demonstrating the existence and understanding the nature of the feedback presents the greatest challenge.

Altered desaturase expression during cold acclimation of carp

Strong evidence points to lipid desaturases as important components of the enzymatic response to cold challenge. First, the introduction of the first unsaturation bond is thought to have the greatest effect on membrane physical properties (Stubbs *et al.*, 1981; Coolbear *et al.*, 1983) and incorporation of this double bond at the central 9–10 position from the carboxyl terminus has maximal effects on lipid physical properties (Barton & Gunstone, 1975). These two observations indicate that membrane physical properties can be controlled most effectively, at least over a certain range of temperatures, by regulating the proportion of saturated fatty acids compared with unsaturated fatty acids. This can be achieved by modifying the activity of the enzyme, the Δ^9-desaturase, which incorporates the first double bond into stearic or palmitic acid. This conclusion needs to be qualified to the extent that changes in the proportions and composition of polyunsaturated fatty acid are often an important component of homeoviscous adaptation. Variations in the activity of desaturases and elongases which modify the composition of polyunsaturates may also form part of the compositional response to cooling.

Second, desaturase activity has been shown to be greatly enhanced during the period following cold-transfer of organisms; examples include the bacterium *Bacillus megaterium* (Fujii & Fulco, 1977), the cyanobacteria *Anabaena variabilis* and *Synechocystis* (Sato & Murata, 1981), and the thermotolerant strain, NT-1, of the protozoan *Tetrahymena pyriformis* (Nozawa & Kasai, 1978). In fish, the cold-induced change in fatty acid composition of membrane phospholipids is more complex and it is likely that coordinated changes in the expression of several and probably many enzymes is involved. Schünke & Wodtke (1983) demonstrated that the progressive cooling of carp from 30 to 10 °C led to a pronounced 14 to 32-fold increase in Δ^9-desaturase specific activity. The induction was transient but correlated closely with changes in membrane physical order as detected by fluorescence polarisation spectroscopy (Wodtke & Cossins, 1991) supporting its role in homeoviscous adaptation. The carp were fed a high-fat trout diet, so the low hepatic desaturase activity of warm-acclimated carp indicates that the dietary input was sufficient for the needs of the animals. Wodtke has previously shown that feeding a more saturated diet caused the upregulation of desaturase activity even in warm-acclimated carp (Wodtke, 1986). Cooling of these carp did not lead to a desaturase induction even though similar cold-induced changes in membrane lipids were observed.

We are currently following a molecular biology approach to understand the mechanism underlying the cold-induced increase in desaturase activity using the carp liver system. Recent developments in the molecular analysis of mammalian Δ^9-desaturases have provided the appropriate immunological and genetic tools for the analysis of desaturase expression, given an adequate degree of homology between mammals and fish. The cDNA for the rat desaturase encodes a protein of 358 amino acids (SCD1) with a predicted molecular weight of 41 400 Two Δ^9-desaturase genes (*SCD1* and *SCD2*) have been identified in mouse and rat that encode two isoforms (Ntambi *et al.*, 1988; Kaestner *et al.*, 1989; Mihara, 1990) which display a tissue-specific distribution. Mouse SCD1 exhibits 92% homology with rat SCD1 at the amino acid level while SCD2 shows 86 and 87% homology to rat and mouse SCD1, respectively. The Δ^9-desaturase gene (*OLE1*) of the yeast *Saccharomyces cerevisiae* codes for a protein of 510 amino acids with a calculated mass of 57.4 kDa (Stukey *et al.*, 1989, 1990). At the amino acid level it shows 36% identity and 60% similarity with the rat SCD and there are three highly conserved regions including one with 11 out of 12 perfect residue match. Significantly, the yeast protein can be functionally replaced by the product of the rat *SCD* gene (Stukey *et al.*, 1990) and a similar experiment has been performed to complement a desaturase mutation in tobacco plants (Grayburn *et al.*, 1992). Thus, Δ^9-desaturase structure and function appears to be conserved across a wide range of organisms and this encourages an approach based on heterologous cDNA hybridisation.

Our strategy has been to define the changes in lipid composition, desaturase activity and protein levels and mRNA levels in carp liver to define the level at which control was exerted. Thus, an increase in desaturase protein during the activity induction implies a role for protein turnover while constant protein levels indicate a post-translational mechanism. Similarly, increasing levels of desaturase mRNA would indicate a transcriptional or RNA turnover mechanism.

We have attempted to raise PCR products from carp liver cDNA using primers corresponding to the conserved segments found in the rat liver and yeast gene Δ^9-desaturase cDNA. This approach proved unsuccessful, possibly because the degeneracy built into the primers reduced the stringency for selective amplification of desaturase to levels that prevented production of any useful DNA. An alternative approach was based on screening a commercial carp liver cDNA library with the rat cDNA. Early experiments indicated that the homology at the nucleotide level between carp and rat was too low to allow detection of a desaturase mRNA by conventional, stringent Northern analysis. Subsequent work on isolated carp hepatocytes, however, indicated that continued long-term culture led to a large isothermal desaturase induction (Macartney *et al.*, 1996) and this was related to the increasing quantities of a transcript that could be detected on moderate stringency Northern analysis using the rat cDNA as probe. The apparently large quantities of mRNA on hepatocyte incubation, together with very long autoradiography exposure periods, allowed refinement of the conditions for selective heterologous hybridisation under conditions of moderate stringency.

The level of this transcript revealed by this technique mirrored the increasing enzymatic activity of the desaturase during hepatocyte culture, as well as the change in membrane phospholipid fatty acid composition. These observations encouraged a belief that the rat cDNA probe could be useful. Successive heterologous screening of a commercial carp hepatocyte cDNA library yielded three clones which tested positive in Northern analysis of mRNA extracts from cold-exposed carp liver under conditions of very high stringency (Fig. 3). All three clones detected the same sized transcript (approximately 2700 bp). Restriction analyses of the clones revealed that two clones (pcDsL5 and L6) were identical, but were different from the third (pcDsL7). The difference in cDNA size (approximately 5500 bp and 2700 bp for L6 and L7, respectively) suggested that they might be different desaturase proteins, perhaps from a desaturase multigene family. Partial DNA sequencing of these two clones indicates differences in the 5′ and 3′ untranslated regions (UTRs).

Clone pcDsL7 has been fully sequenced, giving an open reading frame (ORF) of 879 bp, with a fairly short (approximately 520 bp) 5′ UTR and a long (1271 bp) 3′ UTR (GenBank accession no. CC U31864). Translation of this ORF indicated a polypeptide of 292 amino acid residues and a calculated molecular weight of 33 750. The predicted protein showed a high degree of sequence similarity with other published Δ^9-desaturase sequences, including those from rat liver (55% identity), mouse liver (53% identity), tick (47% identity) and yeast (20% identity). Figure 4*a* compares the predicted hydropathicity plots for the carp desaturase with rat, tick and yeast enzymes, revealing extensive regions of secondary structure similarity and this, together with the near identity of carp, rat and yeast amino acid sequences

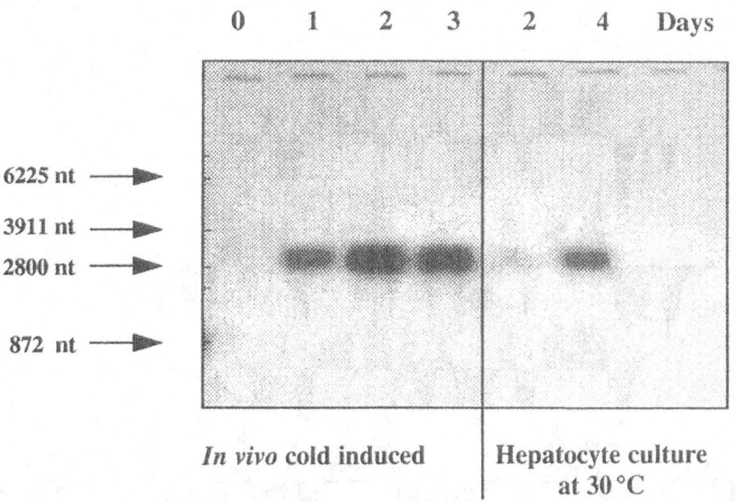

Fig. 3. Northern analysis of hepatic RNA samples using a probe derived from the carp liver desaturase clone, pcDsL7. Samples were extracted from carp maintained for several months at 30 °C (day 0) or at the indicated period following stepped transfer of these fish to 10 °C. Extracts from hepatocyte culture were obtained at the indicated times following initiation of culture of cells from 30 °C acclimated fish at 30 °C.

over several consensus boxes (Fig. 4*b*), establishes that pcDsL7 is indeed the desaturase.

Clone pcDsL7 has been used to design an antisense mRNA probe for ribonuclease protection assays (RPAs), this being a 50- to 100-fold more sensitive assay for homologous mRNA detection than conventional Northern analyses. The RPA has allowed us to investigate levels of liver Δ^9-desaturase transcript during thermal acclimation (Tiku *et al.*, 1996). RPA probes have also been prepared against carp β-actin mRNA and 18S ribosomal RNA and the induction of desaturase has been judged relative to these. This not only corrects by internal calibration for variations in loading and running of gels but also allows the induction of desaturase transcript levels to be measured relative to that of a housekeeping gene not thought to be involved in temperature adaptation. Figure 5 shows the variation in Δ^9-desaturase transcript levels relative to 18S ribosomal RNA in liver RNA extracts from carp acclimated to 30 °C and at different periods after a progressive 3-day cooling regimen to 10 °C. Transcript levels were low in warm-acclimated

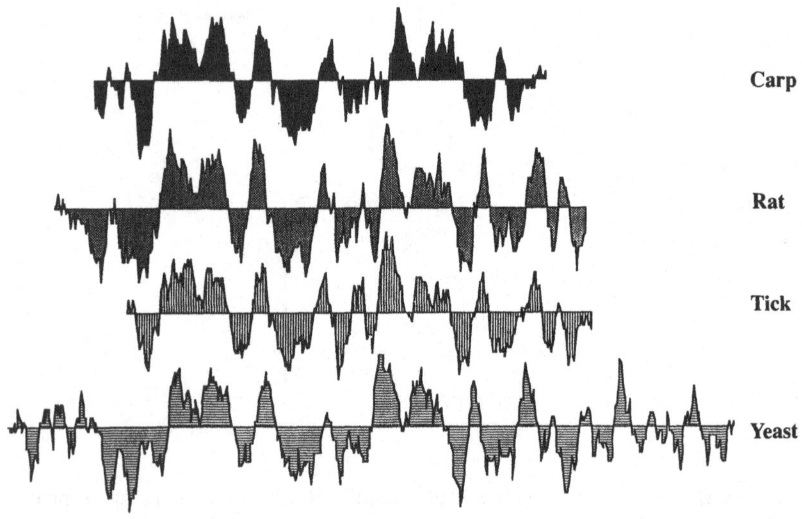
Carp
Rat
Tick
Yeast

carp HNRSYKEKEGPKPPTVIV--------------WRNVILMS 26
rat HDPSYQDEEGPPPKLEYV--------------WRNIILMA 26
tick --------------MEIV--------------WRNVILMG 12
yeast LEKDNQEKEEAKTKIHISEQPWTLNNWHQHLNWLNMVLVC 40
carp LLHLGALYGLFLFP-SARALTWIWFFGCLLFSALGITAGA 65
rat LLHVGALYGITLIP-SSKVYTLLWGIFYYLISALGITAGA 65
tick SLHLISIYGFYLIFFAAQWKTVLAAYIFYTISGIGVTAGS 52
yeast GMPMIGWYFALSGKVPLHLNVFLFSVFYYAVGGVSITAGY 80
carp HRLWSHRSYKASLPLQIFLALGNSMAFQNDIYEWSRDHRV 105
rat HRLWSHRTYKARLPLRIFLIIANTMAFQNDVYEWARDHRA 105
tick HRLWSHRSYKAKLPYRIMLMIFQIMAFQNDIYDWARDHRM 92
yeast HRLWSHRSYSAHWPLRLFYAIFGCASVEGSAKWWGHSHRI 120
carp HHKYSETDADPHNAVRGFFFSHVGWLLVRKHPDVIEKGRK 145
rat HHKFSETHADPHNSRRGFFFSHVGWLLVRKHPAVKEKGGK 145
tick HHKFSETTADPHDATRGFFFSHVGWLLVRKHPDVRNKGKS 132
yeast HHRYTDTLRDPYDARRGLWYSHMGWMLLKPNPKYKARA-- 159
carp LELSDLKADKVVMFQRRFYKPSVLLMCFFVPTFVPWYVWG 185
rat LDMSDLKAEKLVMFQRRYYKPGLLLMCFILPTLVPWYCWG 185
tick IDLSDVLADPVVRFQRRYYLPLMVTICFIVPALLPWWLWG 172
yeast -DITDMTDDWTIRFQHRHYILLMLLTAFVIPTLICGYFF- 197
carp ESLWVAYFVPALLRYALVLNATWLVNSAAHMWGNRPYDSS 225
rat ETFLHSLFVSTFLRYTLVLNATWLVNSAAHLYGYRPYDKN 225
tick ETLWNSFVVCSLTRYCFTLNMTWLVNSAAHIWGNRPYDRH 212
yeast NDYMGGLIYAGFIRVFVIQQATFCINSMAHYIGTQPFDDR 236
carp INPRENRFVTFSAIGEGFHNYHHTFPFDYATSEFGCKLNL 265
rat IQSRENILVSLGSVGEGFHNYHHAFPYDYSASEYRWHINF 265
tick ISPRQNLVTIVGAHGEGFHNYHHTFPYDYRTSFLGCRINT 252
yeast RTPRDNWITAIVTFGEGYHNFHHEFPTDYRNAIKWYQYDP 276
carp TT 267
rat TTFFIDCMAALGLAYDRKK 284
tick TTWFIDFFAWLGQVYDRKE 271
yeast TKVIIYLTSLVGLAYDLKK 295

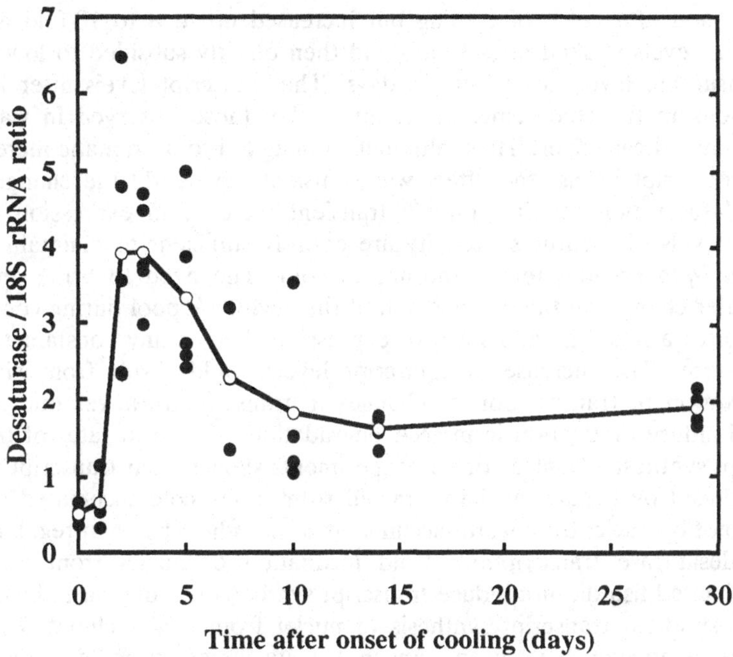

Fig. 5. The effect of cold acclimation of carp upon the levels of hepatic Δ^9-desaturases mRNA transcript. Total RNA was extracted from individual carp subject to indicated periods of the cooling regimen from 30 °C down to 10 °C. Desaturase mRNA levels have been normalised to that determined for the 18S rRNA levels. Filled symbols represent data from individual animals and unfilled symbols represent mean values.

Fig. 4. (*a*) Comparison of the hydrophobicity plots for predicted Δ^9-desaturases from carp, rat, tick and yeast. The published nucleotide sequences from the latter three were obtained from GenBank. Kyte-Doolittle plots with 9 amino acid residue windows were generated with Lasergene software from DNASTAR. (*b*) Comparison of the predicted amino acids sequences for Δ^9-desaturases of carp, rat, tick and yeast. Clustal alignment using a structural residue weight table (DNASTAR Lasergene software). The shaded areas indicate residues which match the consensus within one distance unit.

carp and after 24 h of cooling but increased up to 8 to 10-fold after 48 h. Levels peaked at 2–3 days and then quickly subsided to low but measurable levels for up to 28 days. The transcript levels after long periods in the cold remained about double those observed in warm-acclimated carp liver. Thus, although cooling led to a dramatic increase in transcript levels, the effect was transient. Evidently the change in lipid saturation requires only a transient increase in expression and low levels of desaturase activity are entirely sufficient to maintain the *status quo* on long-term exposure to cold. The need to bring about greater changes in the composition of the acyl-CoA pool during cooling requires a much greater level of expression than at any constant temperature. The increase in transcript levels could result from either activation of transcription or changes in transcript turnover due to a cold-induced reduction in mRNA degradation relative to rates of transcript synthesis. Nuclear run-on experiments showed that transcript was produced by hepatic nuclei extracted from 5-day cold-acclimated carp but not by nuclei from warm-acclimated carp, indicating an upregulation of desaturase transcription. Cold incubation of nuclei from warm-acclimated fish did not induce transcript synthesis nor did warm incubation suppress transcript synthesis in nuclei from cold-acclimated fish. Thus, it appears that chronic and not acute variations in temperature lead to the desaturase transcript induction.

Recently, we have found that a polyclonal antibody raised against purified rat SCD1 by Strittmatter's group (Thiede & Strittmatter, 1985) reacts against a carp liver microsomal protein with a molecular mass of approximately 33 kDa, which corresponds closely to the predicted molecular weight based on the coding sequence (Macartney *et al.*, 1996; Tiku *et al.*, 1996). In early experiments with carp hepatocytes, the increase in specific activity of the Δ^9-desaturase during isothermal culture of carp hepatocytes was matched by an increase in immunodetectable desaturase protein, which suggests that the induction of specific activity was linked to changes in turnover of the protein, rather than to some late post-translational processing of protein already present in the membrane. In more recent experiments using more refined membrane preparations from *in vivo* thermally acclimated carp we have found that the liver of warm-acclimated carp possesses measurable levels of immunodetectable desaturase which increases by an as yet unquantified level during the first 5 days of cooling (Tiku *et al.*, 1996). Thus, while the liver of warm-acclimated carp displayed little or no desaturase activity they did possess considerable quantities of desaturase protein. It therefore appears that these animals possess a latent enzyme and that cooling appears to cause enzymatic activation, possibly by a post-translational mechanism. There seems little doubt that the antibody binds specifically to carp desaturase because the detected proteins from both carp and rat extracts possessed precisely the molecular weights expected from the nucleotide sequence of the respective cDNAs.

We conclude that cold induction of the Δ^9-desaturase in carp liver is the result of two events (Tiku *et al.*, 1996). First, a pre-existing desaturase is activated during the early stages of cold transfer, possibly by a post-translational modification. Second, the levels of desaturase transcript are increased after 48 h possibly due to induced transcription. This presumably leads to an enhanced rate of desaturase synthesis, as evidenced by the increase in immunodetectable protein after 5 days of cooling. Both events correlate in time course and in direction with the change in lipid saturation and membrane lipid order.

Future developments in membrane adaptation

The production of cloned genes and antibodies against the Δ^9-desaturase allows the ultimate mechanisms and adaptive significance of homeoviscous adaptation to be addressed in a more direct manner than previously possible. First, regarding the mechanism of the cold-induced increase in desaturase transcript levels, we have shown that induced desaturase transcription is involved. Transcription is controlled by a series of DNA-binding proteins which by interacting with specific regulatory DNA sequences, influence the transcriptional activity of RNA polymerase II. These sites interact with other enhancer sequences which may lie at a site some distance upstream or downstream from the coding sequence. The number of DNA-binding proteins may be large and their effects and interactions with enhancers may be complex, as in the induction of heat shock proteins (Morimoto *et al.*, 1994). Nevertheless, progress in understanding how cold exposure modulates the activity of DNA-binding proteins requires the identification and characterisation of the regulatory sequences and proteins. We have termed the putative promoter–enhancer complex of carp liver the 'cold-inducible promoter' system or CIP and we are actively seeking to identify its position in relation to the coding sequence for the Δ^9-desaturase gene and its relationship with other regulatory elements, such as the 'fat-specific element' (Mihara, 1990) which controls lipid biosynthetic enzymes in other species.

A second and particularly intriguing question is what is the mechanism that underlies transcriptional upregulation? In other words, what is the regulated variable and what is the temperature sensor in what may be a long and complex feedback pathway? Murata and colleagues have recently shown that catalytic hydrogenation of plasma membrane lipids of cyanobacteria led to an increase in Δ^{12}-desaturase transcript levels (Horvath *et al.*, 1991; Vigh *et al.*, 1993). This implies that, at least in cyanobacteria, the transcriptional activation of the Δ^{12}-desaturase is linked to changes in the fatty acid saturation and/or physical properties of the plasma membrane and that the ultimate sensor is some aspect of membrane physical condition. For events at the plasma membrane to influence transcriptional activity within the nucleus is not without precedence, but for a physical condition of the membrane to be transduced into an appropriate signal is, to our knowledge, a novel expectation. One possibility is that a phosphorylation cascade is involved, one of whose component enzymes binds to the plasma membrane or endoplasmic reticulum membrane in a manner that is sensitive to membrane physical structure.

Third, the identification of enzymes whose induced activity is involved in homeoviscous adaptation and the cloning of the corresponding genes provides more powerful approaches to establishing a causal role for those genes in HA. Expression of the enzyme may become controllable by any one of several methods, including transfection or transgenic manipulation using the cDNA under the control of a constitutive or an inducible promoter. Alternatively, abolition of cold-induced desaturase induction either by gene-targeted mutation or by the introduction of antisense constructs should lead to reduced or totally absent HA. Thus, desaturase expression can be controlled both during cold transitions or even at normal temperatures and the effects on membrane fluidity and lipid composition can be examined. These techniques also allow questions relating to the link between homeoviscous adaptation and acquired cold tolerance of the whole animal or its tissues to be addressed unambiguously (Logue *et al.*, 1995). Functional differences between transgenic individuals lacking cold-inducible desaturase expression and wild-type individuals can be explored and the existence of acquired cold tolerance in genetically-manipulated individuals can establish a causal link. The use of chimaeric or mosaic individuals allows investigation of the temperature tolerance of specific tissues for which the untransformed controls are provided within the same individual.

There is a good precedence for these approaches in more tractable experimental systems. In the cyanobacterium, *Synechocystis* strain PCC6803, the unsaturation of fatty acids in membrane lipids can be modified by genetic manipulation of the various desaturase genes (Sakamoto *et al.*, 1994; Wada *et al.*, 1990) and this influences growth and chilling tolerance (Gombos *et al.*, 1991, 1992, 1994). Elimination of desaturation at the Δ^{12} position of fatty acids has a deleterious effect on the growth of this microorganism at 22 °C (compare normal growth temperature of 34 °C) and transformation of *Synechocystis* sp. PCC7942 with the *desA* gene, produced chill-tolerant cells, clearly demonstrating the role of the gene in acquired chill tolerance (Wada *et al.*, 1990). Genetic manipulation of higher plants has also elegantly demonstrated a similar critical role for lipid biosynthetic enzymes in cold tolerance (Ishizaki *et al.*, 1988; Murata *et al.*, 1992; Nishida *et al.*, 1993). Here the introduction of glycerol-3-phosphate acyltransferase cDNA from squash (which is chill-sensitive) or from *Arabidopsis* (which is chill-resistant) into tobacco (which has intermediate chill-sensitivity) resulted in altered chilling sensitivity and resistance, respectively (Murata *et al.*, 1992). In the former case, the level of unsaturated fatty acids in phosphoglycerolipids fell significantly and chilling sensitivity was markedly increased, whilst in the latter, there was an increase in both the unsaturation of fatty acids in phosphoglycerolipids and chilling tolerance.

Conclusions

Membranes play a key role in the temperature adaptation of animals, both over the seasonal (phenotypic) and evolutionary (genotypic) time-scales and there is good reason to believe that these responses underlie both resistance and capacity adaptations of cells, tissues and whole organisms. The cold-induced increase in unsaturation is certainly the most conserved of thermal responses, yet progress in identifying the underlying mechanisms has been slow, at least in poikilothermic animals. Part of the problem in identifying the enzymatic basis of homeoviscous response has been the technical difficulty of measuring enzymatic activities and the need to determine variations in expression over the full time-course of the response. The availability of newer and more sensitive immunological and molecular genetic methods for exploring the expression of these enzymes allows a more convenient and direct assessment of their role in membrane adaptation. These techniques also allows new approaches to establishing the causal role of these differentially expressed enzymes in the process of membrane adaptation and in due course to the manipulation of whole organism thermotolerance. This knowledge would not come from continued comparisons of lipid composition or of physical structure of membranes isolated from animals subject to different thermal environments.

References

Barton, P.G. & Gunstone, F.D. (1975). Hydrocarbon chain packing and molecular motion in phospholipid bilayers formed from unsaturated lecithins. *Journal of Biological Chemistry* **250**, 4470–6.

Behan-Martin, M., Bolwer, K., Jones, G. & Cossins, A.R. (1993). A near perfect temperature adaptation of bilayer order in vertebrate brain membranes. *Biochimica et Biophysica Acta* **1151**, 216–22.

Coolbear, K.P., Berde, C.P. & Keogh, K.M.W. (1983). Gel to liquid–crystalline phase transitions of aqueous dispersions of polyunsaturated mixed acid phosphatidylcholines. *Biochemistry* **22**, 1466–73.

Cossins, A.R. (1977). Adaptation of biological membranes to temperature: the effect of temperature acclimation of goldfish upon the viscosity of synaptosomal membranes. *Biochimica et Biophysica Acta* **470**, 395–411.

Cossins, A.R. (1994). Homeoviscous adaptation and its functional significance. In *Temperature Adaptation of Biological Membranes* (ed. by A.R. Cossins), pp. 63–75. London: Portland Press.

Cossins, A.R. & Prosser, C.L. (1977). Evolutionary adaptations of membranes to temperature. *Proceedings of the National Academy of Sciences of the USA* **75**, 2040–3.

Fujii, D.J. & Fulco, A.J. (1977). Biosynthesis of unsaturated fatty acids by bacilli: hyperinduction of desaturase synthesis. *Journal of Biological Chemistry* **252**, 3660–70.

Gennis, R.B. (1989). *Biomembranes. Molecular Structure and Function*. New York: Springer-Verlag.

Gombos, Z., Wada, H. & Murata, N. (1991). Direct evaluation of effects of fatty acid unsaturation on the thermal properties of photosynthetic activities, as studied by mutation and transformation of *Synechocystis* PCC6803. *Plant Cell Physiology* **32**, 205–11.

Gombos, Z., Wada, H. & Murata, N. (1992). Unsaturation of fatty acids in membrane lipids enhances tolerance of the cyanobacterium *Synechocystis* PCC6803 to low temperature photoinhibition. *Proceedings of the National Academy of Sciences of the USA* **89**, 9959–63.

Gombos, Z., Wada, H. & Murata, N. (1994). The recovery of photosynthesis from low temperature photoinhibition is accelerated by the unsaturation of membrane lipids: a mechanism of chilling tolerance. *Proceedings of the National Academy of Sciences of the USA* **91**, 8787–91.

Grayburn, W.S., Collins, G.B. & Hildebrand, D.F. (1992). Fatty acid alteration by a Delta 9 desaturase in transgenic tobacco tissue. *Biotechnology* **10**, 675–8.

Hazel, J.R. (1989) Cold adaptation in ectotherms: regulation of membrane function and cellular metabolism. In *Advances in Comparative and Environmental Physiology* (ed. L.C.H. Wang), vol. 4, pp. 1–30. Berlin. Springer Verlag.

Hazel, J.R. (1995). Thermal adaptation in biological membranes: is homeoviscous adaptation the explanation? *Annual Reviews of Physiology* **57**, 19–42.

Horvath, I., Torok, Z., Vigh, L. & Kates, M. (1991). Lipid hydrogenation induces elevated 18 : 1-CoA desaturase activity in *Candida lipolytica* microsomes. *Biochimica et Biophysica Acta* **1085**, 126–30.

Ishizaki, O., Nishida, I., Agata, K., Eguchi, G. & Murata, N. (1988). Cloning and nucleotide sequence of cDNA for the plastid glycerol-3-phosphate acyltransferase from squash. *Federation of European Biochemical Societies Letters* **238**, 424–30.

Johnston, P.V. & Roots, B.I. (1961). Brain lipid fatty acids and temperature acclimation. *Comparative Biochemistry and Physiology* **11**, 303–10.

Kaestner, K.H., Ntambi, J.M., Kelly, T.J. & Lane, M.D. (1989). Differentiation-induced gene expression in 3T3-L1 preadipocytes. A second differentially expressed gene encoding stearoyl-CoA desaturase. *Journal of Biological Chemistry* **264**, 14 755–61.

Logue, J., Tiku, P. & Cossins, A.R. (1995). Cold injury and resistance adaptation in fish. *Journal of Thermal Biology* **20**, 191–7.

Macartney, A.I., Tiku, P.E. & Cossins, A.R. (1996). An isothermal induction of Δ^9-desaturase in cultured carp hepatocytes. *Biochimica et Biophysica Acta* (in press).

McElhaney, R.N. (1974). The effect of alterations in the physical state of the membrane lipids on the ability of *Acholeplasma laidlawii* to grow at various temperatures. *Journal of Molecular Biology* **84**, 145–57.

McElhaney, R.N. (1994). Techniques for measuring lipid phase state and fluidity in biological membranes. In *Temperature Adaptation of Biological Membranes* (ed. by A.R. Cossins), pp. 31–62. London: Portland Press.

Mihara, K. (1990). Structure and regulation of rat liver microsomal stearoyl-CoA desaturase gene. *Journal of Biochemistry, Tokyo* **108**, 1022–9.

Morimoto, R., Tissieres, A. & Georgopoulos, C. (1994). *The Biology of Heat Shock Proteins and Molecular Chaperones*, pp. 1739–40. New York: Cold Spring Harbor Press.

Murata, N., Ishizaki-Nishizawa, O., Higashi, S., Hayashi, H., Tasaki, Y. & Nichida, I. (1992). Genetically engineered alteration in the chilling sensitivity of plants. *Nature* **356**, 710–13.

Nishida, I., Tasaka, Y., Shiraishi, H. & Murata, N. (1993). The gene and the RNA for the precursor to the plastid-located glycerol-3-phosphate acyltransferase of *Arabidopsis thaliana*. *Plant Molecular Biology* **21**, 267–77.

Nozawa, Y., Iida, H., Fukushima, H., Ohki, K. & Ohnishi, S. (1974). Studies on *Tetrahymena* membranes. Temperature-induced alterations in fatty acid composition of various membrane fractions in *Tetrahymena pyriformis* and its effect on membrane fluidity as inferred by spin-label study. *Biochemica et Biophysica Acta* **367**, 134–47.

Nozawa, Y. & Kasai, R. (1978). Mechanism of thermal adaptation of membrane lipids in *Tetrahymena pyriformis* NT-1. Possible evidence for temperature-mediated induction of palmitoyl-CoA desaturase. *Biochimica et Biophysica Acta* **529**, 54–66.

Ntambi, J.M., Bhrow, S.A., Kaestner, K.I., Christy, R.I., Sibley, E., Kelly, T.J. & Lane, M.D. (1988). Differentiation-induced gene expression in 3T3-L1 preadipocytes. Characterization of a differentially expressed gene encoding stearoyl-CoA desaturase. *Journal of Biological Chemistry* **263**, 17 291–300.

Sakamoto, T., Los, D.A., Higashi, S., Wada, H., Nishida, I., Ohmori, M.Z. & Murata, N. (1994). Cloning of omega 3 desaturase from cyanobacteria and its use in altering the degree of membrane-lipid unsaturation. *Plant Molecular Biology* **26**, 249–63.

Sato, N. & Murata, N. (1981). Studies on the temperature shift-induced desaturation of fatty acids in monogalactosyl diacylglycerol in blue green alga (Cyanobacterium) *Anabaena variabililis. Plant Cell Physiology* **22**, 1043–50.

Schünke, M. & Wodtke, E. (1983). Cold-induced increase of Δ^9 and Δ^6-desaturase activities in endoplasmic reticulum of carp liver. *Biochim. Biophys. Acta* **734**, 70–5.

Sinensky, M. (1974). Homeoviscous adaptation: a homeostatic process that regulates the viscosity of membrane lipids in *E. coli. Proceedings of the National Academy of Sciences of the USA* **71**, 522–6.

Stubbs, C.D., Kouyama, T., Kinosita, K. & Ikegami, A. (1981). Effects of double bonds on the dynamic properties of the hydrocarbon region of lecithin bilayers. *Biochemistry* **20**, 2800–10.

Stukey, J. E., McDonough, V. M. & Martin, C. E. (1989). Isolation and characterization of OLE1, a gene affecting fatty acid desaturation from *Saccharomyces cerevisiae. Journal of Biological Chemistry* **264**, 16437–44.

Stukey, J.E., McDonough, V.M. & Martin, C.E. (1990). The *OLE1* gene of *Saccharomyces cerevisiae* encodes the delta 9 fatty acid desaturase and can be functionally replaced by the rat stearoyl-CoA desaturase gene. *Journal of Biological Chemistry* **265**, 20144–9.

Thiede, M.A. & Strittmatter, P. (1985). The induction and characterisation of rat liver stearyl-CoA desaturase mRNA. *Journal of Biological Chemistry* **260**, 14 459–63.

Tiku, P.E., Gracey, A.Y., Macartney, A.I., Beyon, R.B. & Cossins, A.R. (1996). Cold-induced expression of Δ^9-desaturase in carp by transcriptional and post-translation mechanisms. *Science* **271**, 815–18.

Van der Meer, W. (1984). Physical aspects of membrane fluidity. In *Physiology of Membrane Fluidity* (ed. Shinitsky, M.), vol. 1, pp. 53–72. Boca Raton: CRC Press.

Vigh, L., Los, D.A., Horvath, I. & Murata, N. (1993). The primary signal in the biological perception of temperature: Pd-catalysed hydrogenation of membrane lipids stimulated the expression of the *desA* in *Synechocystis* PCC6803. *Proceedings of the National Academy of Sciences of the USA* **90**, 9090–4.

Wada, H., Gombos, Z. & Murata, N. (1990). Enhancement of chilling tolerance of a cyanobacterium by genetic manipulation of fatty acid desaturation. *Nature* **347**, 200–3.

Williams, W.P. (1990). Cold-induced lipid phase transitions. *Philosophical Transactions of the Society of London Biology* **326**, 555–67.

Wodtke, E. (1986). Adaptation of biological membranes to temperature: modifications and their mechanisms in the eurythymic carp. *Biona Reports (Gustav Fischer, Stuttgart)* **4**, 129–38.

Wodtke, E. & Cossins, A.R. (1991). Rapid cold-induced changes of membrane order and delta 9-desaturase activity in endoplasmic reticulum of carp liver: a time-course study of thermal acclimation. *Biochimica et Biophysica Acta* **1064,** 343–50.

G. DI PRISCO and B. GIARDINA

Temperature adaptation: molecular aspects

Introduction

More than any other habitat on earth, Antarctica is a unique natural laboratory, ideal for studying temperature adaptations. Hence special attention will be given to its paleogeography and to the adaptive mechanisms of Antarctic marine organisms. For example some adaptations (freezing avoidance, efficient enzymatic catalysis and cytoskeletal polymer assembly, decreased blood viscosity through reduction or elimination of erythrocytes and haemoglobin) represent a unique character of Antarctic fish and will be examined in detail. Specialisations in haematology and in the oxygen transport system were also developed by other polar and temperate organisms: Arctic mammals (reindeer, musk ox, whale), birds (penguin), reptiles (turtle), crustaceans (krill), cephalopods (squid). We describe the molecular mechanisms of the oxygen transport system in relation to requirements for function at low temperature.

It is pertinent to mention the difficulty in establishing consensus on objective criteria to identify a phenotypic trait as an adaptation. Thus, adaptation remains 'a slippery concept'. The reader will find extensive discussion on this and other issues in two recent reviews (Reeve & Sherman, 1993; Garland & Carter, 1994).

The Antarctic

In the late Precambrian, 590 million years ago (Ma), Antarctica was the central part of the supercontinent Gondwana, which remained intact for 400 million years, during the Paleozoic and part of the Mesozoic, through the Jurassic; fragmentation began and continued during the Cretaceous. The continental drift took Antarctica to its present position about 65 Ma, at the beginning of the Cenozoic. Final separation of East Antarctica from Australia occurred 38 Ma, in the Eocene–Oligocene transition; that of West Antarctica from South

America occurred 22–25 Ma, in the Oligocene–Miocene transition. The opening of the Drake passage produced the Circum-Antarctic Current and the development of the Antarctic Convergence (or Antarctic Polar Front): a well-defined, roughly circular oceanic frontal system, now running between 50 °S and 60 °S, where the surface layers of the north-flowing Antarctic waters sink beneath the less cold, less dense sub-Antarctic waters. With the reduction of heat exchange from northern latitudes, cooling of the environment proceeded to the present extreme conditions. The Antarctic ocean became gradually colder and seasonally ice-covered; to date it is surrounded during winter by a continuous belt of sea ice 1500 km wide, which recedes during the summer, when the migration and reproduction of birds takes place. Although sea ice may have already been present at the end of the Eocene (40 Ma), extensive ice sheets have probably formed only after the middle Miocene (14 Ma) every 1–3 million years. The latest ice sheet expansion, with progressive cooling leading to the current climatic conditions, began 2.5 Ma in the Pliocene. Antarctica is now a continent, almost fully coated with an ice sheet with average thickness of about 2000 m. It had enjoyed a much warmer climate than that of the dry, bitterly cold desert of the current times. The fossil records reveal that such variations produced many diversified forms of terrestrial and aquatic life (Eastman, 1991). Detailed information and pertinent references can be found in the comprehensive book by Eastman (1993).

Antarctic fish fauna

Although in this environment fish from temperate water would rapidly freeze, the oxygen-rich Antarctic waters support a wealth of marine life.

The origin of the first teleosts can be traced in the Jurassic, approximately 200 Ma. These fish continued to evolve through the Cretaceous and, until the late Eocene, were quite cosmopolitan. In contrast, the modern Antarctic fish fauna is largely endemic and, unlike the populations of the other continental shelves, is dominated by a single group: the suborder Notothenioidei with 120 species in total. Ninety-five of the 174 species living on the shelf or upper slope of the Antarctic continent are notothenioids. Fifty-three of the remaining 79 species belong to families (Liparididae and Zoarcidae) originating in the northern Pacific Ocean; other fish typical of the boreal hemisphere are virtually absent (Gon & Heemstra, 1990).

No fossil record of Notothenioidei is available, leaving a void of 38 million years from the Eocene to the present. There is a lack of information on their site of origin, on the existence of a transition fauna and on the time of their radiation in the Antarctic. Indirect indications, however, suggest that notothenioids appeared in the early Tertiary, filling the ecological void on the shelf left by most of the other fish fauna (which experienced local extinction during maximal glaciation), and began to diversity in the middle Tertiary. Notothenioids fill a varied range of ecological niches normally occupied by taxonomically diverse fish communities in temperate waters. The suborder comprises six families (Table 1). Only one of the 11 species of Bovichtidae, the most primitive family, lives south of the Antarctic Polar Front; 15 out of 49 Nototheniidae species are non-Antarctic. Harpagiferidae, Artedidraconidae, Bathydraconidae and Channichthyidae are all Antarctic (Gon & Heemstra, 1990; Eastman, 1993). Notothenioids are red-blooded, with the exception of Channichthyidae, the only known vertebrates (Ruud, 1954) whose pale whitish blood is devoid of haemoglobin (Hb).

The evolutionary adaptation of Antarctic fish includes physiological and biochemical specialisations, some of which characterise these organisms as unique. The ensemble of these specialisations was developed in the last 20–30 million years, during increasing isolation in the cooling seas. By virtue of evolutionary responses to the many environmental constraints, the Antarctic ocean is now the ideal habitat for the fish fauna. Fish are finely adjusted to the environment and intolerant of warmer temperatures, so that an increase in temperature of only a few centigrades has lethal effects (Somero & DeVries, 1967). They live in isolation south of the Polar Front, a natural barrier to migration in both directions, and thus a key factor for fish evolution. The physiological/biochemical adaptations aimed at over-

Table 1. *The families of the suborder Notothenioidei (Gon & Heemstra, 1990; Eastman, 1993)*

Family	Antarctic species	Non-Antarctic species	Total
Bovichtidae	1	10	11
Nototheniidae	34	15	49
Harpagiferidae	6	0	6
Artedidraconidae	24	0	24
Bathydraconidae	15	0	15
Channichthyidae	15	0	15
Total	95	25	120

coming the temperature effects include freezing avoidance, aglomerular kidneys, preferred metabolic pathways (lipid versus carbohydrate), specialisations in cytoskeletal protein polymerisation, in enzymatic, haematological, muscular and nervous systems, and in membrane structure. Adaptive responses have occurred at all levels of organisation (organismal, organic, cellular, molecular). Although the examples described here involve protein structure and function, a key role is fulfilled by all macromolecules (Clarke, 1983; Hochachka & Somero, 1984; Macdonald *et al.*, 1987; di Prisco *et al.*, 1988*a*, 1991*a*; Gon & Heemstra, 1990; di Prisco, 1991; Eastman, 1993).

Some physiological specialisations in Antarctic fish

Freezing avoidance

The average year-round water temperature in the coastal part of the Antarctic Ocean is −1.87 °C, the equilibrium temperature of ice and seawater, well below the freezing temperature (−0.8 °C) of a typical marine teleost hyposmotic to sea water. A few species avoid freezing by supercooling. This is a dangerous strategy. The metastable supercooled state does not allow contact with ice, and even partial freezing invariably causes death.

In Notothenioidei, freezing is avoided by lowering the freezing point of blood and other tissue fluids. About one-half of the freezing point depression is caused mostly by NaCl; the other half is provided by solutes in the colloidal fraction of the fluid, which exert their effect non-colligatively. In most Antarctic fish species these solutes are glycopeptides with molecular mass ranging from 2600 to 33 700 Da, which contain a repeating unit of three amino acid residues in the sequence $[\text{Ala-Ala-Thr}]_n$; a disaccharide is linked with each Thr. In the lighter glycopeptides, Pro periodically substitutes Ala at position 1 of the tripeptide. Antifreeze glycopeptides are synthesised year-round in the liver, secreted into the circulatory system and then distributed into the extracellular fluids, where their concentration approaches 3.5%.

In the Arctic, some fish species have antifreeze peptides and others have glycopeptides similar to the Antarctic ones. Biosynthesis mostly occurs only during winter, as isolation is not so strict as in the Antarctic.

Most of our knowledge of the antifreeze molecules comes from the work of DeVries' group (Raymond & DeVries, 1977; DeVries, 1980, 1988; Eastman & DeVries, 1986; Ahlgren *et al.*, 1988; Hsiao *et al.*, 1990; Cheng & DeVries, 1991; see also Eastman, 1993).

Cytoskeletal polymers: tubulins and actins

The polymerisation of these proteins at low temperatures is regulated by at least two adaptive strategies (Detrich, 1991*b*). (i) Tubulins and microtubule-associated proteins (MAPs) assemble and form microtubules, a major component of the cytoskeleton of eucaryotic cells, participating in many processes (mitosis, nerve growth, intracellular transport of organelles). The *in vitro* assembly of microtubules is temperature-sensitive. In temperate fish, mammals and birds, tubulins associate near 37 °C, but disassemble at temperatures as low as 4 °C. In Antarctic fish, however, adaptive changes in the tubulin molecular structure have made these proteins able to stay polymerised at temperatures as low as −2 °C. Thermodynamic analysis of polymerisation showed large positive enthalpy and entropy changes and indicated that the reaction is entropically driven (Detrich, 1991*a*, *b*) and that, in organisms with widely different evolution histories, the entropic control increases with decreasing body temperature. Interspecific differences in polymerisation thermodynamics have an important adaptive consequence: the critical concentration for microtubule assembly is conserved within a narrow range of tubulin concentrations, and each organism efficiently assembles microtubules within the limits of its body temperature. Polymerisation of Antarctic fish tubulins relies on entropy-generating interactions: more specifically, on an increase in hydrophobic interactions (rather than in exothermic electrostatic bonds) between the molecular domains involved in the process (Detrich & Overton, 1986, 1988; Detrich *et al.*, 1987). (ii) In skeletal muscle actins (Sweezey & Somero, 1982; Hochachka & Somero, 1984; Somero, 1991) subunit self-assembly is also temperature-dependent. Unlike tubulins, in 14 vertebrate species the standard entropy and enthalpy of the globular to filamentous transformation of actins increase as average body temperature increases; stabilisation of actin filaments in cold-adapted fish is suggested to depend on exothermic polar bonds, rather than on endothermic hydrophobic interactions. The low enthalpy change is interpreted as adaptive because a lower heat input is needed to drive polymerisation.

Thus, in the regulation of cytoskeletal polymer assembly at low temperatures, (i) modification of the bond types at subunit contacts may give an entropy of association sufficient to overcome unfavourable enthalpy changes; or (ii) a preponderant role of bonds making negative contributions to the overall enthalpy of polymerisation may limit the destabilising enthalpy changes (Detrich, 1991*b*).

Enzyme catalysis

Thermodynamic analysis of enzyme stability and activity shows a variety of patterns. Some Antarctic fish enzymes are more labile than the corresponding mammalian ones (Clarke, 1983; Macdonald *et al.*, 1987). In others, the negligible differences in heat denaturation (Genicot *et al.*, 1988; Ciardiello *et al.*, 1995, 1996) suggest high molecular structure conservation during evolution. At least two types of adaptation can account for high catalytic rates at low temperatures (Hochachka & Somero, 1984; Somero, 1991): (i) a higher intracellular enzyme concentration (an increased number of catalytic sites compensates for the temperature-induced lower rate per site); (ii) a higher inherent catalytic activity per active site (namely fewer molecules of a more efficient enzyme). In the majority of cases, both mechanisms are likely to be in function at the same time. The temperature dependence of K_m (a measure of the substrate or coenzyme affinity), k_{cat} (the catalytic efficiency), k_{cat}/K_m (the physiological efficiency), often offers meaningful indications. Thermodynamic analysis (Ciardiello *et al.*, 1995, 1996) does suggest that glucose-6-phosphate dehydrogenase (G6PD) from the blood of two notothenioids is a 'better enzyme' than G6PD from temperate fish; however, the highly increased amount of G6PD in the few cells of the Hb-less blood of the channichthyid *Chionodraco hamatus* indicates temperature compensation also via the synthesis of a larger number of G6PD molecules.

Somero (1991) has discussed some general criteria for defining thermal optima for biochemical functions. An optimal situation *in vitro* needs not be optimal in a physiological sense; thus it would be incorrect to identify enzyme thermal optima with the temperatures at which reaction rate, or substrate affinity is highest. K_m values under physiological conditions of temperature, hydrostatic pressure, osmotic concentration and pH, which all strongly interfere with enzyme-ligand interactions, are strongly conserved among species. This conservation was interpreted by Hochachka & Somero (1984) as a mechanism adopted by enzymes to retain optimal maximal responsiveness to changes in substrate concentration and to allosteric effectors in response to variations of the above-mentioned factors. A physiologically appropriate definition of thermal optima is the temperatures at which values for enzymatic properties are held within the range strongly conserved among the species (Somero, 1991). Conservation of K_m in the range of physiological temperatures (generally accompanied by a sharp increase above this range) in many enzymes of cold-adapted and temperate species supports this view (Somero, 1991). Acetylcholinesterase (which is critical for synaptic transmission) shows a very sharp increase in K_m at 3–4 °C in warm-acclimated *Pagothenia borchgrevinki* (Baldwin & Hochachka, 1970; Baldwin, 1971). Also, the release of acetylcholine sharply increases with temperature (Macdonald & Montgomery, 1982); the enzyme may become unable to bind its substrate, and accumulation of large amounts of acetylcholine at synapses can disrupt synaptic transmission. This control breakdown could partially account for the heat death of notothenioids at 5–8 °C (Somero & DeVries, 1967; Somero, 1991).

Blood and the oxygen transport system

Oxygen carriers are one of the most interesting systems for studying the interrelationships between environmental conditions and molecular evolution. Considering the variety of species that have Hb, it is easy to imagine that this molecule, being a direct link between the exterior and body requirements and exploiting its primary function under extremely variable conditions, may have experienced a major evolutionary pressure to adapt and modify its functional features.

Before describing the details of temperature adaptations, it is worthwhile to briefly summarise the molecular aspects of the function modulation of respiratory proteins. To ensure an adequate supply of oxygen to the entire organism, oxygen-carrying proteins have developed a common molecular mechanism based on ligand-linked conformational change in a multi-subunit structure. They generally exhibit a marked degree of cooperativity between oxygen-binding sites (homotropic interactions) that enables maximum oxygen unloading at relatively high oxygen tension. In the simplest model, cooperativity in oxygen binding is achieved through conformational transition between a low-affinity T state and a high-affinity R state, which accounts for the sigmoidal shape of the ligand-binding curve. These two extreme conformational states are also involved in the modulation of the oxygen affinity brought about by several effectors (heterotropic interactions). Under physiological conditions a given effector may preferentially bind to the low-affinity or to the high-affinity state of the protein, thereby lowering or enhancing the overall oxygen affinity of the molecule.

Within the framework of this common mechanism, however, respiratory proteins have acquired special features to meet special needs. Thus, they not only illustrate the variability possible within the same overall mechanism, but also represent a type-case of molecular adaptation to different physiological requirements. For instance, oxygen binding is generally exothermic, so that a decrease in temperature induces an increase of oxygen affinity. The role of temperature and its interplay with heterotropic ligands in modifying the oxygenation–deoxygenation cycle in the respiring tissues is, therefore, of the utmost importance. In gaining deeper insight into the physiology of a specific organism, thermodynamic analysis may thus indeed become the tool of choice. At the molecular level, heat absorption and release can be considered a physiologically relevant modulating factor, similar to hetero- and homotropic ligands.

Antarctic fish

Antarctic fish differ from temperate and tropical species in having a reduced erythrocyte number and Hb concentration in the blood. The subzero seawater temperature would greatly increase the blood viscosity, but the potentially negative physiological effects caused by this increase are counterbalanced by reducing or eliminating erythrocytes and Hb (Wells *et al.* 1990). This adaptation reduces the energy needed for circulation. Channichthyidae, the phylogenetically most highly developed notothenioids, have attained the extreme of this trend. Their colourless blood lacks Hb (Ruud, 1954). It has a very small number of erythrocyte-like cells (Hureau *et al.*, 1977), containing enzymes with key metabolic functions (Ciardiello *et al.*, 1995, 1996) that may explain their physiological significance. Hb has not been replaced by another oxygen carrier and the oxygen-carrying capacity of blood is only 10% of that of red-blooded fish; however, these fish are not at all disadvantaged by the lack of Hb. The physiological adaptations enabling channichthyids to prosper without Hb include: low metabolic rate; large, well-perfused gills; large blood volume, heart, stroke volume and capillary diameter; and cutaneous respiration. In addition to reducing the metabolic demand for oxygen, low temperatures increase its solubility in the plasma, so that more oxygen can be carried in physical solution.

The co-existence of Hb-less and red-blooded species suggests that the need for an oxygen-carrier in a stable, cold habitat is reduced in both groups. The functional incapacitation of Hb and induced reduction of the haematocrit to 1–2% in *Pagothenia bernacchii* (Di Prisco *et al.*, 1992; Wells *et al.*, 1990) caused no discernible harm to the fish, at least in the absence of metabolic challenges. Like channichthyids, red-blooded Antarctic fish can carry oxygen dissolved in plasma. Indeed, the most important adaptation in the haematological features of Antarctic fish appears to be the decrease (or elimination) of Hb and erythrocytes; however, molecular studies are necessary to characterise the Hb functional role in the red-blooded species, and to look for correlations with lifestyle.

What happened to the globin genes in channichthyids? Globin DNA sequences from several Hb-less species have indicated retention of α-globin-related DNA sequences in their genomes, and apparent loss (or rapid mutation) of β-globin genes (Cocca *et al.*, 1995). Deletion of the β-globin locus in the ancestral channichthyid may have been the primary event leading to the Hb-less phenotype; the α-globin gene(s), no longer under selective pressure for expression, would then have accumulated mutations leading to loss of function, without (as yet) complete loss of sequence information.

Another common feature of endemic Antarctic fish is the markedly reduced Hb multiplicity. This may have a phylogenetic origin; however, multiplicity is also linked with the variability of the environment (Riggs, 1970) and the Antarctic waters are a stable habitat. Among notothenioids (Table 2), 34 sedentary bottom dwellers (*A. mitopteryx* and *D. mawsoni* are bentho-pelagic, but the former is very sluggish and the latter is very moderately active) have a single major Hb (Hb 1) and often have a second, minor component (Hb 2, about 5% of the total). Oxygen binding is generally strongly pH- and organophosphate-regulated (D'Avino & di Prisco, 1988, 1989; di Prisco *et al.*, 1988*b*, 1991*b*, 1994; D'Avino *et al.*, 1989, 1991; Kunzmann *et al.*, 1991; Caruso *et al.*, 1991, 1992; Camardella *et al.*, 1992). Another component (Hb C) is present at less than 1% in all species.

In the last decade, many papers have been published on the molecular structure and biological function of the Hbs of many Antarctic species (D'Avino *et al.*, 1991, 1992, 1994; see di Prisco *et al.*, 1991*b*). The effect of pH and endogenous organophosphates on oxygen equilibria and saturation (the Bohr and Root effects) have been investigated in 30 (plus an additional one) of the 34 notothenioids of Table 2 (Table 3). The Hb system has also been studied in non-Antarctic nototheniid and bovichtid species (Fago *et al.*, 1992; R. D'Avino *et al.*, unpublished data).

In comparison with the other notothenioids, higher Hb multiplicity is shown by three Nototheniidae. *Trematomus newnesi*, *Pleuragramma antarcticum* and *P. borchgrevinki* have three to five functionally distinct Hbs (Table 4). Their lifestyle is different from that of the other sluggish bottom dweller species.

T. newnesi actively swims and feeds near the surface (Eastman, 1988). It is the only species (D'Avino *et al.*, 1994) in which Hb C is not present in traces, but reaches 20–25%. Unlike the other noto-

Table 2. *Haemoglobins of Antarctic and non-Antarctic (*Notothenia angustata *and* Pseudaphritis urvillii*) Notothenioideia*

Family	Species	% of haemoglobin components
Bovichtidae	*P. urvillii*	Hb 1 (95), Hb 2 (5)
Nototheniidae	*N. coriiceps*	Hb 1 (95), Hb 2 (5)
	N. rossii	Hb 1 (95), Hb 2 (5)
	N. angustata	Hb 1 (95), Hb 2 (5)
	N. nudifrons	Hb 1 (95), Hb 2 (5)
	N. larseni	Hb 1 (95), Hb 2 (5)
	G. gibberifrons	Hb 1 (90), Hb 2 (10)
	P. hansoni	Hb 1 (95), Hb 2 (5)
	P. bernacchii	Hb 1 (98), (Hb 2?)
	D. mawsoni	Hb 1 (98), (Hb 2?)
	A. mitopteryx	one Hb (99)
	T. nicolai	Hb 1 (95), Hb 2 (5)
	T. pennellii	Hb 1 (95), Hb 2 (5)
	T. loennbergi	Hb 1 (95), Hb 2 (5)
	T. eulepidotus	Hb 1 (95), Hb 2 (5)
	T. lepidorhinus	Hb 1 (95), Hb 2 (5)
	T. scotti	Hb 1 (95), Hb 2 (5)
Bathydraconidae	*C. mawsoni*	Hb 1 (97), Hb 2 (5)
	R. glacialis	Hb 1 (90), Hb 2 (10)
	P. charcoti	one Hb (99)
	G. acuticeps	one Hb (99)
	B. marri	one Hb (99)
	B. macrolepis	one Hb (99)
	A. nudiceps	one Hb (99)
	G. australis	one Hb (99)
Artedidraconidae	*A. skottsbergi*	one Hb (99)
	A. orianae	one Hb (99)
	A. shackletoni	one Hb (99)
	H. velifer	one Hb (99)
	P. scotti	one Hb (99)
	Pogonophryne sp. 1	one Hb (99)
	Pogonophryne sp. 2	one Hb (99)
	Pogonophryne sp. 3	one Hb (99)
Harpagiferidae	*H. antarcticus*	one Hb (99)

[a]The blood of all species contains traces (less than 1%) of Hb C.
Hb, haemoglobin.

Table 3. *Regulation by pH and heterotropic physiological ligands of oxygen binding of Hbs of Antarctic and non-Antarctic (*N. angustata *and* P. urvillii*) Notothenioidei*

Family	Species	Bohr and Root effects; effect of organophosphates
Bovichtidae	*P. urvillii*	Strong in Hb 1, Hb 2
Nototheniidae	*N. coriiceps*	Strong in Hb 1, Hb 2
	N. rossii	Strong in Hb 1, Hb 2
	N. angustata	Strong in Hb 1, Hb 2
	G. gibberifrons	Strong in Hb 1, Hb 2
	P. hansoni	Strong in Hb 1, Hb 2
	P. bernacchii	Strong in Hb 1
	D. mawsoni	Strong in Hb 1
	A. mitopteryx	Root, absent; Bohr, weak
	T. nicolai	Strong in Hb 1, Hb 2
	T. pennellii	Strong in Hb 1, Hb 2
	T. loennbergi	Strong in Hb 1, Hb 2
	T. eulepidotus	Strong in Hb 1, Hb 2
	T. lepidorhinus	Strong in Hb 1, Hb 2
	T. scotti[a]	Strong in Hb 1, Hb 2
Bathydraconidae	*C. mawsoni*	Strong in Hb 1, Hb 2
	R. glacialis	Strong (in haemolysate)
	P. charcoti	Strong
	G. acuticeps	Absent
	B. marri	Strong
	B. macrolepis[a]	Strong, only with ATP
	A. nudiceps[a]	Strong
	G. australis[a]	Strong
Artedidraconidae	*A. orianae*	Weak (Root only with ATP)
	A. shackletoni[a]	Weak, only with ATP
	H. velifer	Weak
	P. scotti	Weak (Root only with ATP)
	D. longedorsalis[a]	Weak (in haemolysate)
	Pogonophryne sp. 1	Weak (Root only with ATP)
	Pogonophryne sp. 2	Weak (Root only with ATP)
	Pogonophryne sp. 3	Weak (Root only with ATP)

[a]The Bohr effect was not measured.

ATP, adenosine triphosphate; Hb, haemoglobin.

Table 4. *Antarctic Notothenioidei (family Nototheniidae) with higher haemoglobin multiplicity and functionally distinct components*

Species	% Hb components	Bohr effect	Root effect	Effect of organophosphates
T. newnesi				
(2 major Hbs)	Hb C (20)	Strong	Strong	Strong
	Hb 1 (75)	Weak	Absent	Absent
	Hb 2 (5)	Weak	Absent	Absent
P. antarcticum[a]				
(3 major Hbs)	Hb C (traces)	Strong	Strong	Strong
	Hb 1 (30)	Strong	Strong	Strong
	Hb 2 (20)	Strong	Strong	Strong
	Hb 3 (50)	Strong	Strong	Strong
P. borchgrevinki				
(1 major Hb)	Hb C (traces)	Strong	Strong	Strong
	Hb 0 (10)	Strong	Strong	Strong
	Hb 1 (70)	Weak	Weak	Absent
	Hb 2 (10)	Weak	Weak	Weak
	Hb 3 (10)	Weak	Weak	Weak

[a]The Hbs of *P. antarcticum* differ thermodynamically (they have different values of heat of oxygenation).

thenioids, the oxygen binding of Hb 1 and Hb 2 is not regulated by pH and organophosphates; on the other hand, Hb C displays effector-enhanced Bohr and Root effects. Thus *T. newnesi* has two functionally distinct major Hbs, possibly required by this more active fish to ensure oxygen binding at the gills and delivery to tissues in conditions of acidosis. To balance the lack of regulation in two Hbs by physiological ligands, evolution may have preserved the expression of high levels of Hb C, conceivably redundant in the other notothenioids.

P. antarcticum is the most abundant species in high-Antarctic shelf areas and has an important role in the pelagic system. Among notothenioids, it has the highest multiplicity of major components (three). These Hbs display effector-enhanced Bohr and Root effects, but differ thermodynamically in the heats of oxygenation (Tamburrini *et al.*, 1994*b*). Temperature-regulated oxygen affinity may reflect molecular adaptation to the pelagic lifestyle of this fish, allowing migration across water regions with significant temperature differences and fluctuations.

P. borchgrevinki is an active cryopelagic species. Of the five Hbs (M. Tamburrini & G. di Prisco, unpublished results), Hb 1 accounts for 70–80% of the total. Besides Hb C, present in traces, also Hb 0 displays strong, effector-enhanced Bohr and Root effects. Hb 1, Hb 2 and Hb 3 bind oxygen with weak pH-dependence and weak or no influence of organophosphates. In view of this high multiplicity of functionally distinct Hbs, the oxygen-transport system of *P. borchgrevinki* is the most specialised among notothenioids.

Arctic mammals

Specialisations have been developed by Hb from Arctic and sub-Arctic mammals (Giardina *et al.*, 1989, 1991*a*; Brix *et al.*, 1989*a*). Hbs from ruminants (reindeer, musk ox, cervus), under physiological conditions, are characterised by an overall oxygenation enthalpy (ΔH) that is much less exothermic than that of human Hb and other mammalian Hbs (Table 5). The physiological implication of this result may become apparent considering the very low habitat temperature (down to −40 °C) experienced by the animals during the year. In fact, as deoxygenation is an endothermic process, in the peripheral tissues, where the temperature may be as much as 10 °C lower than in the lungs and the deep core of the organism, oxygen delivery would be drastically impaired if the molecule were not characterised by a small ΔH, namely by a small temperature dependence of its oxygen binding.

That the protein molecule possesses peculiar features is clearly outlined by musk ox (*Ovibos muschatus*) Hb, whose ΔH is reported as a function of pH in Fig. 1. It should be recalled that in human Hb A, the more exothermic value is observed at very alkaline pH values,

Table 5. *Overall heat of oxygenation of some Arctic ruminant haemoglobins*[a]

Species	ΔH (kcal mol^{-1})
Reindeer	−3.4 ± 0.17
Musk ox	−3.8 ± 0.19
Cervus	−3.2 ± 0.16
Horse	−8.1 ± 0.40
Human	−8.8 ± 0.44

[a]In 0.1 M Hepes plus 0.1 M NaCl at pH 7.4. For comparison, ΔH of human and horse haemoglobins in the presence of 3 mM 2,3-DPG.
ΔH values, corrected for the heat contribution of oxygen in solution (−3.0 kcal mol^{-1}), were calculated from the van't Hoff equation by using the oxygen equilibrium data.

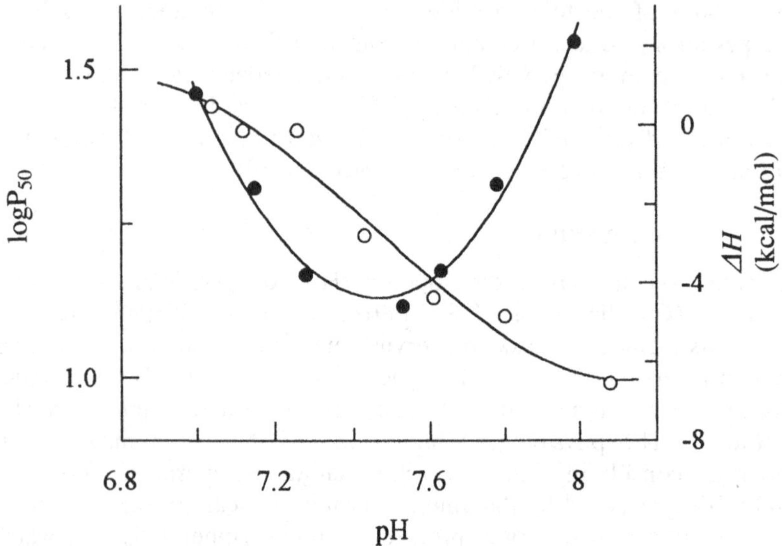

Fig. 1. Oxygen Bohr effect of musk ox Hb at 20 °C (open symbols) and apparent heat of oxygenation as a function of pH (closed symbols). Values of ΔH are corrected for the heat contribution of oxygen in solution (-3.0 kcal mol^{-1}). 0.1 M Bis-Tris or Tris buffer plus 0.1 M NaCl. (Modified from Brix *et al.*, 1989*a*.)

where the Bohr effect is over and the contribution of the Bohr protons (endothermic) is abolished. In musk ox Hb we are confronted by a completely different situation because the apparent heat of oxygenation is at its maximum value (in absolute terms) just within the physiological pH range and tends to zero or even positive values going towards both more acid and more alkaline pH regions. Hence, these positive and unusual ΔH values are obtained under conditions in which the Bohr effect is almost over (Fig.1) and the Bohr protons cannot give any thermal contribution. We are left either with an intrinsic property of the molecule or with the effect of some other ions whose presence *in vivo* could be important for the modulation of the Hb function.

Another special case is that of reindeer (*Rangifer tarandus tarandus*). The shape of the oxygen-binding curve is markedly temperature-dependent, a phenomenon linked to the unusual temperature independence of the upper asymptote which represents the high-affinity R state of the molecule (Giardina *et al.*, 1990). By contrast, the lower asymptote (the low-affinity T state) is strongly exothermic in nature, much like the effect observed in Hb A. This large difference in the thermodynamics of the two forms of reindeer Hb results in a particular dependence of the temperature effect on the degree of oxygen saturation (Y) of the protein; for values of Y higher than 0.6, which are within the range of oxygen saturation at which the protein works *in vivo*, the overall heat of oxygenation increases from -4.2 kcal mol^{-1} of oxygen (at Y = 0.6) to almost zero as Y tends to 1. On the whole, also in this case, the overall heat of oxygen binding is very small and may be useful for organisms that have to face temperatures down to -40 °C.

A very interesting example of the interplay of the effects of organophosphates, carbon dioxide and temperature is shown by Hb from the whale *Balaenoptera acutorostrata* (Brix *et al.*, 1990). In fact, although this Hb has a higher intrinsic temperature sensitivity (ΔH = -14.2 kcal mol^{-1} of oxygen) when the physiological cofactors are added to the system, the overall heat required for oxygenation falls to -2.5 kcal mol^{-1}. This feature brings whale Hb into the same category as Hb from Arctic ruminants. Most of the whale's body is covered by a thick insulating layer of blubber, but the active muscular parts (such as the fins and the large tail), not so well insulated, are kept at a lower temperature by a counter-current heat exchanger that reduces heat loss. Hence, although quantitative information on the temperature gradient within the body is lacking, unloading of oxygen in these tissues seems to recall the situation occurring in the cold leg muscles of Arctic ruminants, the main difference lying in the molecular mechanisms used to achieve this low temperature sensitivity. On the basis of this assumption, the lower temperature that blood finds in the fins and tail, in comparison with the rest of the body, may also explain the temperature dependence of the effect of carbon dioxide (Fig. 2). Thus, the data indicate that, within the core of the organism, carbon dioxide does not display any allosteric effect, because at 37 °C the differential binding of this ligand with respect to oxy- and deoxy- structure is completely abolished. Carbon dioxide facilitates oxygen unloading just at the level of the fins and tail, where great muscular activity takes place and the temperature is well below 37 °C, due to the counter-current heat exchanger which, in order to save energy, allows temperature to decrease in these less insulated parts of the body. In conclusion, in whale Hb the combined effects of organophosphates, carbon dioxide and temperature optimise oxygen delivery at all tissues despite great local heterothermia.

Reptiles and birds

Other examples of adaptive mechanisms are shown by Hbs from sea turtle, *Caretta caretta*, and Emperor penguin, *Aptenodytes forsteri* (Giardina *et al.*, 1992; Tamburrini *et al.*, 1994*a*). Turtles and penguins

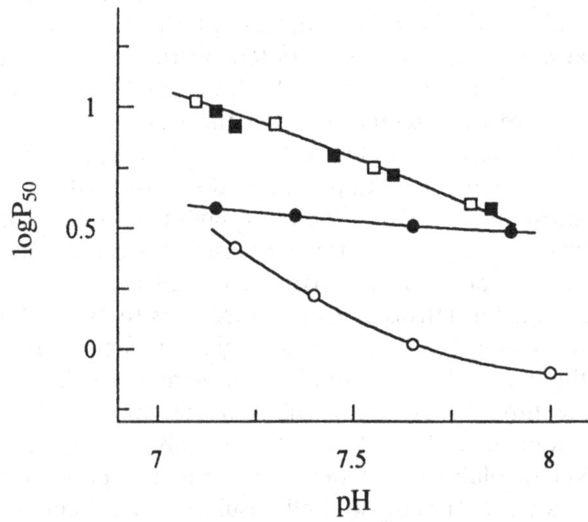

Fig. 2. Effect of carbon dioxide on whale Hb. Oxygen affinity at 20 °C (circles) and 37 °C (squares) in 0.1 M Tris buffer plus 0.1 M NaCl, in the absence (open symbols) and presence (closed symbols) of 2% carbon dioxide. (Modified from di Prisco *et al.*, 1991*c*.)

are fully committed to aquatic life; being accomplished divers and spending much of their lives submerged, they have developed suitable mechanisms for the maintenance of an adequate oxygen supply to tissues under hypoxic conditions. The Bohr curve shape appears well adapted for gas exchange during very prolonged dives, because the amplitude of the effect, as indicated by the Bohr coefficient (Δlog P50/ΔpH), is 50% smaller in the presence of the physiological effector (ATP for turtle and IPP for penguin) than that of human Hb in the presence of 2,3-DPG (Table 6). Moreover, the strongly reduced Bohr effect shows a substantial shift of the midpoint of the transition towards acidic pH values. Thus, the increase of lactic acid and the concomitant decrease in pH accompanying prolonged dives would not affect the oxygen affinity to any great extent, preserving Hb from an uncontrolled and sudden stripping of oxygen. In other words, during diving, oxygen delivery from penguin and turtle Hbs would be essentially modulated by the oxygen partial pressure in the specific tissues.

The effect of temperature on oxygen binding of both Hbs deserves special attention. In turtle Hb (Fig. 3) at pH 7.3 the overall heat of oxygenation is -1.8 kcal mol^{-1}, four times lower than the value gener-

Table 6. *Bohr coefficients and overall pK for the pH dependence of the oxygen affinity of turtle, penguin and human haemoglobins*

Species	Bohr coefficient	overall pK
Turtle	−0.35	7.1
Penguin	−0.35	7.3
Human	−0.73	7.7

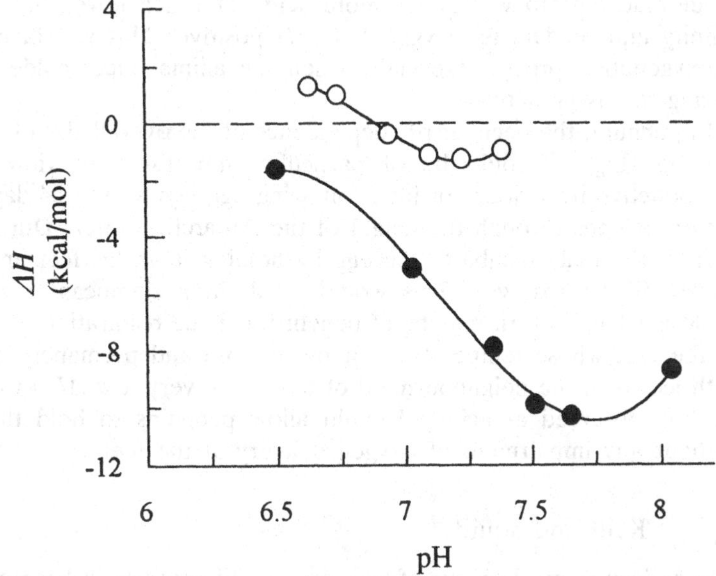

Fig. 3. Overall heat of oxygenation for *Caretta caretta* (open symbols) and Emperor penguin (closed symbols) Hbs as a function of pH. Oxygen equilibria experiments were performed in 0.1 M Hepes plus 0.1 M NaCl and 3 mM ATP (turtle) or IHP (penguin). ΔH values are corrected for the heat contribution of oxygen in solution. (Modified from Giardina *et al.*, 1992; Tamburrini *et al.*, 1994*a*.)

ally observed in vertebrate Hbs and even slightly lower than that observed in Arctic mammals. It should also be noted that the very small ΔH value, negative at neutral and alkaline pH, becomes positive at acid pH (+1.2 kcal mol^{-1} at pH 6.5). Hence, oxygenation is exothermic at the extreme alkaline limit of the Bohr effect and becomes endothermic at the other extreme, being zero at pH near 6.9.

The very low and even positive ΔH is such that, at the flippers and feet, which experience a lower temperature and a great muscular activity, oxygen delivery is not impaired, allowing the animals to endure more prolonged periods of anaerobiosis. In fact, through the very minor enthalpy change observed at acid pH, oxygen delivery becomes essentially independent of the water temperature to which the animal is exposed. Moreover it is particularly suggestive that the overall heat involved in oxygen binding is positive just at those pH values (lower than 6.9) which may be reached by turtle tissues during prolonged dives as a consequence of lactic acid production. Under these conditions, tissue-reaching Hb will find a more acid pH which lowers its oxygen affinity and renders its oxygenation ΔH positive. This will favour the deoxygenation process especially when the animal faces colder water during its diving activity.

In penguin, the peculiar pH dependence of the overall ΔH of oxygen binding (Fig. 3) could be of particular importance in view of its reproductive behaviour. In fact, following egg laying, the 64-day incubation extends through the height of the Antarctic winter. During this period, the male incubates the egg by holding it on his feet and lives on stored fat reserves. This would result in a significant metabolic acidosis which in turn may be of benefit for tissue respiration especially at the feet whose temperature, being in close and permanent contact with ice, is in the neighbourhood of 0 °C. The very low ΔH of oxygen binding observed at acid pH could allow penguins to hold the eggs without any impairment of oxygen delivery at the feet.

Krill and squid

The molecular mechanisms of temperature adaptations in haemocyanin are strongly similar to those seen in Hb. This is clearly demonstrated by krill and squid haemocyanins (Brix *et al.*, 1989*b*; Giardina *et al.*, 1991*b*).

Krill (*Meganyctiphanes norvegica*), a crustacean epipelagic organism, lives in northern Atlantic waters and is abundant throughout the year in Norwegian fjords. Its haemocyanin is characterised by a relatively low oxygen affinity and a high cooperativity of oxygen binding. Moreover, at pH higher than 7.3, the oxygen affinity increases markedly as a function of temperature, thereby reflecting a strong endothermic overall heat of oxygenation (Fig. 4). This is a novel characteristic for haemocyanin and has never been previously reported. The ecophysiological significance of this effect may be related to the feeding excursions of krill towards the warmer and phytoplankton-rich surface layers at night. It

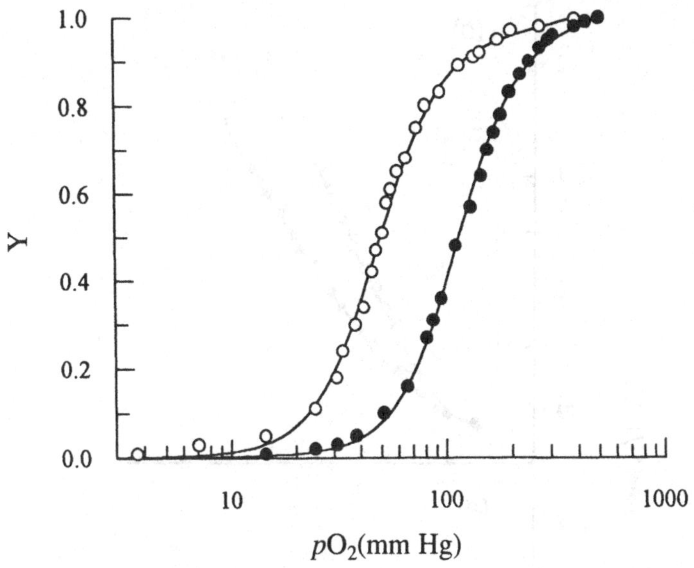

Fig. 4. Oxygen binding isotherms of the haemolymph of krill at two temperatures (○, 10 °C; ●, 5 °C) and at pH 7.7. (Modified from Brix *et al.*, 1989*b*.)

is known that the organism descends at daytime to 300–400 m, and ascends at night to between 100 m depth and the surface. The strong endothermic character of oxygen binding should be of great benefit when krill ascend towards the upper layers, where oxygen availability may be reduced owing to lower solubility at higher temperatures and to the respiratory activity of plankton. On the whole, these thermodynamic features may be interpreted in terms of adaptation to high levels of activity in a particular environment. It should be mentioned that ascending speeds up to 173 m h^{-1} and corresponding to about 1.4 bodylengths s^{-1} have been reported.

Another interesting but more complex example is haemocyanin from the squid *Todarodes sagittatus*, whose functional properties have been carefully characterised as a function of temperature and proton concentration. Within the physiological pH range, the concentration of protons mainly affects the high-affinity R state without significantly influencing the low-affinity T state (Fig. 5). On the other hand, the effect of temperature is just the opposite, because the ligand affinity of the T state is greatly affected by temperature changes whereas that of the R state is almost completely independent of this parameter. In other

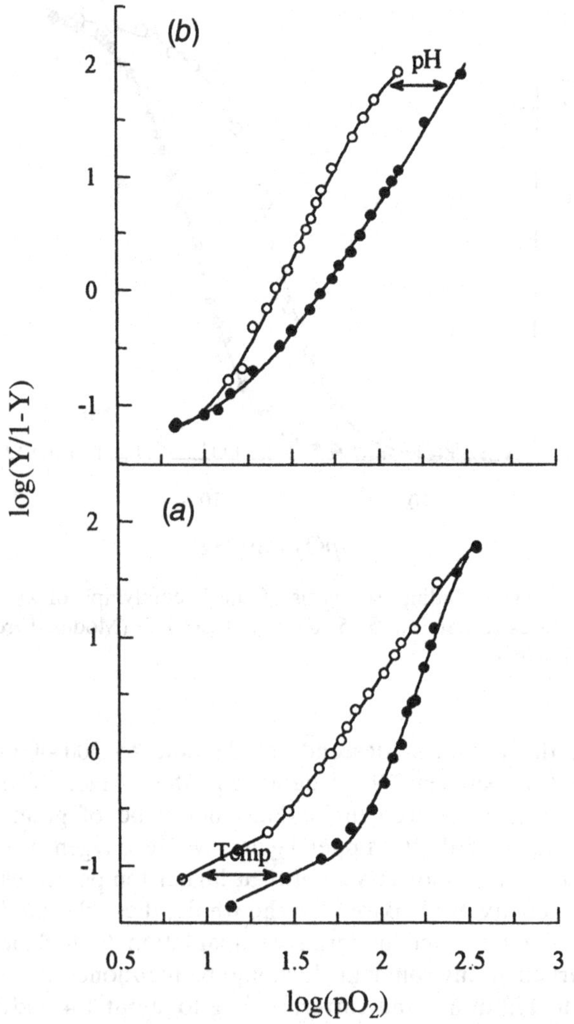

Fig. 5. Oxygen binding curves, expressed by Hill plots, for squid blood as a function of pH and temperature. (*a*) pH 7.15; 6 °C (left) and 20 °C (right). (*b*) 6 °C; pH 7.4 (left) and pH 7.15 (right). (Modified from Giardina *et al.*, 1991*b*).

words, the overall heat of oxygen binding is strongly exothermic for the low-affinity state and almost zero for the high-affinity state.

In the light of the following considerations, the interplay of pH and temperature effects seems to be useful to squid. This organism, like most cephalopods, is characterised by a high aerobic metabolism which has to be satisfied by very efficient respiration. In this connection it should be recalled that, for a functional Haldane coefficient of about 1, the Bohr shift would not come into operation by means of an arteriovenous pH difference but only by means of a change in pH brought about by temperature in both arterial and venous blood. The differential and opposite effects of protons and temperature will minimise the handicap of oxygen unloading induced by a decrease in temperature and a subsequent increase in pH. In fact, as far as the concentration of protons is concerned, the position of the lower asymptote is fixed and varies only through a temperature change, which will increase the oxygen affinity of the lower asymptote of the binding curve. This should cause a decrease of cooperativity because of the temperature independence of the higher asymptote. This decrease, however, is not observed, being almost completely counterbalanced by the Bohr effect which is operative only at high levels of saturations (R state). In fact, the increased affinity of the upper asymptote brought about by pH would increase cooperativity and counteract the effect of temperature on the lower asymptote.

Concluding remarks

Adaptive evolution of Antarctic fish occurred within a simplified framework (reduced number of variables in an isolated environment, dominated by a taxonomically uniform group). Although correlations of molecular data with physiological and biochemical adaptations and with ecology and lifestyle are difficult to identify, this objective is increasingly attracting the interest of scientists. Haematology and Hb (whose molecular features have been conserved throughout evolution) have been analysed as a potential tool. Investigations on a highly representative number of fish species (38 of 80 red-blooded Antarctic, and two non-Antarctic, notothenioids) allow two conclusions of general bearing to be reached: (i) the more phyletically-derived notothenioid families have lower erythrocyte number and Hb concentration and multiplicity, and (ii) a correlation often appears between lifestyle and Hb multiplicity and functional features. Bottom dwellers have a single major Hb with high affinity for oxygen; in the most inactive species, Hb is essentially not regulated by pH and effectors (di Prisco *et al.*, 1991*b*; D'Avino *et al.*, 1992; Kunzman *et al.*, 1992; Tamburrini *et al.*, 1992). The constant physicochemical features of the ocean may have reduced the need for multiple Hbs. The observation that three pelagic and active notothenioids have multiple, functionally distinct Hbs supports this interpretation. In these cases, a link with adaptation to habitat and with lifestyle becomes possible (di Prisco & Tamburrini, 1992), as the selective advantage of multiple Hb genes is clear.

The thermodynamics of oxygen binding and release show that the enthalpy change for oxygenation in species facing low temperature is very low when compared with temperate organisms (di Prisco *et al.*, 1991*c*). The reduced thermal sensitivity of Hbs of all organisms herewith examined is one example of the different strategies adopted during evolution to solve the problem of oxygen transport to respiring tissues. In fact, no significant amount of energy appears to be involved during both oxygenation at gills or lungs and deoxygenation in the tissues. In this regard, the Bohr effect of the single Hb of *P. bernacchii*, a sedentary benthic fish, is almost temperature insensitive, and the overall ΔH of oxygen binding in the pH range 7–8 is positive, being approximately 2 kcal mol^{-1} (di Prisco *et al.*, 1991*c*). It is worth noting that the temperature dependence of oxygenation of tuna Hb in the physiological pH range is also close to zero (Rossi-Fanelli & Antonini, 1960). The extremely active behaviour of tuna, which swims in waters of widely different temperatures, does not resemble the lifestyle of *P. bernacchii* at all. It is astonishing that the Hb systems of species facing such different situations and needs are thermodynamically so similar. Further investigations are needed to address this apparent contradiction. It should be considered that tuna is not entirely poikilothermic and maintains a partial regulation of body temperature, which can reach 14 °C above that of water (Carey & Teal, 1966); this temperature independence may help in maintaining the body heterothermicity.

On the whole, from all the examples discussed (including krill and squid haemocyanins) it is clear that the overall thermodynamics of a biological macromolecule may vary in order to cope with special circumstances. In Hb this is achieved by linking the basic reaction with the binding of different ions and effectors whose thermodynamics contribute to the overall effect of temperature. It is worth recalling the different contributions to the thermal effects measured when oxygen binds to Hb. These may be summarised as: (i) intrinsic heat of oxygenation, namely the heat involved in the binding of oxygen to the haem iron, (ii) heat of ionisation of oxygen-linked ionisable groups (Bohr groups), which is always endothermic, (iii) heat of oxygen solubilisation (exothermic), (iv) heat associated with the T → R allosteric transition, and (v) heat of binding of other ions such as organophosphates and chloride. An overall example of these aspects is whale Hb, which well illustrates how temperature and heterotropic ligands can synergetically act to modulate the basic function and overall thermodynamic characteristics of the protein. In fact, the presence of carbon dioxide and organophosphates brings about a roughly 8-fold decrease in ΔH while temperature controls the regulatory effect of carbon dioxide in switching the differential binding of this ligand on and off. By means of this unusual mechanism, the oxygen concentration throughout the whole body can satisfy the metabolic needs of the fins and tail, which must generate strong forward propulsion.

Finally, functional adaptations may have been produced either by the gradual accumulation of minor mutations or by substitutions in key positions of the amino acid sequence, the two alternatives not being mutually exclusive (Perutz, 1987). Amino acid sequencing and other structural studies are therefore an obligatory step for the molecular approach to adaptive evolution. In fish, further studies of non-Antarctic species of non-endemic Bovichtidae and Nototheniidae (the most ancient Notothenioidei) will be a logical development. In fact, recent experimental evidence suggests that *Notothenia angustata* (currently found in New Zealand waters) was cold-adapted before migrating from Antarctica, thus being an ideal evolutionary link between cold-adapted and temperate species of the same family.

Acknowledgements

This work was partially supported by the Italian National Programme for Antarctic Research. The outstanding contributions of L. Camardella, V. Carratore, C. Caruso, M.A. Ciardiello, M.E. Clementi, E. Cocca, S.G. Condò, M. Corda, R. D'Avino, A. Fago, G. Galtieri, A. Lania, M.G. Pellegrini, M. Romano, A. Riccio, M. Tamburrini and the late B. Rutigliano are gratefully acknowledged. The authors thank the editors and one anonymous referee for very helpful comments and suggestions.

References

Ahlgren, J.A., Cheng, C.C., Schrag, J.D. & DeVries, A.L. (1988). Freezing avoidance and the distribution of antifreeze glycopeptides in body fluids and tissues of Antarctic fish. *Journal of Experimental Biology* **137**, 549–63.

Baldwin, J. (1971). Adaptation of enzymes to temperature: acetylcholinesterase in the central nervous system of fishes. *Comparative Biochemistry and Physiology* **40B**, 181–7.

Baldwin, J. & Hochachka, P.W. (1970). Functional significance of isoenzymes in thermal acclimatization: acetylcholinesterase from trout brain. *Biochemical Journal* **116**, 883–7.

Brix, O., Bardgard, A., Mathisen, S., el-Sherbini, S., Condò, S.G. & Giardina, B. (1989*a*). Arctic life adaptation. II. The function of musk ox hemoglobin. *Comparative Biochemistry and Physiology* **94B**, 135–8.

Brix, O., Bougund, S., Barunung, T., Colosimo, A. & Giardina, B. (1989*b*). Endothermic oxygenation of hemocyanin in the krill. *FEBS Letters* **247**, 177–80.

Brix, O., Condò, S.G., Ekker, M., Tavazzi, B. & Giardina, B. (1990). Temperature modulation of oxygen transport in a diving mammal *(Balaenoptera acutorostrata)*. *Biochemical Journal* **271**, 509–13.

Camardella, L., Caruso, C., D'Avino, R., di Prisco, G., Rutigliano, B., Tamburrini, M., Fermi, G. & Pertuz, M.F. (1992). Haemoglobin of the Antarctic fish *Pagothenia bernacchii*. Amino acid sequence, oxygen equilibria and crystal structure of its carbonmonoxy derivative. *Journal of Molecular Biology* **224**, 449–60.

Carey, F. & Teal, J. (1966). Heat conservation in tuna fish muscle. *Proceedings of the National Academy of Sciences of the USA* **56**, 1464–9.

Caruso, C., Rutigliano, B., Riccio, A., Kunzmann, A. & di Prisco, G. (1992). The amino acid sequence of the single hemoglobin of the high-Antarctic fish *Bathydraco marri* Norman. *Comparative Biochemistry and Physiology* **102B**, 941–6.

Caruso, C., Rutigliano, B., Romano, M. & di Prisco, G. (1991). The hemoglobins of the cold-adapted Antarctic teleost *Cygnodraco mawsoni*. *Biochimica et Biophysica Acta* **1078**, 273–82.

Cheng, C.C. & DeVries, A.L. (1991). The role of antifreeze glycopeptides and peptides in the freezing avoidance of cold-water fish. In *Life Under Extreme Conditions. Biochemical Adaptations* (ed. G. di Prisco), pp. 1–14. Berlin, Heidelberg, New York: Springer-Verlag.

Ciardiello, M.A., Camardella, L. & di Prisco, G. (1995). Glucose-6-phosphate dehydrogenase from the blood cells of two Antarctic teleosts: correlation with cold adaptation. *Biochimica et Biophysica Acta* **1250**, 76–82.

Ciardiello, M. A., Camardella, L. & di Prisco, G. (1996). Enzymes in cold-adapted Antarctic fish: glucose-6-phosphate dehydrogenase. In *Proceedings of the SCAR 6th Biology Symposium* (ed. B. Battaglia, J. Valencia & D. W. H. Walton). Cambridge: Cambridge University Press (in press).

Clarke, A. (1983). Life in cold water: the physiological ecology of polar marine ectotherms. *Annual Review of Oceanography and Marine Biology* **21**, 341–453.

Cocca, E., Ratnayake Lecamwasam, M., Parker, S.K., Camardella, L., Ciaramella, M., di Prisco, G. & Detrich, H.W. III. (1995). Genomic remnants of α-globin genes in the hemoglobinless Antarctic icefishes. *Proceedings of the National Academy of Sciences of the USA* **92**, 1817–21.

D'Avino, R., Caruso, C., Camardella, L., Schininà, M.E., Rutigliano, B., Romano, M., Carratore, V., Barra, D. & di Prisco, G. (1991). An overview of the molecular structure and functional properties of the hemoglobins of a cold-adapted Antarctic teleost. In *Life Under Extreme Conditions. Biochemical Adaptations* (ed. G. di Prisco), pp. 15–33. Berlin, Heidelberg, New York: Springer-Verlag.

D'Avino, R., Caruso, C., Romano, M., Camardella, L., Rutigliano, B. & di Prisco, G. (1989). Hemoglobin from the Antarctic fish *Notothenia coriiceps neglecta*. 2. Amino acid sequence of the α-chain of Hb 1. *European Journal of Biochemistry* **179**, 707–13.

D'Avino, R., Caruso, C., Tamburrini, M., Romano, M., Rutigliano, B., Polverino de Laureto, P., Camardella, L., Carratore, V. & di Prisco, G. (1994). Molecular characterization of the functionally distinct hemoglobins of the Antarctic fish *Trematomus newnesi*. *Journal of Biological Chemistry* **269**, 9675–81.

D'Avino, R. & di Prisco, G. (1988). Antarctic fish hemoglobin: An outline of the molecular structure and oxygen binding properties. 1. Molecular structure. *Comparative Biochemistry and Physiology* **90B**, 579–84.

D'Avino, R. & di Prisco, G. (1989). Hemoglobin from the Antarctic fish *Notothenia coriiceps neglecta* 1. Purification and characterisation. *European Journal of Biochemistry* **179**, 699–705.

D'Avino, R., Fago, A., Kunzmann, A. & di Prisco, G. (1992). The primary structure and oxygen-binding properties of the high-Antarctic fish *Aethotaxis mitopteryx* DeWitt. *Polar Biology* **12**, 135–40.

Detrich, H.W., III. (1991*a*). Cold-stable microtubules from Antarctic fish. In *Life Under Extreme Conditions. Biochemical Adaptations* (ed. G. di Prisco), pp. 35–49. Berlin, Heidelberg, New York: Springer-Verlag.

Detrich, H.W., III. (1991*b*). Polymerization of microtubule proteins from Antarctic fish. In *Biology of Antarctic Fish* (ed. G. di Prisco, B. Maresca & B. Tota), pp. 248–62. Berlin, Heidelberg, New York: Springer-Verlag.

Detrich, H.W., III. & Overton, S.A. (1986). Heterogeneity and structure of brain tubulins from cold-adapted Antarctic fishes: comparison to brain tubulins from a temperate fish and a mammal. *Journal of Biological Chemistry* **261**, 10 922–30.

Detrich, H.W., III. & Overton, S.A. (1988). Antarctic fish tubulins: heterogeneity, structure, amino acid compositions and charge. *Comparative Biochemistry and Physiology* **90B**, 593–600.

Detrich, H.W., III. Prasad, V. & Ludueña, R.F. (1987). Cold-stable microtubules from Antarctic fishes contain unique α-tubulins. *Journal of Biological Chemistry* **262**, 8360–6.

DeVries, A.L. (1980). Biological antifreezes and survival in freezing environments. In *Animals and Environmental Fitness* (ed. R. Gilles), pp. 583–607. Oxford: Pergamon Press.

DeVries, A.L. (1988). The role of antifreeze glycopeptides and peptides in the freezing avoidance of Antarctic fishes. *Comparative Biochemistry and Physiology* **90B**, 611–21.

Di Prisco, G. (ed.) (1991). *Life Under Extreme Conditions. Biochemical Adaptations*. Berlin, Heidelberg, New York: Springer-Verlag.

Di Prisco, G., Camardella, L., Carratore, V., Caruso, C., Ciardiello, M.A., D'Avino, R., Fago, A., Riccio, A., Romano, M., Rutigliano, B. & Tamburrini, M. (1994). Structure and function of hemoglobins, enzymes and other proteins from Antarctic marine and terrestrial organisms. In *Proceedings of the 2nd Meeting 'Biology in Antarctica'* (ed. B. Battaglia, P.M. Bisol & V. Varotto), pp. 157–77. Padova: Edizioni Universitarie Patavine.

Di Prisco, G., Condò, S.G., Tamburrini, M. & Giardina, B. (1991*c*). Oxygen transport in extreme environments. *Trends in Biochemical Sciences* **16**, 471–4.

Di Prisco, G., D'Avino, R., Caruso, C., Tamburrini, M., Camardella, L., Rutigliano, B., Carratore, V. & Romano, M. (1991*b*). The biochemistry of oxygen transport in red-blooded Antarctic fish. In *Biology of Antarctic Fish* (ed. G. di Prisco, B. Maresca & B. Tota), pp. 263–81. Berlin, Heidelberg, New York: Springer-Verlag.

Di Prisco, G., Giardina, B., D'Avino, R., Condò, S.G., Bellelli, A. & Brunori, M. (1988*b*). Antarctic fish hemoglobin: an outline of the molecular structure and oxygen binding properties. II. Oxygen binding properties. *Comparative Biochemistry and Physiology* **90B**, 585–91.

Di Prisco, G., Macdonald, J.A. & Brunori, M. (1992). Antarctic fishes survive exposure to carbon monoxide. *Experientia* **48**, 473–5.

Di Prisco, G., Maresca, B. & Tota, B. (eds). (1988*a*). *Marine Biology of Antarctic Organisms. Comparative Biochemistry and Physiology*, pp. 459–637.

Di Prisco, G., Maresca, B. & Tota, B. (eds). (1991*a*). *Biology of Antarctic Fish*. Berlin, Heidelberg, New York: Springer-Verlag.

Di Prisco, G. & Tamburrini, M. (1992). The hemoglobins of marine and freshwater fish: the search for correlations with physiological adaptation. *Comparative Biochemistry and Physiology* **102B**, 661–71.

Eastman, J.T. (1988). Ocular morphology in Antarctic notothenioid fishes. *Journal of Morphology* **196**, 283–306.

Eastman, J.T. (1991). The fossil and modern fish faunas of Antarctica: Evolution and diversity. In *Biology of Antarctic Fish* (ed. G. di Prisco, B. Maresca & B. Tota), pp. 116–130. Berlin, Heidelberg, New York: Springer-Verlag.

Eastman, J.T. (1993). *Antarctic Fish Biology. Evolution in a Unique Environment*. San Diego, CA: Academic Press.

Eastman, J.T. & DeVries, A.L. (1986). Renal glomerular evolution in Antarctic notothenioid fishes. *Journal of Fish Biology* **29**, 649–62.

Fago, A., D'Avino, R. & di Prisco, G. (1992). The hemoglobins of *Notothenia angustata*, a temperate fish belonging to a family largely endemic to the Antarctic Ocean. *European Journal of Biochemistry* **210**, 963–70.

Garland, T., Jr. & Carter, P.A. (1994). Evolutionary physiology. *Annual Review of Physiology* **56**, 579–621.

Genicot, S., Feller, G. & Gerday, C. (1988). Trypsin from Antarctic fish (*Paranotothenia magellanica* Forster) as compared with trout (*Salmo gairdneri*) trypsin. *Comparative Biochemistry and Physiology* **90B**, 601–9.

Giardina, B., Condò, S.G., Bardgard, A. & Brix, O. (1991*a*). Life in arctic environments: molecular adaptation of oxygen carrying proteins. In *Life Under Extreme Conditions* (ed. G. di Prisco), pp. 51–60. Berlin, Heidelberg, New York: Springer-Verlag.

Giardina, B., Condò, S.G. & Brix, O. (1991*b*). Modulation of oxygen binding in squid blood. In *Structure and Function of Invertebrate Oxygen Carriers* (ed. S.N. Vinogradov & O.H. Kapp), pp. 333–9. Berlin, Heidelberg, New York: Springer-Verlag.

Giardina, B., Condò, S.G., Petruzzelli, R., Bardgard, A. & Brix, O. (1990). Thermodynamics of oxygen binding to arctic hemoglobins: the case of reindeer. *Biophysical Chemistry* **37**, 281–6.

Giardina, B., el-Sherbini, S., Mathiesen, S., Tylar, N., Nuutinen, M., Bardgard, A., Condò, S.G. & Brix, O. (1989). Arctic life adaptation. I. The function of reindeer hemoglobin. *Comparative Biochemistry and Physiology* **94B**, 129–33.

Giardina, B., Galtieri, A., Lania, A., Ascenzi, P., Desideri, A., Cerroni, L. & Condò, S.G. (1992). Reduced sensitivity of oxygen transport to allosteric effectors and temperature in loggerhead sea turtle hemoglobin: functional and spectroscopic study. *Biochimica et Biophysica Acta* **1159**, 129–33.

Gon, O. & Heemstra, P.C. (eds) (1990). *Fishes of the Southern Ocean*. Grahamstown, South Africa: JLB Smith Institute of Ichthyology.

Hochachka, P.W. & Somero, G.N. (1984). *Biochemical Adaptation*. Princeton, NJ: Princeton University Press.

Hsiao, K., Cheng, C.C., Fernandes, I.E., Detrich, H.W. & DeVries, A.L. (1990). An antifreeze glycopeptide gene from the Antarctic cod *Notothenia coriiceps neglecta* encodes a protein of high peptide copy number. *Proceedings of the National Academy of Sciences of the USA* **87**, 9265–9.

Hureau, J.-C., Petit, D., Fine, J.M. & Marneux, M. (1977). New cytological, biochemical and physiological data on the colorless blood of the Channichthyidae (Pisces, Teleosteans, Perciformes). In *Adaptations within Antarctic Ecosystems* (ed. G.A. Llano), pp. 459–77. Washington: Smithsonian Institution.

Kunzman, A., Caruso, C. & di Prisco, G. (1991). Haematological studies on a high-Antarctic fish: *Bathydraco marri* Norman. *Journal of Experimental Marine Biology and Ecology* **152**, 243–55.

Kunzmann, A., Fago, A., D'Avino, R. & di Prisco, G. (1992). Haematological studies on *Aethotaxis mitopteryx* DeWitt, a high-Antarctic fish with a single haemoglobin. *Polar Biology* **12**, 141–5.

Macdonald, J.A. & Montgomery, J.C. (1982). Thermal limits of neuromuscular function in an antarctic fish. *Journal of Comparative Physiology* **147**, 237–50.

Macdonald, J.A., Montgomery, J.C. & Wells, R.M.G. (1987). Comparative physiology of Antarctic fishes. *Advances in Marine Biology* **24**, 321–88.

Perutz, M.F. (1987). Species adaptation in a protein molecule. *Advances in Protein Chemistry* **36**, 213–44.

Raymond, J.A. & DeVries, A.L. (1977). Adsorption inhibition as a mechanism of freezing resistance in polar fishes. *Proceedings of the National Academy of Sciences of the USA* **74**, 2589–93.

Reeve, H.K. & Sherman, P.W. (1993). Adaptation and the goals of evolutionary research. *The Quarterly Review of Biology* **68**, 1–32.

Riggs, A. (1970). Properties of fish hemoglobins. In *Fish Physiology*, Vol. 4 (ed. W.S. Hoar & D.J. Randall), pp. 209–52. New York: Academic Press.

Rossi-Fanelli, A. & Antonini, E. (1960). Oxygen equilibrium of haemoglobin from *Thunnus thynnus*. *Nature* **186**, 895–6.

Ruud, J.T. (1954). Vertebrates without erythrocytes and blood pigment. *Nature* **173**, 848–50.

Somero, G.N. (1991). Biochemical mechanisms of cold adaptation and stenothermality in Antarctic fish. In *Biology of Antarctic Fish* (ed. G. di Prisco, B. Maresca & B. Tota), pp. 232–47. Berlin, Heidelberg, New York: Springer-Verlag.

Somero, G.N. & DeVries, A.L. (1967). Temperature tolerance of some Antarctic fishes. *Science* **156**, 257–8.

Sweezey, R.R. & Somero, G.N. (1982). Polymerization thermodynamics and structural stabilities of skeletal muscle actins from vertebrates adapted to different temperatures and hydrostatic pressures. *Biochemistry* **21**, 4496–503.

Tamburrini, M., Brancaccio, A., Ippoliti, R. & di Prisco, G. (1992). The amino acid sequence and oxygen-binding properties of the single hemoglobin of the cold adapted Antarctic teleost *Gymnodraco acuticeps*. *Archives of Biochemistry and Biophysics* **292**, 295–302.

Tamburrini, M., Condò, S.G., di Prisco, G. & Giardina, B. (1994*a*). Adaptation to extreme environments: structure–function relationships in Emperor penguin hemoglobin. *Journal of Molecular Biology* **237**, 615–21.

Tamburrini, M., D'Avino, R., Fago, A., Carratore, V., Kunzmann, A. & di Prisco, G. (1994*b*). The unique hemoglobin system of

Pleuragramma antarcticum, a high-Antarctic fish with holopelagic mode of life. *Presented at the SCAR 6th Biology Symposium,* Poster Abstracts 261.

Wells, R.M.G., Macdonald, J.A. & di Prisco, G. (1990). Thin-blooded Antarctic fishes: a rheological comparison of the haemoglobin-free icefishes, *Chionodraco kathleenae* and *Cryodraco antarcticus*, with a red-blooded nototheniid, *Pagothenia bernacchii. Journal of Fish Biology* **36**, 595–609.

G. N. SOMERO, E. DAHLHOFF
and J.J. LIN

Stenotherms and eurytherms: mechanisms establishing thermal optima and tolerance ranges

Introduction

Enormous differences exist among ectothermic animals in optimal body temperatures and breadth of thermal tolerance ranges. Extreme stenothermy, coupled with cold tolerance, is exemplified by highly cold-adapted notothenioid fishes of Antarctica, that have a thermal tolerance range of only about 6 °C (from the freezing point of seawater, −1.86 °C, to approximately 4 °C; Somero & DeVries, 1967; Eastman, 1993). In contrast, extreme eurythermy and heat tolerance is exhibited by fishes such as the intertidal goby *Gillichthys seta*, whose body temperature may range from approximately 8 °C to 40 °C, as a function of both seasonal and diurnal changes in water temperature (Dietz & Somero, 1992). The physiological, biochemical and molecular mechanisms that distinguish stenotherms and eurytherms are likely to play critical roles in establishing biogeographical patterning and in establishing the susceptibility of animals to shifts in ambient temperature, such as are predicted as a consequence of global warming.

This review compares homologous biochemical and physiological systems in stenotherms and eurytherms, and relates interspecific differences in these systems to the thermal optima and tolerance ranges characteristic of the whole organism. In keeping with a central theme of this symposium, namely, the similarities and differences found between evolutionary adaptation to temperature and short-term phenotypic acclimatisation, this review contrasts genetically-fixed traits that are important in setting thermal limits and thermal optima, with more 'plastic' traits that provide significantly different phenotypes under different thermal conditions. Two principal types of biochemical systems will be discussed in this review: enzymatic proteins and membranes. In both types of systems, inherent, genetically-based differences in eurythermy at both the physiological and biochemical levels distinguish stenotherms from eurytherms. Phenotypic plasticity is evident in protein isoform systems and in membrane-based systems, such that considerable rebuilding of the biochemical machinery of eurytherms is accomplished in response to seasonal and, in some cases, diurnal (Carey & Hazel, 1989) changes in temperature. This phenotypic plasticity, which is of such importance to eurytherms, may be greatly attenuated in highly stenothermal species.

The data and analysis presented in this review will lead to several overall conclusions, which are briefly summarised as follows. First, the homologous proteins of stenothermal and eurythermal species are themselves tolerant of narrow or wide ranges of temperature, respectively, and these genetically-fixed differences may play key roles in establishing thermal tolerance ranges. A second and related conclusion is that proteins of warm-adapted species may be inherently more eurythermal than the homologous proteins of cold-adapted species. A decrease in body temperature for a warm-adapted ectotherm may be less stressful than a comparable increase in body temperature for a cold-adapted stenotherm. A third conclusion is that all life stages of a species must be considered when one attempts to interpret the thermal characteristics of biochemical systems. Thermal stress in early larval or juvenile stages may be the dominant selective influence on the temperature sensitivities of physiological and biochemical systems. Fourth, the ability to restructure protein and membrane systems in a thermally-adaptive manner in response to seasonal temperature changes differs significantly between stenotherms and eurytherms. Stenotherms may have lost genetic information or the capability of expressing certain genes during evolution in thermally stable environments. Finally, through reviewing the effects of high temperature on both protein and membrane function, potential mechanisms that determine upper lethal temperatures of species will be discussed. Synaptic transmission appears to be an especially critical physiological process in this regard.

Protein systems

Orthologous homologues: eurythermal and stenothermal characteristics of enzymes

Studies of orthologous homologues of proteins, defined as variants on a protein theme that are encoded by the same gene locus in different populations or species, have provided important data for understanding how the genetically-fixed biochemical properties of organisms may help to establish thermal optima and tolerance limits. Studies of protein stability and of temperature effects on enzyme kinetic properties such as Michaelis–Menten constants (K_m values) have revealed regular differences among orthologous homologues of several classes of proteins. These differences at the protein level reflect the absolute temperatures and ranges of temperatures to which the species are adapted (for a review, see Somero, 1995).

Thermal denaturation temperatures correlate with adaptation temperature

Figure 1 presents data on several orthologous homologues of the A-type (muscle-type) isozyme of the glycolytic enzyme lactate dehydrogenase (LDH-A). Figure 1*a* shows data on the thermal stabilities of LDH-A homologues from vertebrates with body temperatures spanning a range of almost 50 °C, from −1.86 °C for the Antarctic notothenioid fish *Pagothenia borchgrevinki* to approximately 47 °C for the desert iguana, *Dipsosaurus dorsalis*. Thermal perturbation of LDH-A structure was indexed by monitoring changes in the intensity of fluorescence by tryptophyl residues in the presence of the fluorescence quenching agent acrylamide (Donahue, 1982). Increases in relative fluorescence intensity with increasing measurement temperature indicate an unfolding of LDH-A higher order structure (see legend to Fig. 1 for details; Donahue, 1982). In keeping with numerous other studies of variations in thermal stability among orthologous homologues from differently thermally-adapted species (Hochachka & Somero, 1984; Somero, 1995), the protein homologue from the most cold-adapted species, *P. borchgrevinki*, unfolds at the lowest temperature and the homologue from the most thermophilic species, the desert iguana, is most heat resistant. Unfolding of the other LDH-A homologues also clearly reflects the species' environmental temperatures.

Note that the LDH-As of all five species appear to be thermally stable, under the *in vitro* conditions employed in these experiments, at temperatures well above the upper lethal temperatures of the organism. The disparity between the upper lethal temperature of the whole organism and the temperature of heat denaturation of LDH-A is approximately 40 °C in the case of *P. borchgrevinki*. This disparity between the upper lethal temperature of the organism and the heat denaturation temperature of an isolated protein may at first seem to indicate that protein thermal stability is not a relevant factor in establishing thermal relationships, despite the trend shown for LDH-A and many other enzymes and structural proteins (Johnston & Walesby, 1977; Swezey & Somero, 1982; McFall-Ngai & Horwitz, 1990; Jaenicke, 1991; Dahlhoff & Somero, 1993*a*; Somero, 1995). In fact, as shown

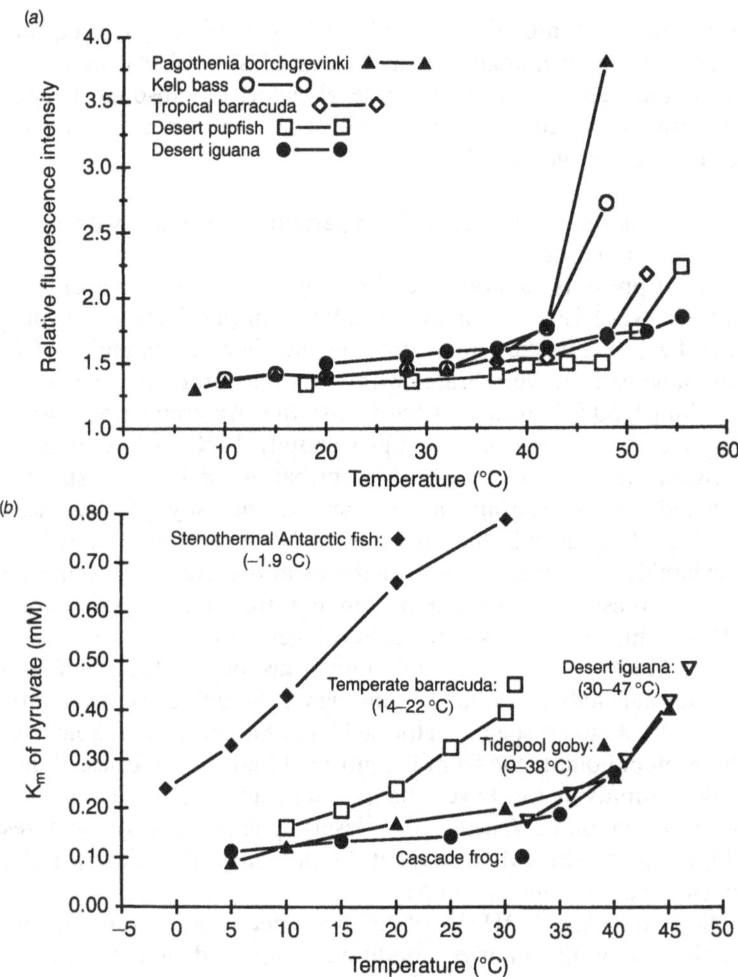

Fig. 1. Adaptive differences in thermal stability and kinetic properties among orthologous homologues of lactate dehydrogenase-A (LDH-A). (*a*) Differences in thermal stability as indexed by quenching of tryptophyl residues by acrylamide. Relative fluorescence intensity is the ratio of fluorescence in the absence of acrylamide to the value in the presence of acrylamide. Thus, a high relative fluorescence intensity indicates effective penetration by acrylamide into the protein structure, i.e. indicates unfolding of the native enzyme structure (data from Donahue, 1982). (*b*) Effect of temperature on the Michaelis–Menten constant (K_m) of pyruvate for orthologous homologues of LDH-A. (After Donahue (1982), Graves & Somero (1982), Somero (1995) and the unpublished results of E. Winter, J. Podrabsky & G. Somero.)

below, the disruption of protein function by elevated temperature is evident at temperatures far below those at which total denaturation of structure and full loss of enzymatic activity are observed.

Heat denaturation data such as those in Fig. 1 are, however, an indication that the stabilities of proteins are finely adjusted during evolution to maintain the appropriate stabilities at physiological temperatures (Jaenicke, 1991; Somero, 1995). Selection favours retention of a marginal degree of stability at physiological temperatures. The protein must retain adequate conformational flexibility to permit the changes in shape required for function, yet at the same time the protein must not be so labile that normal body temperatures can distort the protein's geometry and thereby disrupt function. This evolutionary 'balancing act' between structure-stabilising forces (such as the hydrophobic effect) and structure-destabilising forces (such as configurational entropy) is a critical feature of protein evolution, and has led to net free energies of stabilisation of protein structure that correspond to the energies of only a few non-covalent ('weak') chemical bonds (Jaenicke, 1991). It follows that the differences in stabilisation free energy among orthologous homologues of a protein from species adapted to different body temperatures are extremely small and, as discussed later, are the result of very minor changes in primary structure (Matthews, 1987; Jaenicke, 1991).

Temperature effects on ligand binding

The ability of temperatures that are far lower than heat denaturation temperatures to perturb enzyme function is shown in Fig. 1*b*. These data illustrate how temperature affects the Michaelis–Menten constant of pyruvate for LDH-A homologues from animals adapted to different body temperatures. Included in this comparison are data for a highly stenothermal Antarctic species, *Trematomus penellii*; a highly eurythermal tidepool goby, *Gillichthys seta*; and the desert iguana, one of the most heat tolerant ectothermic vertebrates. Temperature-induced increases in K_m of substrate are indicative of perturbation of substrate binding ability, and may reflect significant disruption of controlled enzymatic activity in the cell (Hochachka & Somero, 1984).

The differences seen in this interspecific comparison of temperature effects on K_m of pyruvate for LDH-A demonstrate how orthologous enzyme homologues of stenotherms and eurytherms vary in their responses to temperature. For example, the effect of variation in temperature on the K_m of pyruvate is extremely small for the LDH-A of the tidepool goby, the most eurythermal species in this comparison. In contrast, the LDH-A of the stenothermal Antarctic notothenioid fish exhibits a rapid rise in K_m of pyruvate as the temperature is increased only slightly above 5 °C. Despite differences among species in patterns of thermal dependence of K_m of pyruvate, a high degree of conservation in K_m of pyruvate is maintained at the respective environmental temperatures of the different species. This conservation is maintained over a wide range of temperatures for eurytherms, but only over a narrow range of temperatures for stenotherms, as shown clearly by the comparison of the Antarctic fish with the eurythermal tidepool goby.

On the basis of these and many other data examining temperature effects on K_m values of different enzymes (e.g. Baldwin, 1971; Hochachka & Somero, 1984; Yancey & Siebenaller, 1987; Dahlhoff & Somero, 1991, 1993*a*), it is clear that the degree of eurythermy or stenothermy found at the whole organism level is mirrored in an important process at the biochemical level, the formation of enzyme-ligand (substrate and cofactor) complexes.

Enzyme adaptation and biogeographical patterning: lessons from studies of congeneric species

Interspecific temperature-adaptive differences in enzyme thermal stability and K_m versus temperature relationships may play important roles in establishing and maintaining biogeographical patterning. This important aspect of temperature adaptation is especially clear in comparisons of closely-related congeneric species from different thermal gradients. Similarities among congeners in life-history characteristics and overall ecological relationships may allow the effects of temperature to be revealed more easily because of the absence of extraneous complicating effects associated with comparisons involving diverse phylogeny.

Dahlhoff & Somero (1993*a*) studied five species of abalone (Mollusca; genus *Haliotis*) found at different depths and latitudes along the Pacific Coast of North America (Fig. 2). One comparison involved the cytosolic isozyme of malate dehydrogenase (cMDH). Orthologous homologues of cMDH of the five species differed between cold, stenothermal species (*Haliotis kamtschatkana kamtschatkana* (pinto abalone) and *H. rufescens* (red abalone)) and more warm-adapted, eurythermal species (*H. cracherodii* (black abalone), *H. corregata* (pink abalone), and *H. fulgens* (green abalone)). cMDHs of the former two species were significantly more heat labile and exhibited larger changes in K_m of cofactor (NADH) with changing temperature, relative to the homologues from the more warm-adapted, eurythermal species. At the different species' normal environmental temperatures, however, a high degree of conservation in K_m of NADH was observed.

Similar patterns of interspecific variation in enzyme kinetic responses have been observed among the LDH-As of congeneric barracuda fishes (Graves & Somero, 1982) and confamilial goby fishes (Fields, 1995). These studies of LDH-As and cMDHs all demonstrate that differences in maximal habitat temperature of only 3–10 °C are sufficient to favour selection for temperature-adaptive differences in protein structure and function. Thus, these studies support the hypothesis that environmental temperature changes of only a few degrees Celsius resulting from global warming could have important, direct effects on the physiological functions of ectothermic organisms (Fields et al., 1993; Somero & Hofmann, 1996) as well as more complex, indirect effects from ecosystem-level interactions (Lubchenco *et al.*, 1993).

Thermal properties of enzymes may reflect thermal selection at early life stages

The thermal relationships discussed above reflect strong correlations between the structural and functional properties of enzymes and the habitat temperatures encountered by adults of the species examined. The effects of temperature on proteins, however, may not always accord with the thermal relationships of adult stages of organisms. For example, note the very flat K_m versus temperature function for LDH-A of the frog *Rana cascadae* (Fig. 1). This frog lives at high elevations in the Cascade Mountains of the Pacific Northwest region of the United States, and is unlikely to experience body temperatures much in excess of 20 °C as an adult (A. Blaustein, personal communication). The LDH-A of this species, however, exhibits a temperature versus K_m pattern similar to that of the heat tolerant, eurythermal tidepool goby, *G. seta*. Thus, the LDH-A of this temperate frog would appear to be adapted for function at temperatures as high as 30–35 °C. Although adult *R. cascadae* are unlikely to experience these high temperatures, tadpoles are known to be subjected to considerably warmer temperatures than 20 °C. Tadpoles of this species live in shallow ponds, and congregate near the surface during the day, where temperatures can be very high. The thermal properties of the LDH-A of this amphibian, therefore, may reflect temperature conditions encountered by tadpoles, rather than those commonly experienced by adult frogs. LDH-A homologues of amphibians with tadpole stages that do not encounter high temperatures do not exhibit the flat K_m versus temperature response shown by the LDH-A of *R. cascadae* (J. Podrabsky & G. Somero, unpublished results). Interpreting the thermal responses of biochemical systems therefore requires a perspective that encompasses the temperature regimes of the entire life-history of an organism.

'Physiological denaturation' of proteins

What is the linkage, if any, between thermal stability of enzyme structure and thermal sensitivity of kinetic function? Two considerations are important in this context. First, as emphasised above, functionally important effects of temperature on enzyme function, e.g. ligand binding (Figs. 1 and 2), can occur at temperatures well below those at which denaturation and full loss of activity take place. Impairment of function, which can be regarded as 'physiological denaturation' of an enzyme, may occur well before the protein is largely unfolded and, thus, catalytically inactive. Although irreversible heat denaturation followed by proteolysis appears to occur under extremes of natural thermal stress (Hofmann & Somero, 1995; Somero & Hofmann, 1996), disruption of physiological function may result from only a relatively minor and fully reversible alteration of enzyme conformation.

A second link between structural stability of proteins and sensitivity of enzymatic function to changes in temperature involves interspecific differences in thermal perturbation of ligand binding. It is proposed that differences among orthologous homologues in temperature effects on K_m values result from different thermal sensitivities of protein conformation, rather than from differences in active site sequences. Thus, because the amino acid residues involved in ligand binding are fully conserved among all LDH-A homologues sequenced to date (Tsuji *et al.*, 1994; Fields, 1995; L.Z. Holland, M. McFall-Ngai & G.N. Somero, unpublished results), interspecific differences in K_m values and in the effect of temperature on K_m are likely to derive from differences in the fine-scale conformations of the proteins, i.e. on the precise geometries of the residues involved in establishing the energies of ligand binding events (Somero, 1995). Even minor perturbations of protein conformation by temperature may be sufficient to distort sufficiently the geometry of the enzyme's active site to weaken enzyme–substrate interactions. For example, even though the inherently thermally labile LDH-As of Antarctic fishes may not undergo gross denaturation until 45 °C (Fig. 1), minor temperature-induced changes in conformation at temperatures only slightly above 0 °C may be enough to impact kinetic function. In contrast, thermally stable LDH-As of warm-adapted, eurythermal ectotherms like *G. seta* may retain the necessary geometry for strong ligand binding over wide ranges of temperature. The importance of temperature-induced losses in ligand binding ability will be considered again below, in the context of lethal effects of high temperature.

Fig. 2(*a*)

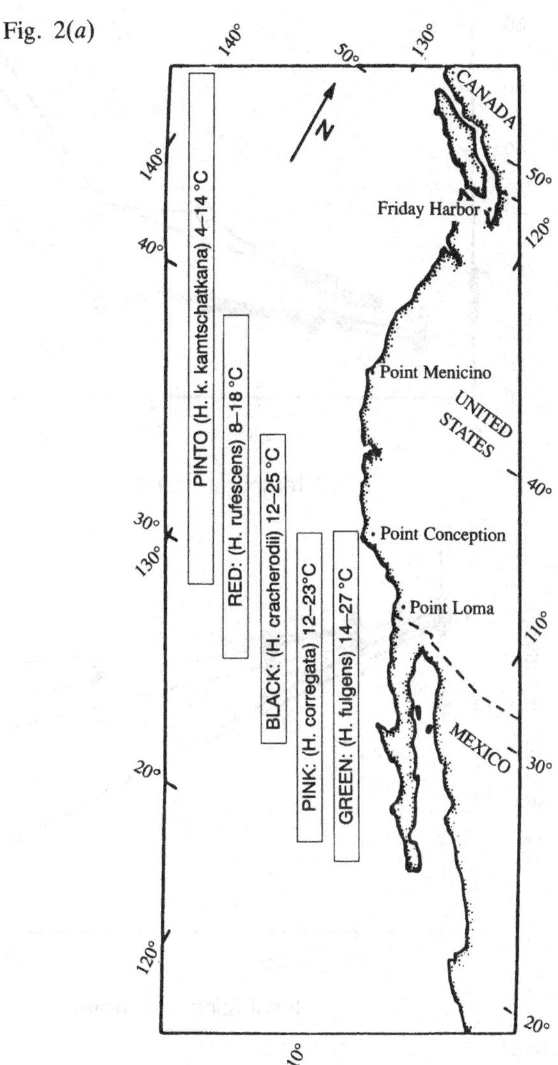

Fig. 2. Temperature-adaptive differences in orthologous homologues of cytosolic malate dehydrogenases (cMDHs) of abalone (genus *Haliotis*). (*a*) Distribution limits and habitat temperature ranges of five *Haliotis* species found on the Pacific Coast of North America. (*b*) Effect of temperature on the Michaelis–Menten constant (K_m) of cofactor (NADH) for orthologous homologues of cMDH. The thick

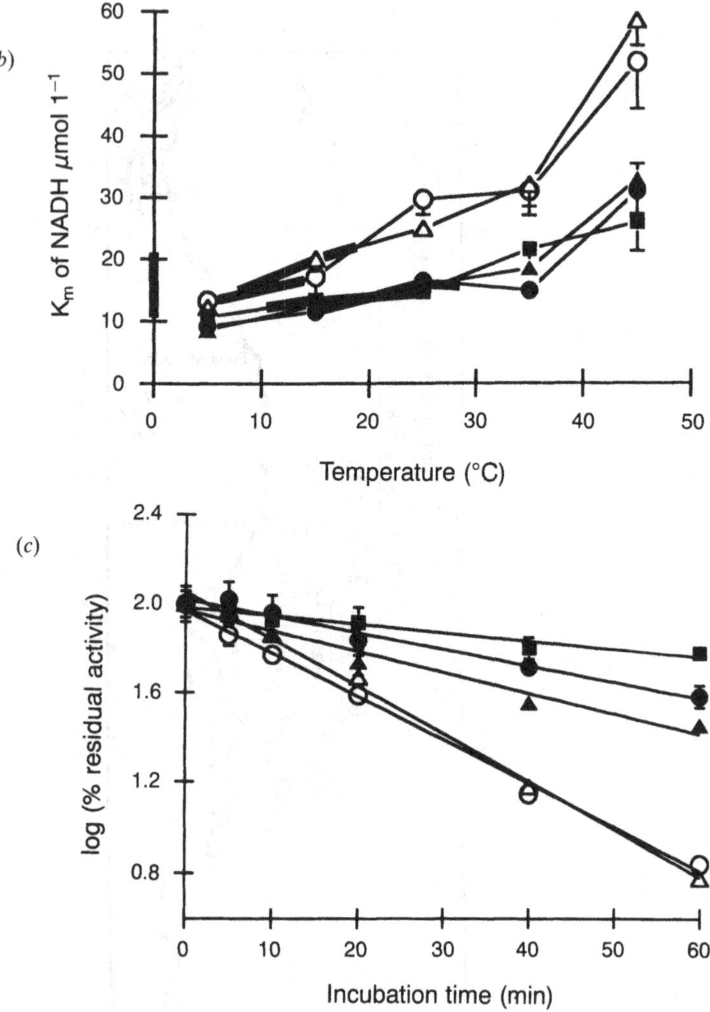

Fig. 2. *contd.*
line segments indicate the habitat temperature ranges of the species. (*c*) Differences in thermal stability among cMDHs of five species of abalone. Activity remaining after different periods of incubation at 40 °C is shown. (●, Green; ▲, pink; ■, black; △, red; ○, pinto. (After Dahlhoff & Somero (1993*a*).)

Only minor changes in protein sequence may be required to effect adaptive change

To conclude this analysis of temperature adaptation of orthologous enzyme homologues, it is appropriate to inquire about the amount of change in protein sequence required to modify the thermal sensitivity of a protein in an adaptive manner. The amount of sequence change is very small. Powers *et al.* (1993) have found that a single amino acid substitution in heart-type LDH (LDH-B) is sufficient to alter thermal stability. LDH-A homologues of barracuda fishes with different environmental temperatures differ by only a few residues (L.Z. Holland, M. McFall-Ngai & G.N. Somero, unpublished results). These studies of evolutionary changes in proteins are fully in accord with recent studies in which the techniques of site-directed mutagenesis have been used to alter protein thermal stability and function (Matthews, 1987). Both types of studies have shown that only very minor changes in protein sequence may be sufficient to cause major modifications of enzyme function. The fact that the evolutionary changes in sequence that cause temperature-adaptive changes in proteins do not involve active site residues is consistent with the conjecture that sequence changes in any regions of the protein molecule that influence its conformational flexibility may bring about temperature-adaptive alterations in structure and function (Somero, 1995).

Although emphasis in the above discussion has been placed solely on adaptive changes in primary structure, there is the possibility that post-translational modifications of proteins, including both covalent modifications such as phosphorylation and temperature-dependent folding into two or more conformational states (Somero, 1969), also could contribute to adaptation to temperature. Before differences in deduced protein sequences (obtained from sequencing complementary DNAs) can be accepted as the sole basis for temperature-adaptive differences in function and stability, the occurrence of temperature-adaptive post-translational modifications must be excluded.

Paralogous enzyme homologues as a mechanism for enhancing eurythermy

A second mechanism for fostering eurythermy of protein function involves isozyme variants encoded by multiple gene loci, paralogous homologues of enzymes. If gene duplication gives an organism two or more gene loci which encode a given type of enzyme, then it is conceivable that paralogous homologues with different thermal characteristics could evolve following the gene duplication event. The paralogous isozymes might differ in thermostability and in kinetic properties, e.g. in K_m versus temperature responses. The isozymes with different thermal characteristics might be synthesised constitutively at all temperatures, or only the isozyme appropriate for the current thermal regimen might be produced. The latter mechanism could give the organism a high degree of plasticity and efficiency in dealing with temperature change, in the sense that only the 'right' isozyme for the current thermal regimen would be produced at any given time.

The role of paralogous isozymes in temperature acclimation has been studied in a variety of ectotherms, and several examples of temperature-dependent expression of different protein isoforms are known (for a review, see Johnston, 1983). Although eurythermy of protein function seems to depend more on single protein homologues with flat K_m versus temperature characteristics than on the occurrence of paralogous isozymes with different thermal responses (Somero, 1995), there are cases in which the latter mechanism may contribute to eurythermy.

Recent studies of cytosolic malate dehydrogenases of teleost fishes illustrate the types of adaptive responses that can occur in paralogous isozyme systems during both evolutionary adaptation and seasonal acclimatisation (Lin & Somero, 1995*a*, *b*). Unlike higher vertebrates, most teleost fishes possess two genes for cMDH (Schwantes & Schwantes, 1982). The two cMDH isozymes differ in thermal stability (Schwantes & Schwantes, 1982; Lin & Somero, 1995*a*, *b*) and in kinetic properties (Lin & Somero, 1995*a*, *b*) (Fig. 3). The differences in the K_m versus temperature responses of the cMDH isozymes of the eurythermal goby *Gillichthys mirabilis* mirror the differences noted for orthologous variants of LDH-A in vertebrates (Fig. 1) and cMDHs in abalone (Fig. 2): the K_m of substrate (or cofactor) for the cold-adapted enzyme variant is higher, at any given temperature of measurement, than the K_m of the warm-adapted variant. Thus, at physiological temperatures, there is a strong conservation of K_m seen among species (orthologous homologues) and between differently-acclimatised (or differently-acclimated) individuals of a single species (paralogous homologues).

Differences in cMDH expression as a function of body temperature have been found in two very different temporal contexts. In *G. mirabilis*, cMDH expression varied seasonally and with laboratory acclimation (Fig. 3) (Lin & Somero, 1995*a*). Under all conditions of thermal acclimatisation and acclimation, specimens expressed both cMDH isozymes, but the ratio of the thermolabile to thermostable isozymes differed as a reflection of thermal history. Thus, in this eurythermal

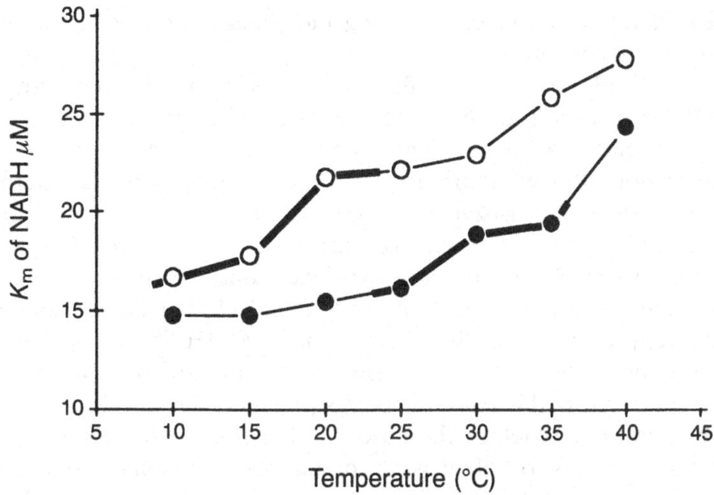

Fig. 3. Effect of temperature on the Michaelis–Menten constant (K_m) of cofactor (NADH) for cMDHs of skeletal muscle of winter (○)- and summer (●)-acclimatised individuals of *Gillichthys mirabilis*. Data are for unfractionated isozymes. During both seasons, muscle contained both the thermolabile and thermostable isozymes (as shown by native gel electrophoresis and activity staining), but the ratio of the two paralogous isozymes differed between seasons. Changes in the ratio of thermostable to thermolabile cMDH were correlated with the differences in K_m shown in the figure. The thicker line segments show K_m values at physiological temperatures in winter and summer, and illustrate the conservation of K_m of NADH at different seasons. (After Lin & Somero (1995*a*).)

species, regulation of expression of cMDHs isozymes is not an 'on–off' response, but rather an adjustment of the ratio of expression of the two isozymes. Fish collected in winter and fish held at 10 °C had a higher percentage of the thermolabile cMDH isozyme than fish collected in summer or held at 30 °C. Acclimation studies revealed that the rate of change in the ratio of cMDH isozymes was faster in the case of warm acclimation (shift from 10 to 30 °C) than during the reciprocal transfer from 30 to 10 °C. This difference between the rates of warm- and cold-acclimation is likely a Q_{10} effect, i.e. a more rapid rate of synthesis and/or degradation of the enzymes at higher temperature. The difference in time course of cMDH isozyme expression is in accord with other studies of the time course of thermal acclimation (e.g. Sidell *et al.*, 1973; Cossins *et al.*, 1977). The possible roles of transcriptional and translational regulatory events, as well as protein degradative activity, in governing the changes in cMDH isozyme ratio are not yet resolved.

The study of interspecific differences in paralogous isozymes of cMDHs in barracuda fishes (genus *Sphyraena*) from different latitudes has revealed a more dichotomous, i.e. 'on–off', form of cMDH expression. Lin & Somero (1995*b*) found that both the thermolabile and thermostable isozymes of cMDH were present in two temperate and one subtropical species of barracuda. In the warm-adapted equatorial species *Sphyraena ensis*, however, only the thermostable cMDH isozyme could be detected in kinetic and electrophoretic studies. The apparent absence of the thermolabile cMDH in this warm-adapted stenothermal fish suggests that life at the continually high temperatures encountered by this tropical species precludes the need for the thermolabile cMDH. Whether the absence of thermolabile cMDH is a consequence of a lost or silent gene, or the result of some other regulatory mechanism, is not known.

Changing the milieu to conserve protein structure and function

There is a third common and basic mechanism by which the structural integrity and kinetic properties of an enzyme can be stabilised in the face of temperature change. Adjustments of the milieu in which a protein occurs, as can be achieved by changing the composition of the solution bathing a protein or the biophysical properties of the lipid environment in which a membrane-based protein occurs (Cossins, 1983), can effect a high degree of stabilisation of both structural and kinetic properties. In principle, changes in the milieu can influence the thermal responses of a large number of proteins, which could make this type of adaptation a highly efficient means for offsetting thermal stress.

The broad influences of milieu effects are illustrated by temperature-dependent changes in intracellular pH (pH_i), which can stabilise the function and structure of enzymes that rely on histidyl residues for activity or maintenance of structural integrity (Reeves, 1977; Somero, 1986). Eurythermy of protein structure and function is thus enhanced for many proteins by the temperature-dependent pH regulatory pattern termed 'alphastat regulation' (Reeves, 1977), which is found in virtually all types of organisms. The conservation of K_m values at different temperatures is strongly dependent on the alphastat pH_i regulatory pattern in the case of enzymes that utilise histidyl residues for ligand binding (Yancey & Somero, 1978). pH-dependent subunit assembly processes likewise are stabilised by alphastat pH regulation (Hand & Somero, 1983).

Milieu influences on protein thermal stability also involve low molecular weight organic solutes that are capable of stabilising proteins (Yancey *et al.*, 1982; Somero, 1992). Heat shock of yeast cells leads to the accumulation of high concentrations of trehalose, a powerful stabiliser of protein and membrane structure (Hottiger *et al.*, 1987). Hensel & Konig (1988) found that the accumulation of high concentrations of cyclical 2,3-diphosphoglycerate raised the heat tolerance of proteins of certain thermophilic archaebacteria to permit function at temperatures in excess of 90 °C. Cossins (1983) has shown that the lipid microenvironment of a membrane-bound protein can also influence the protein's thermal stability, an important point we return to below. All of these data support the conjecture that adaptive change in the medium in which proteins function can play an important role in adaptation to temperature.

Milieu adaptations may be especially important in extending the temperature range over which protein structure and function can be maintained, as manifested by the stabilisation under alphastat pH_i conditions of K_m values for enzymes that employ histidyl residues in the binding or catalytic mechanism. Eurythermy of protein function and structure thus can be significantly enhanced through adjustments of the aqueous or lipid microenvironments of proteins. The extent to which stenotherms and eurytherms differ in their abilities to modify adaptively the milieu of proteins is not well understood. Recent studies of acclimation in membrane-based systems suggest that eurytherms may surpass stenotherms in their ability to modify adaptively the milieu of membrane-associated proteins (Dahlhoff & Somero, 1993*b*).

Membrane-based systems: critical loci for thermal sensitivity

Membrane systems of animals are a critical locus of thermal sensitivity (Cossins, 1983; Cossins & Bowler, 1987; Hazel & Williams, 1990; Hazel, 1995). Disruption of membrane function at high temperature has been invoked as a primary factor in heat death, and even sublethal temperatures may cause significant impairment of diverse behavioural characteristics dependent on the integrity of neural membranes (Cossins & Bowler, 1987). Temperature effects at the membrane level also provide important examples for distinguishing the different modes of adaptation noted in short-term phenotypic acclimation or acclimatisation versus long-term evolutionary adaptation and, as in the case of protein systems, the responses of membrane systems to temperature frequently mirror the degree of eurythermy or stenothermy noted at the level of the whole organism. The two membrane-based systems discussed below illustrate these various relationships very strikingly.

Impairment of synaptic transmission: a potential mechanism of thermal death

Synaptic transmission is a highly temperature-sensitive component of neural function; effects of temperature at the synapse appear to be greater than at subsequent events involving impulse conduction (Cossins & Bowler, 1987). Disruption of synaptic function by changes in body temperature would be expected to lead to numerous behavioural effects. Likewise, temperature-compensatory acclimation of the properties of synaptic membranes would be predicted to correlate with recovery of behavioural capacities. Relationships of this nature have been observed in studies of thermal acclimation in fishes. For example, Cossins *et al.* (1977) showed that the time course of acclimation of chill coma temperature and temperature of loss of equilibrium closely paralleled the time course of homeoviscous acclimation of the membranes of synaptic vesicles, as indexed by fluorescence polarisation of the membrane probe diphenylhexatriene (DPH). Membrane fluidity and behavioural traits changed more quickly during warm- than cold-acclimation, illustrating the type of Q_{10} effect discussed earlier in the context of differential expression of cMDH isozymes.

The importance of maintaining satisfactory synaptic transmission, and the consequences of loss of this ability for survival, are suggested by three independent studies of stenothermal Antarctic notothenioid fishes, including the cryopelagic species *P. borchgrevinki*. Investigations of the thermal tolerance ranges of notothenioid species belonging to two genera demonstrated that the upper incipient lethal temperature of *P. borchgrevinki* is approximately 4 °C, the lowest temperature of heat death known for any animal (Somero & DeVries, 1967). The involvement of neurological damage in their heat death is suggested by the finding that, at temperatures near 10 °C, the fish exhibited rapid and erratic swimming motions and irreversible flaring of the opercula occurred. The involvement in heat death of a membrane-associated enzyme of importance in synaptic function was suggested by work of Baldwin (1971), who showed that the K_m of acetylcholine (ACh) of the acetylcholine esterase (AChE) of *P. borchgrevinki* increased extremely rapidly at temperatures slightly above 0 °C (Fig. 4*a*). The response of

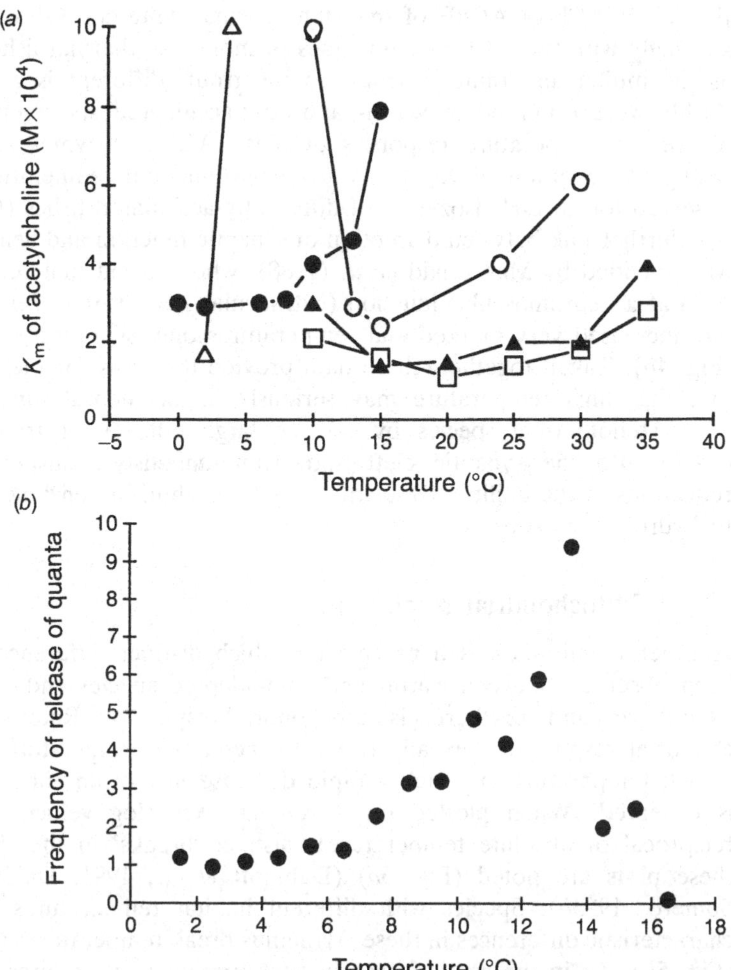

Fig. 4. Effects of temperature on brain acetylcholine esterase of fishes adapted to different temperatures, and on acetylcholine release at synapses of *Pagothenia borchgrevinki*. (*a*) Effect of temperature on the K_m of acetylcholine for acetylcholine esterases of fishes adapted to different temperatures (data from Baldwin, 1971). △, *Pagothenia*; ○, Rainbow trout at 18 °C; ●, Rainbow Trout at 2 °C; ▲, Mullet at 25 °C; □, Ladyfish at 25 °C. (*b*) Effect of temperature on release of quanta of acetylcholine at a neuromuscular junction (extraocular nerve) in *P. borchgrevinki* (data from Macdonald *et al.*, 1988).

the K_m of ACh of AChE of this stenothermal Antarctic fish contrasts strikingly with the AChE homologues of more eurythermal fishes such as the mullet and rainbow trout. In the trout, different isozymes of AChE were observed in warm- and cold-acclimated specimens. The K_m versus temperature responses of these AChE isozymes reflect a strong conservation of K_m at normal environmental temperatures, as observed for cMDHs isozymes of differently-acclimated fishes (Fig. 3).

A further link between disruption of synaptic function and heat death was provided by Macdonald *et al.* (1988), who showed that release of ACh at a neuromuscular junction (extraocular nerve) of *P. borchgrevinki* increased very markedly at temperatures only slightly above 0 °C (Fig. 4*b*). Taken together, these data provide the basis for the conjecture that high temperature may seriously impair neural function in these stenothermal species by causing large effluxes of transmitter (ACh) into the synaptic cleft and, simultaneously, causing major reductions in the ability of the transmitter metabolising enzyme AChE to hydrolyse its substrate.

Mitochondrial respiration

Another membrane-based process for which distinct differences have been observed between warm- and cold-adapted species and between stenotherms and eurytherms is mitochondrial respiration. Rates of mitochondrial respiration typically rise with increasing temperature, up to a high temperature at which a rapid decrease in oxygen consumption is observed. When plotted on Arrhenius axes (log velocity versus reciprocal of absolute temperature), distinct 'breaks' in the slope of these plots are noted (Fig. 5*a*) (Dahlhoff *et al.*, 1991; Dahlhoff & Somero, 1993*b*). Species with different habitat temperatures exhibit characteristic differences in these Arrhenius break temperatures (ABTs) (Fig. 5*b*). As in the case of protein denaturation temperatures, ABTs lie above the upper lethal temperatures of the organisms. In both cases, however, the correlations between heat inactivation temperature and adaptation temperature are viewed as reflections of temperature adaptive differences among species that are instrumental in setting thermal optima and tolerance limits (Dahlhoff *et al.*, 1991; Somero, 1995).

Thermal acclimation leads to shifts in ABT, as shown for congeneric species of abalone found in different thermal habitats (Fig. 2) and differing in their thermal tolerance ranges (Fig. 6) (Dahlhoff & Somero, 1993*b*). The thermal sensitivity of mitochondrial respiration is seen to be a very plastic trait, in the sense that mitochondria of abalone alter

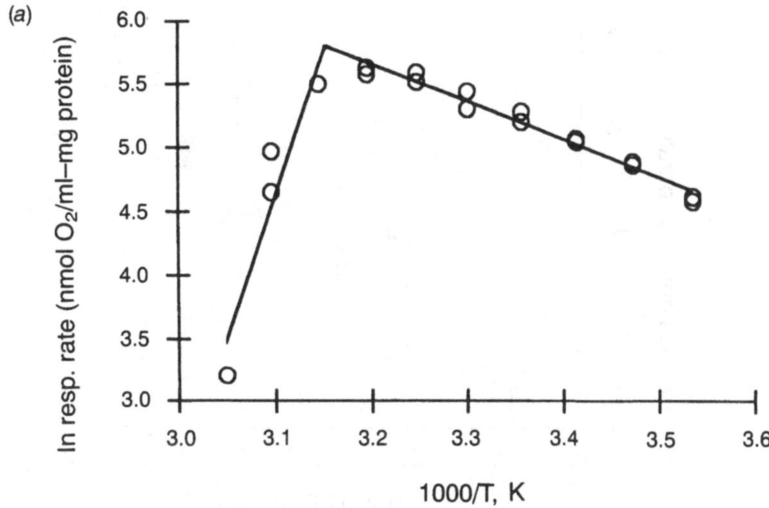

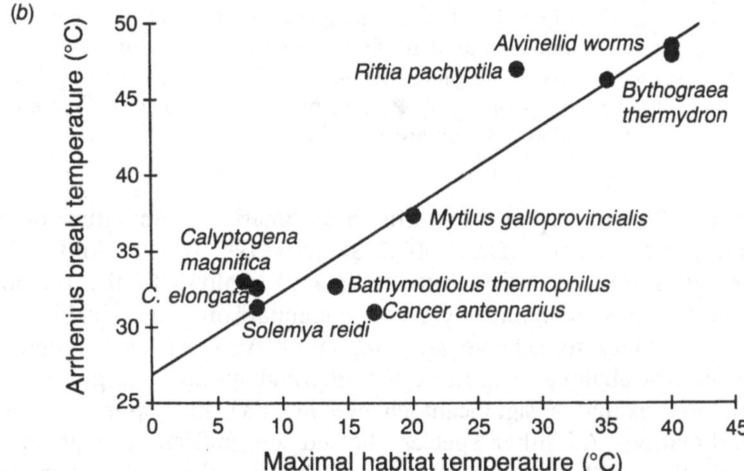

Fig. 5. Effect of temperature on respiration by isolated mitochondria from species adapted to different temperatures. (*a*) Effect of measurement temperature on rate of oxygen consumption by mitochondria of abalone. The Arrhenius break temperature (ABT) is defined as the temperature at which the two regression lines intersect (see Dahlhoff & Somero, 1993*b* for experimental details). (*b*) Relationship between Arrhenius break temperature and adaptation temperature for ten species of marine invertebrates adapted to different temperatures (Dahlhoff *et al.*, 1991).

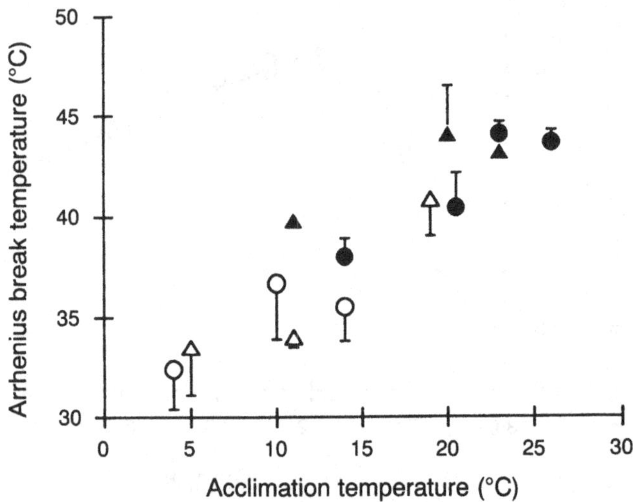

Fig. 6. Effect of acclimation temperature on Arrhenius break temperature for mitochondrial respiration of four species of abalone. For all species except the pinto abalone, acclimation temperature had a significant effect on ABT. ●, Green; ▲, pink; △, red; ○, pinto. Data from Dahlhoff & Somero (1993*b*).

their ABTs in response to shifts in acclimation temperature of only a few degrees Celsius (Dahlhoff & Somero, 1993*b*). The ABTs of mitochondria from the differently-acclimated abalone fit the relationship found in the taxonomically broader comparison shown in Fig. 5.

The ability to achieve acclimation of ABT was not found in all species of abalone. The most stenothermal species, the pinto abalone, did not exhibit a significant change in ABT in response to thermal acclimation. All other species showed a significant change in ABT, with the most eurythermal species showing the widest temperature range over which ABT could be adjusted. It was also demonstrated in these acclimation studies that at temperatures outside of the normal physiological temperature range, acclimation of ABT did not occur, even though the organisms survived these exposures. These data indicate that the ability to acclimate ABT, and possibly other traits, is restricted to a narrower range of temperatures than the full thermal tolerance range of a species. Moreover, as shown for ABT of mitochondria of the pinto abalone, in stenothermal species that encounter only narrow ranges of body temperature the capacity to undergo acclimation may not exist at all.

The biochemical mechanisms underlying the ABT phenomenon *per se*, and acclimation-induced shifts in ABT are not fully understood. Disruption of lipid–protein interactions may be partially responsible for ABTs (O'Brien *et al.*, 1991). During acclimation, changes in membrane fluidity may contribute to adjustments of ABTs. In the abalone species, changes in ABT with thermal acclimation were generally, although not always, correlated with acclimatory shifts in the fluidity of mitochondrial membranes, as indexed by the fluorescence polarisation of the membrane probe DPH (Dahlhoff & Somero, 1993*b*). With rapid growth in our knowledge about the mechanisms that achieve temperature-adaptive restructuring of membrane lipids (Hazel, 1995), it will be important to determine if stenotherms lack some of the plasticity in membrane restructuring systems that occurs in closely-related eurytherms.

What makes a 'good' eurytherm?

Species that exhibit a high degree of eurythermy are seen to differ from stenotherms in several important ways. Homologous gene products like enzymes show responses to temperature that mirror very clearly the degree of eurythermy of the whole organism. Temperature effects on ligand binding by orthologous homologues of enzymes illustrate this type of difference (Figs. 1 and 2). Increases in the number of genes encoding a particular type of protein may provide additional variation, in the form of paralogous isozymes, for selection to act on. Temperature-specific paralogous isoforms of proteins, as seen with cMDHs of teleosts (Fig. 3), thus could facilitate the evolution of a eurytherm. Conversely, for stenothermal species that have existed for substantial periods in thermally stable environments, loss of genetic information may result from the lack of selection for gene products that enhance eurythermy. The absence of the thermolabile isozyme of cMDH in the tropical barracuda *S. ensis* might be an example of this type of effect.

The genetic information present in eurytherms must also support the types of acclimatory changes seen in lipid systems, in which major restructuring of membranes occurs in response to temperature change. It is not known if the complex regulatory systems involved in membrane restructuring are more elaborate and responsive to temperature change in eurytherms than in stenotherms. The observation that the pinto abalone did not shift mitochondrial ABTs during acclimation is consistent with the view that regulatory mechanisms may be attenuated in stenotherms. Similarly, the fact that Antarctic notothenioid fishes, stenotherms *par excellence*, cannot be acclimated to temperatures higher than 4 °C, suggests a loss of mechanisms for regulating acclimatory changes, as well as the evolution of proteins with functional abilities severely limited at temperatures just above zero (Figs. 1 and 4).

Another factor contributing to the degree of eurythermy characteristic of a species may be the absolute temperature to which the species is adapted. The ability to function at high temperatures may confer a significant degree of eurythermy to a species (Coppes & Somero, 1990). In general, reductions in temperature from the upper end of the physiological range down to the lowest temperatures that a species tolerates are less perturbing of enzymes than increases in temperature above the physiological temperature range. Large increases in K_m are almost always seen when temperature is increased above the physiological temperature range, but decreases in temperature through the physiological temperature range seldom lead to large perturbations of K_m. In other words, proteins that are adapted to function at high temperatures are capable of functioning well at lower temperatures, but the converse is not generally true.

In summary, the development of a high degree of eurythermy appears to be based in part on the evolution of proteins that maintain relatively stable values for key functional properties such as ligand binding over wide ranges of temperature. Eurythermy is also fostered by the ability to restructure the phenotype, e.g. isozyme composition and membrane function and fluidity, in response to temperature change. Thus, genetically-fixed, adaptive differences among orthologous homologues of proteins and the degree of phenotypic 'plasticity' that characterises a species both contribute to determining the range of temperatures over which life is possible.

Acknowledgement

This research was supported in part by a National Science Foundation grant, IBN92–06660.

References

Baldwin, J. (1971). Adaptation of enzymes to temperature: acetylcholinesterases in the central nervous system of fishes. *Comparative Biochemistry and Physiology* **40B**, 181–7.

Carey, C. & Hazel, J.R. (1989). Diurnal variation in membrane lipid composition of Sonoran desert teleosts. *Journal of Experimental Biology* **147**, 375–91.

Coppes, Z.L. & Somero, G.N. (1990). Temperature-adaptive differences between the M_4-lactate dehydrogenases of stenothermal and eurythermal Sciaenid fishes. *Journal of Experimental Zoology* **254**, 127–31.

Cossins, A.R. (1983). The adaptation of membrane structure and function to changes in temperature. In *Cellular Acclimatisation to Environmental Change, SEB Seminar Series 17* (ed. A.R. Cossins & P. Sheterline), pp. 3–32. London: Cambridge University Press.

Cossins, A.R. & Bowler, K. (1987). *Temperature Biology of Animals*. London: Chapman and Hall.

Cossins, A.R., Friedlander, M.J. & Prosser, C.L. (1977). Correlations between behavioural temperature adaptations by goldfish and the viscosity and fatty acid composition of their synaptic membranes. *Journal of Comparative Physiology* **120**, 109–21.

Dahlhoff, E., O'Brien, J., Somero, G.N. & Vetter, R.D. (1991). Temperature effects on mitochondria from hydrothermal vent invertebrates: evidence for adaptation to elevated and variable habitat temperatures. *Physiological Zoology* **64**, 1490–508.

Dahlhoff, E. & Somero, G.N. (1991). Pressure and temperature adaptation of cytosolic malate dehydrogenases of shallow- and deep-living marine invertebrates: evidence for high body temperatures in hydrothermal vent animals. *Journal of Experimental Biology* **159**, 473–87.

Dahlhoff, E. & Somero, G.N. (1993*a*). Kinetic and structural adaptations of cytoplasmic malate dehydrogenases of eastern Pacific abalones (genus *Haliotis*) from different thermal habitats: biochemical correlates of biogeographical patterning. *Journal of Experimental Biology* **185**, 137–50.

Dahlhoff, E. & Somero, G.N. (1993*b*). Effects of temperature on mitochondria from abalone (genus *Haliotis*): adaptive plasticity and its limits. *Journal of Experimental Biology* **185**, 151–68.

Dietz, T.J. & Somero, G.N. (1992). The threshold induction temperature of the 90-kDa heat shock protein is subject to acclimatization in eurythermal goby fishes (genus *Gillichthys*). *Proceedings of the National Academy of Sciences of the USA* **89**, 3389–93.

Donahue, V.E. (1982). *Lactate dehydrogenase: structural aspects of environmental adaptation*. Ph.D. dissertation, University of California, San Diego.

Eastman, J. (1993). *Antarctic Fish Biology*. San Diego: Academic Press.

Fields, P. (1995). *Adaptation to temperature in two genera of coastal fishes,* Paralabrax *and* Gillichthys. Ph.D. dissertation, University of California, San Diego .

Fields, P., Graham, J.B., Rosenblatt, R. H. & Somero, G.N. (1993). Effects of expected global change on marine faunas. *Trends in Ecology and Evolution* **8**, 30–7.

Graves, J. E. & Somero, G.N. (1982). Electrophoretic and functional enzymic evolution in four species of eastern Pacific barracudas from different thermal environments. *Evolution* **36**, 97–106.

Hand, S.C. & Somero, G.N. (1983). Phosphofructokinase of the hibernator, *Citellus beecheyi*: temperature and pH regulation of activity via influences on the tetramer-dimer equilibrium. *Physiological Zoology* **56**, 380–8.

Hazel, J.R. (1995). Thermal adaptation in biological membranes: Is homeoviscous adaptation the explanation? *Annual Review of Physiology* **57**, 19–42.

Hazel, J.R. & Williams, E.E. (1990). The role of alterations in membrane lipid composition in enabling physiological adaptations to organisms to their physical environment. *Progress in Lipid Research* **29**, 167–227.

Hensel, R. & Konig, H. (1988). Thermoadaptation of methanogenic bacteria by intracellular ion concentration. *FEMS Microbiological Letters* **49**, 75–9.

Hochachka, P.W. & Somero, G.N. (1984). *Biochemical Adaptation*. New Jersey: Princeton University Press.

Hofmann, G.E. & Somero, G.N. (1995). Evidence for protein damage at environmental temperatures: seasonal changes in levels of ubiquitin conjugates and hsp70 in the intertidal mussel *Mytilus trossulus*. *Journal of Experimental Biology* **198**, 1509–18.

Hottiger, T., Boller, T. & Wiemken, A. (1987). Rapid changes of heat and desiccation tolerance correlated with changes of trehalose content in *Saccharomyces cerevisiae* cells subjected to temperature shifts. *FEBS Letters* **220**, 113–15.

Jaenicke, R. (1991). Protein stability and molecular adaptation to extreme conditions. *European Journal of Biochemistry* **202**, 715–28.

Johnston, I.A. (1983). Cellular responses to an altered body temperature: the role of alterations in the expression of protein isoforms. In *Cellular acclimatisation to environmental change, SEB Seminar Series 17* (ed. A.R. Cossins & P. Sheterline), pp. 121–43. London: Cambridge University Press.

Johnston, I.A. & Walesby, N.J. (1977). Molecular mechanisms of temperature adaptation in fish myofibrillar adenosine triphosphatases. *Journal of Comparative Physiology B* **119**, 195–206.

Lin, J.J. & Somero, G.N. (1995*a*). Temperature-dependent changes in expression of thermostable and thermolabile isozymes of cytosolic malate dehydrogenase in the eurythermal goby fish *Gillichthys mirabilis*. *Physiological Zoology* **68**, 114–28.

Lin, J.J. & Somero, G.N. (1995*b*). Thermal adaptation of cytoplasmic malate dehydrogenases of eastern Pacific barracuda (*Sphyraena* spp.): the role of differential isoenzyme expression. *Journal of Experimental Biology* **198**, 551–60.

Lubchenco, J., Navarette, S.A., Tissot, B.N. & Castilla, J.C. (1993). Possible ecological responses to global climate change: nearshore benthic biota of northeastern Pacific ecosystems. In *Earth Systems*

Responses to Global Change (ed. H. Mooney, E.R. Fuentes & B.I. Kronberg), pp. 147–66. London: Academic Press.

Macdonald, J.A., Montgomery, J.C. & Wells, R.M.G. (1988). The physiology of McMurdo Sound fishes: current New Zealand research. *Comparative Biochemistry and Physiology* **90B**, 567–78.

McFall-Ngai, M.J. & Horwitz, J. (1990). A comparative study of the thermal stability of the vertebrate eye lens: Antarctic ice fish to the desert iguana. *Experimental Eye Research* **50**, 703–9.

Matthews, B.W. (1987). Genetic and structural analysis of the protein stability problem. *Biochemistry* **26**, 6885–8.

O'Brien, J., Dahlhoff, E. & Somero, G.N. (1991). Thermal resistance of mitochondrial respiration: hydrophobic interactions of membrane proteins may limit thermal tolerance. *Physiological Zoology* **64**, 1509–26.

Powers, D.A., Smith, M., Gonzalez-Villasenor, I., DiMichele, L., Crawford, D., Bernardi, G. & Lauerman, T. (1993). A multidisciplinary approach to the selectionist/neutralist controversy using the model teleost, *Fundulus heteroclitus*. In *Oxford Surveys in Evolutionary Biology*, vol. 9 (ed. D. Futuyma & J. Antonovics), pp. 43–107. Oxford: Oxford University Press.

Reeves, R.B. (1977). The interaction of body temperature and acid–base balance in ectothermic vertebrates. *Annual Review of Physiology* **39**, 559–86.

Schwantes, M.L.B. & Schwantes, A.R. (1982). Adaptive features of ectothermic enzymes. I. Temperature effects on the malate dehydrogenase from a temperate fish *Leiostomus xanthurus*. *Comparative Biochemistry and Physiology* **72B**, 49–58.

Sidell, B.D., Wilson, F.R., Hazel, J. & Prosser, C.L. (1973). Time course of thermal acclimation in goldfish. *Journal of Comparative Physiology* **84**, 119–27.

Somero, G.N. (1969). Pyruvate kinase variants of the Alaskan king crab. *Biochemical Journal* **114**, 237–41.

Somero, G.N. (1986). Protons, osmolytes, and fitness of the internal milieu for protein function. *American Journal of Physiology* **252** (*Regulatory, Integrative, Comparative Physiology 20*), R197–R213.

Somero, G.N. (1992). Adapting to water stress: convergence on common solutions. In *Water and Life* (ed. G.N. Somero, C.B. Osmond & C. L. Bolis), pp. 3–18. Berlin: Springer-Verlag.

Somero, G.N. (1995). Proteins and temperature. *Annual Review of Physiology* **57**, 43–68.

Somero, G.N. & DeVries, A.L. (1967). Temperature tolerance of some Antarctic fishes. *Science* **156**, 257–8.

Somero, G.N. & Hofmann, G.E. (1996). Temperature thresholds for protein adaptation: when does temperature change start to 'hurt'? In *Global Warming: Implications for Freshwater and Marine Fish,*

SEB Seminar Series 61 (ed. C. Wood & G. McDonald), Cambridge: Cambridge University Press (in press).

Swezey, R.R. & Somero, G.N. (1982). Polymerization thermodynamics and structural stabilities of skeletal muscle actins from vertebrates adapted to different temperatures and pressures. *Biochemistry* **21**, 4496–503.

Tsuji, S., Qureshi, M.A., Hou, E.W., Fitch, W.M. & Li, S.S.-L. (1994). Evolutionary relationships of lactate dehydrogenases (LDHs) from mammals, birds, an amphibian, fish, barley, and bacteria: LDH cDNA sequences from *Xenopus*, pig, and rat. *Proceedings of the National Academy of Sciences of the USA* **91** 9392–6.

Yancey, P.H., Clark, M.E., Hand, S.C., Bowles, R.D. & Somero, G. N. (1982). Living with water stress: evolution of osmolyte systems. *Science* **217**, 1214–22.

Yancey, P.H. & Siebenaller, J.F. (1987). Coenzyme binding ability of homologs of M_4-lactate dehydrogenase in temperature adaptation. *Biochimica et Biophysica Acta* **924**, 483–91.

Yancey, P.H. & Somero, G.N. (1978). Temperature dependence of intracellular pH: its role in the conservation of pyruvate apparent K_m values of vertebrate lactate dehydrogenase. *Journal of Comparative Physiology B* **125**, 129–34.

M.E. FEDER

Ecological and evolutionary physiology of stress proteins and the stress response: the *Drosophila melanogaster* model

A context for molecular studies of the thermal phenotype

The thermal phenotype of an organism comprises hundreds if not thousands of traits. Some of these traits determine the tolerance limits, others determine the thermal sensitivity of physiological performance within the zone of tolerance, and still others underlie responses to changes in temperature such as acclimation, behavioural thermoregulation and physiological thermoregulation. The descriptions or explanations of these traits now constitute an enormous but still-growing literature (Fig. 1*a*).

The traits underlying the thermal phenotype might behave as an ensemble in at least two ways. On the one hand, each trait could play an essential role such that variation in any trait would have discernible if not major effects on physiological performance and evolutionary fitness. (Fig. 1*b* analogises this alternative to a 'house of cards', in which removal of any one card causes the entire structure to collapse.) On the other hand, numerous and redundant traits could underlie each aspect of the thermal phenotype such that variation in any given trait might have negligible consequences for the thermal phenotype as a whole; i.e. variation in numerous traits is necessary to affect the thermal phenotype. (Fig. 1*c* analogises this alternative to a 'roller coaster', in which many struts must be removed before the structure will fail.) These are obviously extreme alternatives, and intermediate states of ensemble behaviour clearly exist in organisms.

In any event, the large number and extraordinary variety of the traits underlying the thermal phenotype pose a general challenge for comparative biologists' understanding of the thermal phenotype as a whole and a particular challenge in deciding where, on the continuum between a 'house of cards' and a 'roller coaster', any given trait falls. The difficulty arises from the evolutionary covariation of traits: as a new thermal phenotype evolves, numerous traits change consecutively

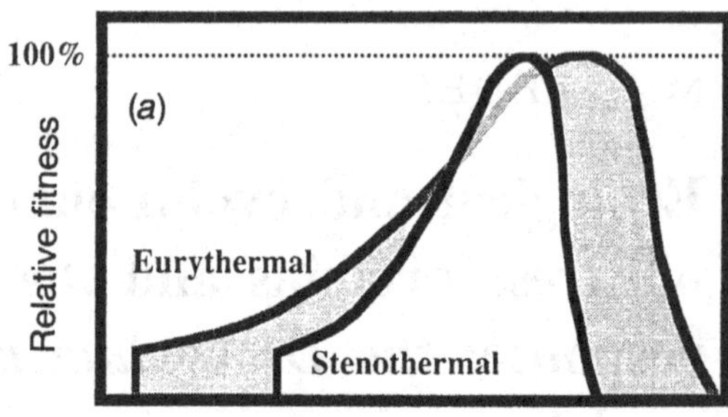

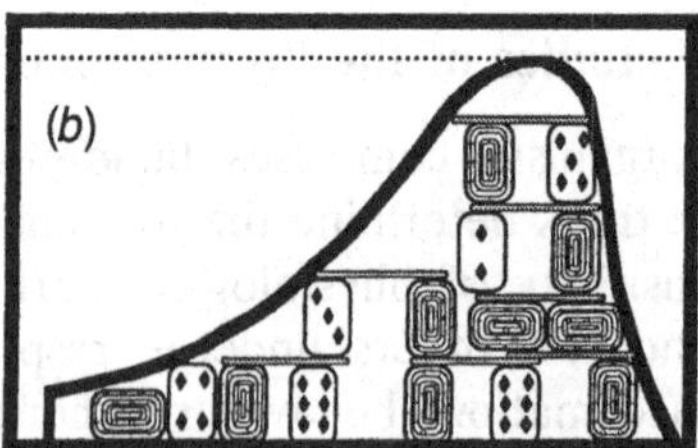

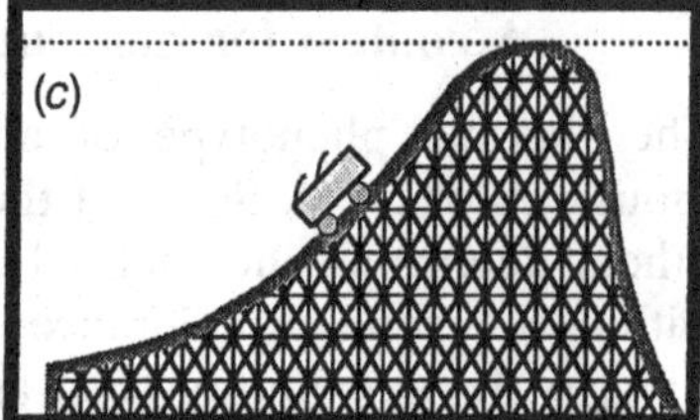

Fig. 1. Views of the thermal phenoptype. (*a*) Representation of two thermal phenotypes, one eurythermal (shaded) and one stenothermal (solid line, no shading). The former is able to tolerate more extreme environments and maintain a higher relative fitness within the zone of tolerance than the latter. These could represent a single individual at different times in its life, different individuals within a population, or individuals from different populations, species, or higher taxa. Many traits comprise each of these thermal phenotypes. (*b*) One alternative view of the ensemble properties of traits comprising the thermal phenotype: a house of cards. According to this view, experimental manipulation of any given trait in isolation is sufficient to alter the thermal phenotype. (*c*) Another alternative view: a roller coaster. According to this view, numerous traits redundantly support the thermal phenoptype such that manipulation of any one trait is likely to have a negligible effect on the thermal phenotype as a whole.

if not simultaneously (Feder, 1987). Thus, in any comparison of species or populations from differing thermal environments or even of individuals before and after thermal acclimation, unambiguous attribution of a difference in thermal phenotype to any trait or group is problematic. For example, a phylogenetically controlled comparison of fish species from high and low latitudes might document a correlation between ability to reproduce at low temperatures (that is, a difference in thermal phenotype) and fatty acid desaturase activity, transcription of the *Ldh-B* gene at low temperatures, and blood oxygen capacity (all traits that might reasonably underlie the difference in thermal phenotype). To what extent is each of these particular underlying traits responsible for the difference in thermal phenotype? This question becomes very difficult to answer because all three underlying traits have changed simultaneously. Moreover, in all likelihood numerous other underlying traits have changed as well.

In theory, one way to address the difficult question of a trait's contribution to an observed difference in thermal phenotype is to manipulate that trait while all other traits are held constant. One of the most exciting developments in the modern biological sciences is that the tool kit of molecular biology and allied techniques enables the organismal and integrative biologist to perform such experimental manipulations (Feder & Block, 1991). In particular, techniques such as site-directed mutagenesis, gene targeting and insertion of antisense or ribozyme constructs afford unprecedented power to the organismal physiologist. Even so, the techniques of molecular biology are not without their drawbacks. A principal problem is that most techniques have been optimised for model systems that lend themselves to experimental manipulation in the laboratory (e.g. bacteria (*Escherichia coli*), yeast (*Saccharomyces cerevisiae*), fruit flies (*Drosophila melanogaster*), mouse (*Mus*) and cultured cell lines). Often, models chosen on this basis either are seemingly irrelevant to understanding the evolution of thermal phenotypes in a natural setting, lack a well-documented natural history, or both. This seemingly leaves the investigator with an unpalatable choice between work with an experimentally tractable but ecologically irrelevant organism, or *vice versa*. Investigators are now responding to this seeming but false dilemma in various ways (see work by Bennett and Huey, Chapters 9 and 10). In my own research programme, the development of a novel technique for genetic engineering in *D. melanogaster* induced me to document the thermal ecology of these insects and thus to provide the ecological context for interpretation of molecular physiological experimentation. The particular trait under investigation is the inducible cytoplasmic–nuclear 70 kDa heat-shock protein (hsp) of *D. melanogaster*, hsp70 (*Drosophila melanogaster* expresses several hsp70 family members, some constitutively. Hereafter, 'hsp70' will refer to the specific family member characterised by Velazquez and colleagues (Velazquez *et al.*, 1980, 1983; Velazquez & Lindquist, 1984). This chapter first provides a general introduction to the heat-shock or stress response, then examines the thermal stress that *D. melanogaster* actually encounters in the field as well as natural stress protein expression in response to this stress, and finally discusses the impact of hsp70 on the thermal phenotype of *D. melanogaster* in an ecological context.

Portions of this work have appeared as an abstract (Feder *et al.*, 1994); the remainder is unpublished results.

Stress response and stress proteins

In 1962, Ritossa reported that heat and the metabolic inhibitor dinitrophenol induced a characteristic pattern of puffing in the chromosomes of *Drosophila*. This discovery eventually led to the identification of the heat-shock or stress proteins (these names will be used interchangeably) whose expression these puffs represented, the cloning of the genes encoding these proteins and elucidation of the regulation of expression of these genes. Beginning in the mid-1980s, investigators recognised that many hsps function as molecular chaperones and thus play a critical role in protein folding, intracellular trafficking of proteins and coping with proteins denatured by heat and other stresses. Accordingly, the study of stress proteins has undergone explosive growth, now accounting for more than 1000 papers per year and numerous monographs, edited volumes and reviews (Morimoto *et al.*, 1990, 1994*a*). Feder *et al.* (1995), provide a general overview for physiologists and cite many recent reviews. Given the pace of progress in the field, that review is necessarily superficial and already dated. Some key points, for which references are provided in that review, are as follows.

The stress proteins themselves can be assigned to several families or varieties of relatively large proteins (hsp100, hsp90, hsp70, hsp60 and Lon) and of smaller proteins (hsp27, DnaJ, GrpE, hsp10 and ubiquitin). Although originally recognised by their induction by heat or other stresses, many of the hsps are now known to be expressed constitutively or have constitutively-expressed cognates. Within a family, hsps can be extraordinarily highly conserved. The hsp70s of humans and *E. coli*, for example, are 50% identical in sequence; those of humans and *Drosophila* are more than 70% identical. The entire stress response (i.e. the induction of the characteristic suite of stress proteins) is itself highly conserved and has been reported in each of the hundreds of species in which it has been sought, save one. Details may differ from species to species, however. For example, the small hsps dominate the stress response in many plants, whereas yeast and mammals express only one small hsp.

Although heat has received more attention than other hsp-inducing stresses, it is by no means the only inducer. Inducing stresses include ethanol, heavy metals, hypoxia, hyperoxia, changes in pH, free radicals, various poisons and toxins, ischaemia, osmotic shock, ionising radiation, and many others. To date, no stresses have been reported not to induce heat shock proteins. Studies examining stress protein expression in the wild or in response to laboratory simulations of natural stress regimens are still few. Nonetheless, even these few studies are sufficient to demonstrate that patterns of stress protein expression can be correlated with the species' natural thermal environments; i.e. cells and species from warm environments undergo a stress response at warmer temperatures than counterparts from cool environments (Lindquist, 1986; Huey & Bennett, 1990; Sarge *et al.*, 1995; Somero, 1995).

The expression of hsps during stress and in the recovery period after stress abates coincides with the induction of improved stress tolerance. In terms of high temperature stress, this induction of thermal tolerance requires minutes to hours, and is thus more rapid than the acclimatory changes typically documented in studies of the critical or lethal limits of animals. Its time course resembles that of heat-hardening (Hutchison & Maness, 1979), although the induction conditions of these two phenomena may differ. Stress protein expression and inducible thermotolerance are often correlated, suggesting that the former is responsible for the latter. Experimental manipulations of stress protein expression have now proven this suggestion. For example, Sanchez & Linquist (1990) engineered a yeast (*S. cerevisiae*) strain in which the *hsp104* gene was deleted. This deletion greatly reduced inducible thermotolerance. Reintroduction of the *hsp104* gene on a plasmid restored inducible thermotolerance to that of wild-type yeast. Accordingly, the yeast *hsp104* conforms to the 'house of cards' model in that deletion of this single protein markedly affects the thermal phenotype. Similar studies have now been performed on yeast, *E. coli* and various cultured cell lines for many of the hsp families, with parallel results. Other work has examined cells or organisms engineered to over-express particular hsps, with resultant improvement in thermotolerance (see Welte *et al.* (1993) and later).

As noted above, the major advance in our understanding of the function of stress proteins came about with the recognition of molecular chaperones (Parsell & Lindquist, 1993, 1994; Feder *et al.*, 1995). The basic thesis is as follows: although cellular proteins are typically folded in their native conformations while functioning in cells, proteins may be unfolded in several contexts: (i) during *de novo* synthesis of polypeptides and assembly of multimeric proteins; (ii) during intracellular transport and organellar import, when a protein must unfold or remain unfolded to cross the boundary of a cellular compartment; and (iii) during or after exposure to a protein-denaturing stress. At these times, unfolded proteins may be susceptible to inappropriate interactions with one another or with other cellular components. For example, hydrophobic side groups that are normally sequestered in the interior of folded proteins may be exposed in an unfolded protein and interact with other such groups that are normally inaccessible. Moreover, once unfolded, a protein can prospectively interact with folded proteins and induce them to unfold. Such interactions can result in aggregates of unfolded protein that at best diminish the pool of functional protein and at worst are cytotoxic. Molecular chaperones are a class of proteins that enable the cell to cope with this problem. Chaperones can recognise and bind to the exposed side groups that characterise unfolded proteins. In so doing, molecular chaperones prevent the bound side groups from engaging in inappropriate interactions with other cellular components, as well as stabilising the bound proteins in an unfolded state. Typically in an ATP-dependent manner, molecular chaperones can then release bound proteins to allow them to refold properly. Alternatively, chaperones can target bound proteins for degradation or removal from the cell. According to this scheme, the heat shock cognates or constitutively expressed hsps perform these roles for nascent polypeptides or proteins that unfold during normal cellular processes, while the inducible hsps function primarily in response to the protein denaturation that occurs during or after stress. Molecular chaperones are not presently thought to assist a protein along its folding pathway; rather, chaperones reduce the probability of off-pathway reactions such as the formation of aggregates.

Some initial support for this hypothesis came from findings that the presence of unfolded proteins in the cell was sufficient to induce a stress response, and from the phenomenon of cross tolerance. In yeast, for example, exposure to a first stress (in this case high temperature) induces tolerance of a second stress (such as ethanol), and *vice versa* (Sanchez *et al.*, 1992). These phenomena are consistent with a role for hsps as a general response to protein denaturation. Subsequent support for this role has come about from direct studies, both *in vitro* and *in vivo*, of reporter proteins. A typical experiment (Fig. 2) involves the denaturation of the reporter protein, typically by heat or a chemical denaturant. Next, the denaturant is removed and some aspect of the structure or function of the reporter protein is assayed periodically in the presence or absence of a prospective chaperone, co-chaperone, or cofactor (ATP, for example). A positive finding is that the reporter protein recovers some or all of the structure or function it exhibited before denaturation. Such findings are now available for at least some representatives of many of the major hsp families, including hsp100,

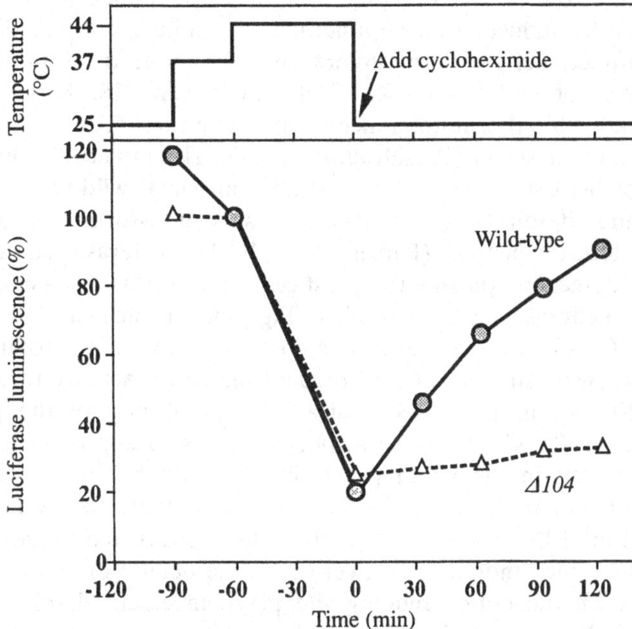

Fig. 2. Hsp104 is necessary for reactivation of luciferase *in vivo* after thermal denaturation. This figure demonstrates the time course, sequence and outcome of a typical experiment to assess the effect of a molecular chaperone on a reporter protein. Luciferase luminescence is standardised as a percentage of activity determined at the end of a 1 h pretreatment at 37 °C. Shaded circles and solid line represent wild-type yeast, which express hsp104. Open triangles and dotted line represent a *hsp104* deletion mutant strain. See Parsell *et al.* (1994) for additional details.

hsp90, hsp70, hsp60 and hsp27, and for a variety of reporter proteins, including dihydrofolate reductase, RUBISCO, citrate synthase, rhodanese and luciferase. Still other work has addressed diverse aspects of the chaperone hypothesis. Flynn *et al.* (1991), for example, have shown that BiP, a hsp70 family member, preferentially binds to polypeptides enriched in hydrophobic amino acids, and such polypeptides stimulate the ATPase activity of BiP much more effectively than polypeptides enriched in hydrophilic amino acids.

Recent work with hsp104 in yeast can exemplify many of these findings. The hsp104 monomer includes two nucleotide-binding sites separated by a spacer region (Parsell *et al.*, 1991). In the cell, hsp104 exists as a hexamer (Parsell *et al.*, 1993). As mentioned above, hsp104 is essential for inducible thermotolerance. Both nucleotide binding sites must be functional for thermotolerance to be induced, as site-directed mutagenesis of either site (K→T at amino acid 218, K→T at 620) reduces inducible thermotolerance to the same level seen in the *hsp104* deletion mutant strain (Parsell *et al.*, 1991). The molecular chaperone activity of hsp104 is evident from studies in which wild-type, deletion mutant and site-directed mutant strains were transformed with a bacterial luciferase construct (Parsell *et al.*, 1994): luciferase luminescence is readily detectable outside the yeast cell, and so provides a convenient assay of luciferase activity *in vivo* (Fig. 2). In each strain, transfer from 25 °C (the normal culture temperature) to 37 °C (to induce a stress response) and then 44 °C reduces luciferase activity to approximately 20% of initial levels, indicating denaturation of the protein. Thereafter at 25 °C, luciferase activity recovers to approximately 80% of initial levels in the wild-type strain (i.e. with hsp104 present) but only to 30% of initial levels in the deletion mutant, clearly indicating the function of hsp104 as a chaperone. The site-directed mutant strains exhibit the same amount of recovery as in the deletion mutant, indicating that each nucleotide binding site plays an essential role. Hsp104 also clearly helps to resolve intracellular aggregates. In all experimental strains, heat shock decreases the fraction of soluble luciferase and increases the proportion found in insoluble aggregates. In the wild-type strain (i.e. with hsp104 present), luciferase resolubilised during recovery at 25 °C, whereas little resolubilisation occurred in the deletion mutant strain. Similar trends in the elimination of protein aggregates are clearly evident in electron micrographs of yeast cells after heat shock.

How molecular chaperones accomplish these effects is now receiving considerable scrutiny. Studies are most advanced for hsp60 (=GroEL in *E. coli*) and its co-chaperone, hsp10 (=GroES in *E. coli*) (Martin *et al.*, 1993; Braig *et al.*, 1994; Schmidt *et al.*, 1994). *In vivo*, 14 GroEL monomers combine to form a barrel-like structure comprising two stacked 7-membered rings. GroES monomers form a 7-membered ring. The emerging consensus is that the GroEL has a high affinity for unfolded protein, which it binds in the interior of the barrel-like structure. One model for GroEL–GroES function (Martin *et al.*, 1993) is that binding of the GroES ring to the GroEL cylinder reduces the affinity of the GroEL cylinder for unfolded protein, which is released from GroEL and is then free to fold. Once the protein is fully folded, GroEL no longer has an affinity for it and its association with the chaperone complex ends. If the protein is not fully folded, it can re-bind GroEL and, in so doing, reduces GroEL–GroES affinity so that GroES dissociates from the complex. Rebinding of GroES can then initiate subsequent rounds of unfolded protein dissociation, folding and rebinding if folding is incomplete. These cycles continue until the protein is fully folded. Many details of this scheme are presently controversial, however, including whether GroES binds on one or both sides of the GroEL cylinder, whether folding occurs within the GroEL cylinder or outside of it, and whether the unfolded protein associates with one or more than one GroEL during a series of folding cycles.

The regulation of the heat shock response is linked to the activity of a specific transcription factor, HSF (heat shock factor) (Morimoto *et al.*, 1994*b*). In the absence of heat shock, HSF exists primarily in the cytoplasm and as a monomer. On heat shock, HSF trimerises and migrates to the nucleus, where the trimer binds to HSEs (heat shock elements, consensus sequences in the promoters of genes encoding heat-shock proteins) and initiates transcription. One heat-shock protein, hsp70, plays a key role in the regulation of HSF trimerisation and hence transcription. Hsp70, either by itself or as part of a protein complex, may interact with HSF monomer to inhibit trimerisation and may interact with HSF trimers to promote their dissociation from HSEs or otherwise inhibit their transcriptional activation. Unfolded proteins apparently compete with HSF monomer for interaction with hsp70. Thus, heat or other cellular stresses may denature proteins and thereby depress hsp70-mediated inhibition of HSF, stimulating transcription of hsps. When unfolded proteins have folded or been removed from the cell, repression of hsp70-mediated inhibition of transcription will decrease, as will transcription of hsps.

The stress response has significant implications for our understanding of not only the regulation of gene expression and the homeostasis of cellular proteins, but also of the ecological and evolutionary physiology of whole organisms and populations (Huey & Bennett, 1990; Feder & Block, 1991; Hoffmann & Parsons, 1991). In this regard, several investigators are now characterising the response in wild animals exposed to natural stresses (Somero, 1995). To date, these studies have yielded few surprises or patterns qualitatively different from those in laboratory studies, but nonetheless pose several critical questions: To what extent do wild organisms experience stresses severe enough to benefit from the protective effects of stress proteins? Can these protective effects, largely demonstrated *in vitro* or in single cells undergoing exposure to prospectively unnatural stresses, actually improve the fitness of entire organisms? If so, are these beneficial effects negligible against the background of the numerous other traits underlying the thermal phenotype (Fig. 1*c*), or are they critical (Fig. 1*b*)?

Thermal ecology of *D. melanogaster*

A necessary first step in the execution of ecologically and evolutionarily relevant experiments on heat shock is the demonstration that the subject species actually experiences temperatures severe enough to induce stress protein expression. Surprisingly, although *D. melanogaster* is one of the most thoroughly studied complex metazoans other than humans, its thermal ecology is almost completely unknown. To be sure, many investigators have observed *D. melanogaster* to occur in diverse climates. For an organism as small as *D. melanogaster*, however, even the most stressful of gross climates offers microhabitats in which *D. melanogaster* could evade thermal stress, and conversely even equable climates contain microhabitats in which thermal stress may exceed lethal limits. Accordingly, what is necessary to establish thermal stress in *D. melanogaster*, and what is so often lacking, are direct measurements of body temperatures or microhabitat temperatures at which *D. melanogaster* occur. The entire relevant literature for high temperatures is as follows: McKenzie & McKenzie (1979) measured larval temperatures in a pile of grape residues, and reported a mode of 25–30 °C and a range of 10–45 °C. By releasing and recapturing a strain in which eye colour of adults is related to pupal temperature, Jones *et al.* (1987) inferred that pupae experienced temperatures of 15–31 °C.

These few reports, while exemplary, are scarcely a sufficient basis for the design of ecologically relevant experiments and thus prompted a systematic investigation of the thermal ecology of *D. melanogaster*. Several considerations guided this investigation: adults are highly mobile and potentially able to move to avoid extreme conditions. Oviposition peaks late in the day (David *et al.*, 1983), and ovipositing females are able to sense and avoid warm oviposition sites (Fogleman, 1979; Schnebel & Grossfield, 1986). Accordingly, eggs and early embryos are likely to escape heat stress. Larvae, by contrast, inhabit rotting fruit that, when insolated, can become quite warm (Barber & Sharpe, 1971; Sampsell, 1977), potentially stressing any larvae within. Thus, larval *D. melanogaster* deserve particular scrutiny as a model of natural thermal stress.

Initial observations sampled rotting fruit (bananas, peaches, plums, apricots and tomatoes) randomly dispersed by the investigators across a second-growth forest–open field ecotone in Cook County, Illinois, in early summer 1994. On a sunny day, fruit temperatures ranged from 25 °C (for fruits in deep shade) to 52 °C (for the upper surface of tomatoes in direct sun). Fruit temperatures varied between these extremes depending on time of day, local shading, orientation and position within the fruit. Subsequent work focused on aspects of rotting fruit that prospectively determined the kinetics of thermal equilibration and the ultimate equilibrium temperatures of fruit in natural environments. Figure 3 exemplifies one of many such studies, this one of two

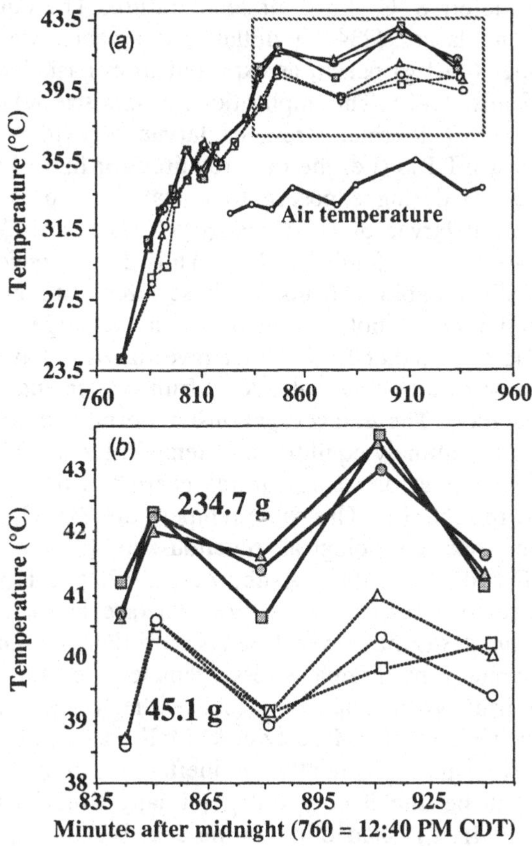

Fig. 3. Effect of time of day and fruit mass on temperatures within rotting peaches, typical larval habitats of *Drosophila melanogaster*. (*a*) Entire record; (*b*) expansion of points within shaded box in (*a*). See text for additional details. For context, the LT_{50} of third-instar Oregon R larvae measured at the time of year at which these measurements were taken is 30 min at 39 °C, 15 min at 40 °C and 8.5 min at 41 °C. Solid line, 234.7 g peach; broken line, 45.1 g peach; circle, top of fruit, 1 cm depth; square, core of fruit; triangle, bottom of fruit, 1 cm depth.

rotting peaches of similar colour but different size, each fitted with indwelling thermocouples at the core, 1 cm beneath the uppermost skin, and 1 cm below the bottom skin. The peaches were transferred from the laboratory to an exposed grass lawn at approximately 1250° Central Daylight Time (CDT) on a sunny day in late June. The temperatures of the peaches required approximately 60 min to equilibrate at approximately 8 °C above air temperature. The equilibrium temperatures of the larger (234.7 g initial mass) peach were above those of the smaller (45.1 g) peach, as expected from basic biophysical considerations (Gates, 1980). The implications of this size-related temperature difference for thermal stress of larvae is evident from a consideration of larval LT_{50}s (i.e. the time required for half of a sample of larvae to succumb during exposure to a stated temperature; see later). For third-instar larvae of *D. melanogaster*, these are: 30 min at 39 °C, 15 min at 40 °C and 8.5 min at 41 °C. Thus, for *D. melanogaster* larvae hypothetically inhabiting fruits of those sizes, size means the difference between rapid if not certain death in the larger fruit and possible survival in the smaller fruit. Other investigations have characterised the effects of evaporative water loss, fruit colour and fruit size in peaches and apples. These investigations uniformly establish that small (<250 g) fruits routinely equilibrate at temperatures >35 °C after 60–90 min of insolation at air temperatures characteristic of summer in midwestern North America. This same temperature, 35 °C, is clearly stressful for numerous physiological functions in *D. melanogaster* (Parsons, 1978; David *et al.*, 1983; Ashburner, 1989) and is sufficient to induce high rates of stress protein expression (see below).

Drosophila melanogaster has several behaviours that seemingly may minimise if not circumvent thermal stress. Females are able to sense and avoid warm fruit when ovipositing (see earlier), and larvae can thermoregulate behaviourally (McKenzie & McKenzie, 1979). These behaviours, however, may frequently be ineffective in nature. Ovipositing females will be unable to avoid potentially stressful fruits on cloudy days or if the fruit is shaded late in the day. Several experiments examined whether ovipositing females could recognise and avoid fruit that had previously been heated but was equithermal with control fruit at the time of oviposition. Females did not avoid such fruit, nor did they avoid fruit laden with heat-killed larvae, another potential cue. Larvae are likely to be no more successful in avoiding heat stress than are ovipositing females. Larvae were able to invade experimental fruit only through wounds in the cuticle or skin, and thereafter were restricted to rotting areas. Insofar as rotting areas were superficial, larvae were unable to burrow into the core of the fruit to avoid heat stress.

Detailed thermal profiles of fruit suggest that larvae might at best realise a 2 °C decrease in temperature by moving from the warmest to the coolest accessible microhabitat around a wound.

The foregoing studies suggest that larvae of *D. melanogaster* are likely to experience stressful temperatures whenever a rotting fruit they inhabit receives sustained insolation, and that behavioural mitigation of such stress may frequently be ineffective. Direct measurements of temperatures of wild *Drosophila* in an apple orchard in LaPorte County, Indiana (Fig. 4) bear out these suggestions. Both *D. melanogaster* and *D. simulans* (a sibling species) occur in such orchards. The apples represented by these temperatures were collected from the orchard substrate, across which they had been dispersed by human fruit harvesters or natural causes; they were untouched by the investigators before sampling. Temperatures of larvae and pupae ranged up to 44.5 °C, and were obtained for both living and dead *Drosophila*. These were single measurements; temperatures preceding the measurements and the survival of sampled larvae are unknown. Nonetheless, these measurements establish an ecological and evolutionary context for experimental studies of heat shock in *D. melanogaster*. Moreover, larvae express hsps in this context. During these field measurements, rotting apples in deep shade were placed in direct sunlight, simulating an acute increase in natural insolation that might occur because of the

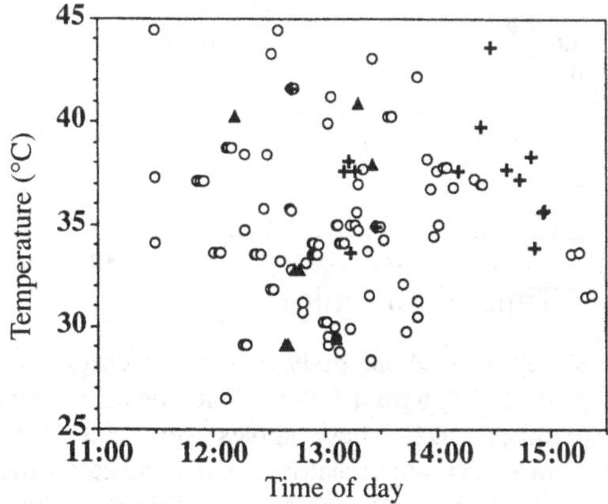

Fig. 4. Temperatures of wild larvae and pupae of *Drosophila* recorded *in situ* from fallen apples in an orchard during August and September 1994. Apples were untouched by the investigators prior to measurement of temperature. ○, Live larvae; ▲, live pupa; +, dead larvae.

earth's rotation, cloud movement or change in weather. At various times, apples were dissected, temperatures of larval microhabitats were determined and larvae were immediately preserved in liquid nitrogen. Subsequently, concentrations of hsp70 were determined via ELISA (see later). Figure 5 shows larval temperatures recorded in the fruit in relation to duration of insolation. Hsp70 concentrations increased during this experiment, eventually equaling the highest of concentrations detectable in laboratory studies with defined strains (Fig. 5; see also later).

Experimental manipulation of hsp70 expression and its consequences

The preceding section reports that larval *D. melanogaster* exploit a habitat that is prone to thermal stress, that at least some larvae experience thermal stress in nature, and that stress proteins are expressed in response. Now that this context has been established, the original question re-emerges (and with it the problematic nature of

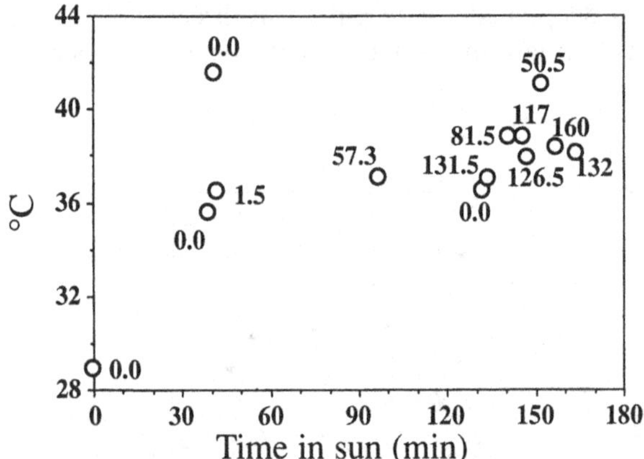

Fig. 5. Temperatures and whole body hsp70 concentrations of wild larvae of *Drosophila* living within fallen apples in an orchard during August and September 1994. These apples were collected in deep shade and placed in direct sun, resulting in the indicated increase in temperature. At various intervals, apples were dissected, the temperatures of fruit next to larvae immediately taken and corresponding larvae immediately frozen in liquid nitrogen for subsequent analysis of hsp70 concentration (see text). Numbers indicate hsp70 concentration as percentage of standard.

complex thermal phenotypes): Do these stress proteins actually improve tolerance of natural stress or recovery from it?

In theory, genetic engineering may allow the manipulation of a specific trait while all others are held constant, and thereby elucidate the impact of that trait on the thermal phenotype. In practice, the execution of this paradigm may be extremely difficult depending on the model system to be studied. The *hsp104* gene of yeast for example, is a single-copy gene in a relatively simple organism for which the techniques of molecular manipulation are considerably advanced. Similar work with the *hsp70* gene of *D. melanogaster* is considerably more problematic; the diploid genome includes at least ten copies (Welte *et al.*, 1993) and the intact multicellular organism is less amenable to manipulation than is single-celled yeast. With respect to the latter difficulty, however, P element-mediated germ-line transformation is an effective technique for inserting transgenes in *D. melanogaster* and could be used to transform *D. melanogaster* with extra copies of *hsp70*. This procedure is not a panacea because germ-line transformation with extra copies of *hsp70* must: (i) incorporate numerous copies to be detectable against the background of ten copies in the wild-type; and (ii) control for insertional mutagenesis, in which integration of transgenes inadvertently disrupts other genes and produces a phenotype solely as a consequence of this disruption. Fortuitously, Golic & Lindquist (1989) have developed a means to circumvent these difficulties; Welte *et al.* (1993) have already used this technique to create strains of *D. melanogaster* that are ideal for studies of hsp70. The technique results in unequal crossing over between homologous chromatids, yielding an 'extra copy' strain with all extra copies of the transgene, and an 'excision' strain whose homologous chromatid has undergone the same insertional disruption as that of the extra copy strain but lacks the extra copies of the transgene. Recombinants are easily recognised by eye colour. Existing extra copy strains created with this technique have 12 inserted copies of *hsp70*, plus the ten pre-existing copies (Welte *et al.*, 1993).

Welte *et al.* (1993) have compared these transgenic strains to examine the impact of variation in hsp70 expression on embryonic thermotolerance. Quantitation of hsp70 concentration established that hsp70 levels increased more rapidly on heat shock in the extra copy strain than in the excision strain. Transgene-specific probes of mRNA from heat-shocked embryos established that expression of the transgenes was specifically elevated, as opposed to a generalised increase in hsp expression.

Thermotolerance is easily assayed in embryos by counting the fraction of egg cases from which surviving larvae have emerged (Welte *et al.*,

1993). The investigation focused on embryos given varying durations of pretreatment at 35–36 °C and then exposed to severe heat stress (41–42 °C) at 6 h of development. With no pretreatment (and, consequently, no induction of stress protein expression before exposure to the severe heat stress), no embryo survived more than 15 min of heat stress. Depending on its duration and the experimental strain, pretreatment dramatically increased embryonic thermotolerance. Twenty minutes at 36 °C, for example, improved the LT_{50} at 41 °C from <10 min to 25 min in several different control strains. This improvement, however, was even more pronounced in the extra copy strain, whose LT_{50} after pretreatment was 35 min. Indeed, in numerous thermotolerance assays, extra copy embryos consistently outperformed excision embryos under identical conditions. The extra copy embryos, moreover, developed more rapidly than excision embryos after a near-lethal heat shock, an added advantage of extra hsp70. These beneficial effects of hsp70, however, are limited in several ways. First, they are temporary; age-matched excision strain embryos, which still have ten copies of the *hsp70* gene, eventually acquire the same level of thermotolerance as in extra copy embryos. Second, extra copies of *hsp70* improve inducible thermotolerance only at certain embryonic stages; their effects on 12- and 18-h embryos are negligible. Finally, and most importantly, the ecological and evolutionary significance of these effects is unclear; because of the timing and thermal preference of oviposition in *D. melanogaster* (see earlier), embryos reach 6 h of development in the middle of the night, a time not noted for thermal stress.

The clear natural thermal context established for studies of larvae (see earlier) prompted a detailed examination of hsp70 levels and thermotolerance in this stage of the life cycle. Hsp70 concentrations were measured by enzyme-linked immunosorbent assay (ELISA) as described by Welte *et al.* (1993). In brief, whole larvae were frozen in liquid nitrogen and stored at −80 °C until sonication in ice-cold phosphate-buffered saline containing 3 mM PMSF. Microwell plates were coated with 2 μg total protein. Bound hsp70 was detected with a hsp70-specific monoclonal antibody (7.FB) coupled, via secondary and tertiary antibodies, to alkaline phosphatase. Each assay included a standard of 2 μg protein in lysate prepared from *Drosophila* Schneider 2 cells that had been maintained at 36 °C for 60 min and then at 25 °C for 60 min before lysis. Comparable measurements of standard and lysate of control cells (at 25 °C constantly before lysis) mixed in various ratios demonstrated that the final signal depended linearly on the quantity of hsp70 present in the lysate.

To characterise hsp70 expression in larvae, third-instar larvae from each of three strains (extra copy, excision and Oregon R (wild-type)) were analysed after 120 min at 32 °C, 34 °C, 36 °C and 38 °C. Hsp70 concentration increased with temperature up to 36 °C (Fig. 6), a temperature found routinely in insolated fruit and experienced by wild larvae (Figs. 3 and 4). The kinetics of this increase at 36 °C were also characterised in all three strains (Fig. 7). Hsp70 concentrations were slightly but not significantly greater in the extra copy strain than in the excision strain for the first 2 h of exposure to 36 °C. Thirty minutes later, hsp70 concentration had increased in the extra copy strain so that it was approximately twice that in the excision strain for the ensuing hour.

The three strains were analysed during recovery at 25 °C after a 60 min exposure to 36 °C (Fig. 8). In excision and Oregon R larvae, hsp70 concentrations decreased or remained approximately constant for the first 2 h of recovery. In extra copy larvae, by contrast, hsp70 concentrations increased dramatically during recovery, peaking at 1 h after the start of the recovery period and eventually subsiding to match the excision and Oregon R larvae by 2 h. Thus, at least at some times and conditions, the expected differences among the *Drosophila* strains

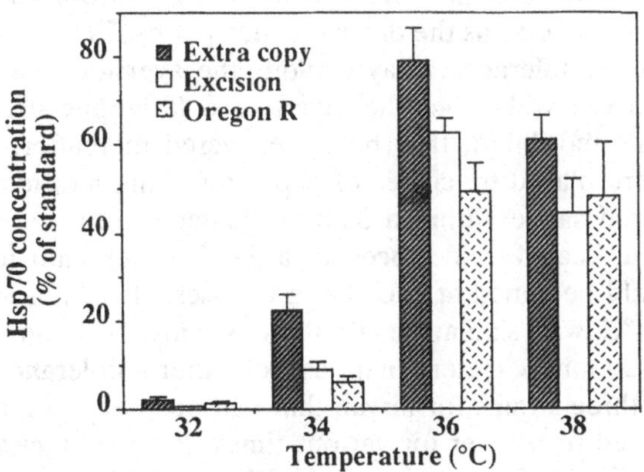

Fig. 6. Effect of temperature on hsp70 accumulation in third-instar larvae. Larvae of each experimental strain were subjected to the indicated temperature for 120 min and then frozen in liquid nitrogen for subsequent analysis of hsp70 concentration. Hsp70 concentration was measured via ELISA and standardised by comparison with the signal detected from a standard lysate of heat-shocked Schneider 2 cells (see text). Means are plotted ± one standard error.

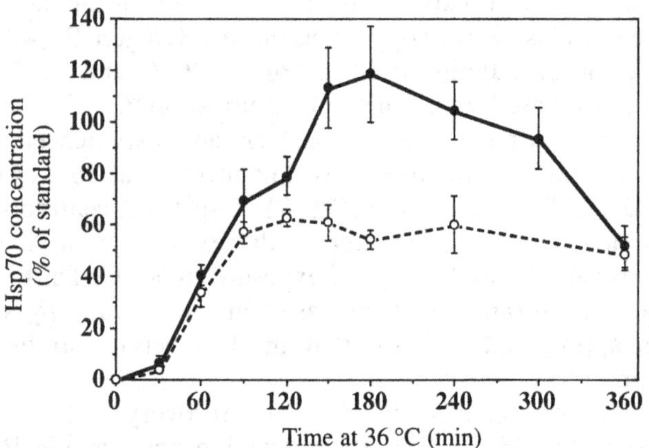

Fig. 7. Effect of duration of heat shock at 36 °C on hsp70 accumulation in third-instar larvae. Data are standardised and plotted as in Fig. 6 for extra copy and excision larvae. ●, Extra copy; ○, excision.

that have previously been demonstrated for embryos also occur in third-instar larvae.

Finally, thermotolerance has been assessed for third-instar larvae along the same time course as the determinations of hsp70 concentration (Fig. 8). The thermotolerance assay exploits the aversion of larvae to intense light. Larvae will avoid the beam of a fibreoptic illuminator until they become moribund; they never recovered thereafter. Groups of 6–9 larvae were placed on circles of paper towelling moistened with phosphate-buffered saline within a 35 mm diameter culture dish. The dish was covered, sealed, and placed in a 39 °C water bath in which a light beam could be aimed at each larva to assess its response. Basal tolerance of 39 °C was similar in all three strains, with an LT_{50} of approximately 20 min. Determinations of thermotolerance were repeated for all three strains for larvae that had been exposed to 36 °C for 1 h and allowed to recover for various times. Pretreatment at 36 °C alone had no effect on tolerance of 39 °C. As it did in embryos, recovery after the pretreatment increased tolerance of 39 °C, with the LT_{50} increasing about 40% in Oregon R larvae and 100% in excision larvae. The improvement in thermotolerance was much greater in the extra copy larvae than in the other strains, with the LT_{50} increasing nearly threefold. This striking increase in thermotolerance, moreover, coincides with the considerable increase in hsp70 concentration seen only in the extra copy strain.

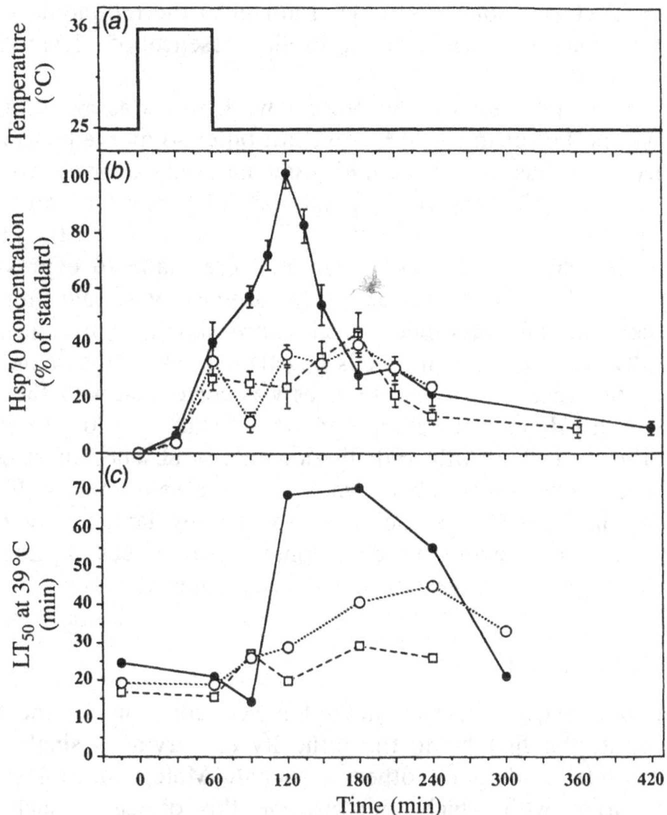

Fig. 8. Effect of heat shock at 36 °C and recovery at 25 °C on hsp70 accumulation and thermotolerance in third-instar larvae. (*a*) Experimental temperature regimen. (*b*) Hsp70 concentrations determined for larvae frozen at indicated time points. Data are standardised and plotted as in Fig. 6. (*c*) At indicated times, samples of larvae were transferred to 39 °C for determination of thermotolerance. Plotted LT_{50}s indicate time after transfer required for half of each sample to succumb. □, Oregon R; ●, extra copy; ○, excision.

Although additional work, now in progress, is necessary to confirm these conclusions, findings to date clearly establish that an increase in hsp70 concentration directly results in improved tolerance of temperatures that *D. melanogaster* larvae might well encounter in the field. The thermal phenotype, in this instance, behaves in part as a 'house of cards' (Fig. 1*b*) in that manipulation of a single trait, hsp70 concentration, is sufficient to alter thermotolerance. Hsp70 concentration, however, is clearly not the sole determinant of thermotolerance in that basal thermotolerance is evident in the absence of elevated hsp70 concentrations.

Like many such studies, the present work raises as many questions as it answers. For example, if hsp70 contributes so markedly to thermotolerance, why has not *D. melanogaster* naturally evolved even more copies of the *hsp70* gene than already occur? Part of the answer may be foreshadowed by a study by J. Feder *et al.* (1992), in which *Drosophila* Schneider 2 cells in culture were made to express hsp70 constitutively (i.e. chronically and in the absence of an inducing stress). These cells initially multiplied much more slowly than control cells. Eventually, the experimental cells began to grow at the same rate as controls; this recovery in growth rates was correlated with the sequestration of hsp70 in intracellular granules. Thus, hsp70 expression is deleterious to cell growth and division in the absence of stress, and these deleterious effects abate when hsp70 is sequestered. If similar effects occur in whole larvae, then extra copy larvae may be at a disadvantage in growth and development (see Krebs & Loeschcke, 1994). Tests of this suggestion are already underway.

Conclusion

This chapter began with two methodological concerns for the thermal physiologist, the first being the difficulty of varying a single trait of interest while holding all others constant. Molecular biology offers effective tools with which to overcome this difficulty, such as the technique of Golic & Lindquist (1989). In the present work, to paraphrase Welte *et al.* (1993: 850), it is the manner in which the experimental strains have been created that allows conclusions to be established with confidence. Both the excision and extra copy strains have had a transgene construct integrated at the same site in the genome, and literally differ solely in hsp70 copy number. If this level of experimental control is insufficient, the entire set of experiments can be repeated with extra copy and excision lines derived from several different transformants at multiple chromosomal sites, as did Welte *et al.* (1993), with each independent derivation yielding similar results.

The second concern was that of identifying model systems that are tractable to molecular manipulation, physiological experimentation, and ecological characterisation (Feder & Watt, 1993). *Drosophila melanogaster* emerges as a particularly promising model in this respect. Not only has it a thermal ecology after all, but it also clearly experiences thermal stress in nature. Thus, of particular importance are the field measurements that establish an ecological and evolutionary context for the interpretation of these measurements. Without such measurements, the findings may contribute to the mechanistic understanding of a molecular chaperone but remain an ecophysiological and evolutionary irrelevance.

Acknowledgements

I thank those numerous individuals who made fruitful contributions to the cultivation and fermentation of the work reported here, including J. Coyne J. Feder, R. Huey, D. Parsell, W. Porter, B. and B. Sampsell, M. Welte, and L. Yue. N. Blair, N. Cartano, H. Figueras and L. Milosz provided technical assistance, the proprietors of Garwood Orchard, LaPorte, IN, kindly allowed fieldwork on their property, and the entire Lindquist laboratory provided remedial training in molecular biology. I am especially indebted to Susan Lindquist, my distinguished colleague and good friend, for graciously accommodating me in her laboratory and for insight on the relative roles and contributions of molecular and organismal approaches to thermal physiology. Research was supported by a Sabbatical Supplement Award in Molecular Studies of Evolution from the Alfred P. Soan Foundation, National Science Foundation grants IBN94-08216 and BIR94-19545, and the Louis Block Fund of the University of Chicago.

References

Ashburner, M. (1989). *Drosophila: A Laboratory Manual.* Cold Spring Harbor, NY: Cold Spring Harbor Laboratory Press.

Barber, H.N. & Sharpe, P.J.H. (1971). Genetics and physiology of sunscald of fruits. *Agricultural Meteorology* **8**, 175–91.

Braig, K., Otwinowski, Z., Hegde, R., Boisvert, D.C., Joachimiak, A., Horwich, A.L. & Sigler, P.B. (1994). The crystal structure of the bacterial chaperonin GroEL at 2.8 *Å*. *Nature* **371**, 578–86.

David, J.R., Allemand, R., Van Herrewege, J. & Cohet, Y. (1983). Ecophysiology: abiotic factors. In *The Genetics and Biology of Drosophila*, 3rd edn (ed. M. Ashburner, H.L. Carson & J.N. Thompson), pp. 105–70. London, Academic Press.

Feder, J.H., Rossi, J.M., Solomon, J., Solomon, N. & Lindquist, S. (1992). The consequences of expressing hsp70 in *Drosophila* cells at normal temperatures. *Genes and Development* **6**, 1402–13.

Feder, M.E. (1987). The analysis of physiological diversity: the future of pattern documentation and general questions in ecological physiology. In *New Directions in Ecological Physiology* (ed. M.E. Feder, A.F. Bennett, W.W. Burggren & R.B. Huey), pp. 38–75. Cambridge: Cambridge University Press.

Feder, M.E., Blair, N. & Figueras, H. (1994). Natural potential body temperatures of non-adult *Drosophila melanogaster* in relation to heat-shock protein expression. *Physiologist* **37**, A88.

Feder, M.E. & Block, B.A. (1991). On the future of physiological ecology. *Functional Ecology* **5**, 136–44.

Feder, M.E., Parsell, D.A. & Lindquist, S.L. (1995). The stress response and stress proteins. In *Cell Biology of Trauma* (ed. J.J. Lemasters & C. Oliver), pp. 177–91. Boca Raton, FL: CRC Press.

Feder, M.E. & Watt, W.B. (1993). Functional biology of adaptation. In *Genes in Ecology* (ed. R.J. Berry, T.J. Crawford & G.M. Hewitt), pp. 365–91. Oxford: Blackwell Scientific Publications.

Flynn, G.C., Pohl, J., Flocco, M.T. & Rothman, J.E. (1991). Peptide-binding specificity of the molecular chaperone BiP. *Nature* **353**, 726–30.

Fogleman, J. (1979). Oviposition site preference for temperature in *D. melanogaster*. *Behavioral Genetics* **9**, 407–412.

Gates, D.M. (1980). *Biophysical Ecology*. New York: Springer-Verlag.

Golic, K.G. & Lindquist, S. (1989). The FLP recombinase of yeast catalyzes site-specific recombination in the *Drosophila* genome. *Cell* **59**, 499–509.

Hoffmann, A.A & Parsons, P.A. (1991). *Evolutionary Genetics and Environmental Stress*. Oxford: Oxford University Press.

Huey, R.B. & Bennett, A.F. (1990). Physiological adjustments to fluctuating thermal environments: an ecological and evolutionary perspective. In *Stress Proteins in Biology and Medicine* (ed. R. I. Morimoto, A. Tissieres & C. Georgopoulus), pp. 37–59. Cold Spring Harbor, NY: Cold Spring Habor Laboratory Press.

Hutchison, V.H. & Maness, J.D. (1979). The role of behavior in temperature acclimation and tolerance in ecotherms. *American Zoologist* **19**, 367–84.

Jones, J.S., Coyne, J.A. & Partridge, L. (1987). Estimation of the thermal niche of *Drosophila melanogaster* using a temperature-sensitive mutation. *American Nauralist* **130**, 83–90.

Krebs, R.A. & Loeschcke, V. (1994). Costs and benefits of activation of the heat-shock response in *Drosophila melanogaster. Functional Ecology* **8**, 730–7.

Lindquist, S. (1986). The heat-shock response. *Annual Review of Biochemistry* **55**, 1151–91.

Martin, J., Mayhew, M., Langer, T. & Hartl, F.U. (1993). The reaction cycle of GroEL and GroES in chaperonin-assisted protein folding. *Nature* **366**, 228–33.

McKenzie, J.A. & McKechnie, S.W. (1979). A comparative study of resource utilization in natural populations of *Drosophila melanogaster* and *D. simulans. Oecologia* **40**, 299–309.

Morimoto, R.I., Jurivich, D.A., Kroeger, P.E., Mathur, S.K., Murphy, S. P., Nakai, A., Sarge, K., Abravaya, K. & Sistonen, L.T. (1994*a*). Regulation of heat shock gene transcription by a family of heat shock factors. In *The Biology of Heat Shock Proteins and Molecular Chaperones* (ed. R.I. Morimoto, A. Tissieres & C. Georgopoulos), pp. 417–56. Cold Spring Harbor, NY: Cold Spring Harbor Laboratory Press.

Morimoto, R.I., Tissieres, A. & Georgopoulos, C. (eds). (1990). *Stress Proteins in Biology and Medicine*. Cold Spring Harbor, NY: Cold Spring Harbor Laboratory Press.

Morimoto, R.I., Tissieres, A. & Georgopoulos, C. (eds.). (1994*b*). *Heat Shock Proteins: Structure, Function and Regulation*. Cold Spring Harbor, NY: Cold Spring Harbor Laboratory Press.

Parsell, D.A., Kowal, A.S. & Lindquist, S. (1993). *Saccharomyces cerevisiae* Hsp104 protein: purification and characterization of ATP-induced structural changes. *Journal of Biological Chemistry* **269**, 4480–7.

Parsell, D.A., Kowal, A.S., Singer, M.A. & Lindquist, S. (1994). Protein disaggregation mediated by heat-shock protein Hsp104. *Nature* **372**, 475–8.

Parsell, D.A. & Lindquist, S. (1993). The function of heat-shock proteins in stress tolerance: degradation and reactivation of damaged proteins. *Annual Review of Genetics* **27**, 437–96.

Parsell, D.A. & Lindquist, S. (1994). Heat shock proteins and stress tolerance. In *The Biology of Heat Shock Proteins and Molecular Chaperones* (ed. R.I. Morimoto, A. Tissieres & C. Georgopoulos), pp. 457–94. Cold Spring Harbor, NY: Cold Spring Harbor Laboratory Press.

Parsell, D.A., Sanchez, Y., Stitzel, J.D. & Lindquist, S. (1991). Hsp104 is highly conserved protein with two essential nucleotide-binding sites. *Nature* **353**, 270–3.

Parsons, P. (1978). Boundary conditions for *Drosophila* resource utilization in temperate regions, especially at low temperatures. *American Naturalist* **112**, 1063–74.

Ritossa, F.M. (1962). A new puffing pattern induced by a temperature shock and DNP in *Drosophila. Experientia* **18**, 571–3.

Sampsell, B.M. (1977). *Alcohol dehydrogenase thermostability variants in natural populations of* Drosophila melanogaster: *genetic basis and adaptive significance*. Ph.D. dissertation, University of Iowa.

Sanchez, Y. & Lindquist, S.L. (1990). HSP104 required for induced thermotolerance. *Science* **248**, 1112–5.

Sanchez, Y., Taulien, J., Borkovich, K.A. & Lindquist, S. (1992). Hsp104 is required for tolerance to many forms of stress. *EMBO Journal* **11**, 2357–64.

Sarge, K.D., Bray, A.E. & Goodson, M.L. (1995). Altered stress response in testis. *Nature* **374**, 126.

Schmidt, M., Rutkat, K., Rachel, R., Pfeifer, G., Jaenicke, R., Viitanen, P., Lorimer, G. & Buchner, J. (1994). Symmetric complexes of GroE chaperonins as part of the functional cycle. *Science* **265**, 656–9.

Schnebel, E.M. & Grossfield, J. (1986). Oviposition temperature range in four *Drosophila* species triads from different ecological backgrounds. *American Midland Naturalist* **116**, 25–35.

Somero, G.N. (1995). Proteins and temperature. *Annual Review of Physiology* **57**, 43–68.

Velazquez, J.M., DiDomenico, B.J. & Lindquist, S. (1980). Intracellular localization of heat shock proteins in *Drosophila. Cell* **20**, 679–89.

Velazquez, J.M. & Lindquist, S. (1984). Hsp70: nuclear concentration during environmental stress and cytoplasmic storage during recovery. *Cell* **36**, 655–62.

Velazquez, J.M., Sonoda, S., Bugaisky, G. & Lindquist, S. (1983). Is the major *Drosophila* heat shock protein present in cells that have not been heat shocked? *Journal of Cell Biology* **96**, 286–90.

Welte, M.A., Tetrault, J.M., Dellavalle, R.P. & Lindquist, S.L. (1993). A new method for manipulating transgenes: engineering heat tolerance in a complex, multicellular organism. *Current Biology* **3**, 842–53.

A.J.S. HAWKINS

Temperature adaptation and genetic polymorphism in aquatic animals

Introduction

Compared with techniques for analysing genetic variation at the DNA level, the older established technique of protein electrophoresis has made by far the largest contribution towards our understanding of population and evolutionary genetics. Protein electrophoresis measures biochemical genetic polymorphism as allozymes, that may be defined as the product of different alleles at a single locus. Numerous studies of allozyme variation have established that genetic diversity is very high within animal species (for a recent review, see van Delden, 1994). This suggests that the consequences of biodiversity, which is often only assessed in terms of species, should also be considered both within and between populations of the same species. An emerging body of evidence indicates that genetic variation does not simply result from the neutral processes of mutation, genetic drift and migration. Positive relations have been predicted on theoretical grounds between fitness and multiple locus heterozygosity, assessed for each individual as the proportion of measured polymorphic loci that are heterozygous (for an example, see Charlesworth, 1991). Such associations have frequently been confirmed in many species of animal (for reviews, see Mitton & Grant, 1984; Zouros & Foltz, 1987). Further investigations have established that genetic differences in the catalytic efficiency of specific enzymes have profound effects on physiological performance, and which may represent the mechanistic basis for long-term adaptation resulting from the functional consequences of natural selection (for reviews, see Nevo, 1983; van Delden, 1994).

Poikilotherms are directly affected by ambient temperature, which is among the most pervasive of environmental factors. Temperature influences metabolism, activity levels, spawning, development and growth; and because of selective pressures associated with these processes, temperature is an ecological resource, influencing the proportion of potential habitat that is suitable for a species. 'Preferred temperatures', temperature optima, the critical thermal limits on growing season and lethal thermal limits all tend to occur within the geographical distribution of a species (for an example, see Wilson & Elkaim, 1991).

Responses to environmental temperature change in a poikilotherm will depend on the history of non-genetic thermal acclimation in that species (for reviews, see Prosser, 1973; Somero & Hochachka, 1976). Active phenotypic adjustments that compensate for the immediate effects of temperature shift include changes in concentration and/or conformation both of enzymes of energy metabolism and of 'heat shock proteins', as well as changes in membrane phospholipid composition. Thermal compensation is complete if the acclimated net energy balance is similar to that before the temperature change, whether following an increase or a decrease in temperature. But if the energy available for growth, reproduction and activity remains diminished, then compensation is not complete, with impaired performance and possible death (Prosser, 1973). Under such circumstances, then natural selection will act on any variation within a population of genetically-influenced traits that include protein thermal stability and thermal optima. Genetic variability within a population thus affords some adaptation to changing temperature, with associated changes in thermal responses that include the limits for ultimate tolerance.

This chapter reviews what is understood of the relations between temperature adaptation and biochemical genetic polymorphism measured as allozyme variation in aquatic poikilotherms. Evidence is presented for evolutionary adaptations to temperature, focusing primarily on intraspecific differences. Then, as the basis of those adaptations, influences of allozyme polymorphism are considered on metabolism and physiological performance, before discussing how these interrelations affect the immediate individual response to temperature change, and provide an adaptive mechanistic explanation for the general advantage of multi-locus heterozygotes exposed to fluctuations in temperature and other environmental variables.

Throughout this review, it has not been possible to reference all relevent literature. Instead, points are illustrated with pertinent examples, and the reader is referred to previous comprehensive reviews of biochemical genetic polymorphism and associated traits (Mitton & Grant, 1984; Zouros & Foltz, 1987; van Delden, 1994; Zouros & Pogson, 1994).

Evidence for intraspecific evolutionary adaptation to temperature

Variations in allozyme frequency and/or multi-locus heterozygosity along geographical clines, between age groups and according to different habitats or lifestyles provide correlative evidence suggesting (rather than confirming) that biochemical genetic polymorphism is of adaptive value in natural populations.

Allozyme frequency

Ecogeographical associations with genetic polymorphism at single loci have been widely documented among animals; and probably the most common of all reported interrelations are between water (body) temperature and clines of allozyme frequency in aquatic species. These include clines in fish (for an example, see Nyman, 1975) and barnacles (Nevo *et al.*, 1977) subjected to heated effluents, as well as among fish subjected to cold water discharge from a hydroelectric dam (Zimmerman & Richmond, 1981). Other clines in shellfish (for an example, see Levinton & Lassen, 1978) and finfish (for examples, see Koehn, 1969; Powers *et al.*, 1986) are associated with macrogeographic temperature gradients.

While allozyme clines may reflect natural selection, it is important to bear in mind that such associations may also result from functionally or structurally associated genes, or through parameters of population structure and history (Lavie & Nevo, 1986). Three complementary lines of evidence confirm that ecogeographical effects of temperature change are influenced by adaptive consequences of enzyme polymorphism.

1. Protein thermal stability and thermal optima for function are under rigid genetic control, and there are several examples of how thermal properties of fish enzymes reflect the physiological limits of different species for temperature adaptation, as well as zonation patterns within pelagic and benthic ecosystems (for a review, see Somero, 1992).
2. Parallel responses in gene frequency to similar conditions but among different species suggest common selective processes. Thermal tolerance has consistently been associated with allozyme variation for phosphoglucose isomerase (PGI) in a sea anemone (Hoffman, 1983) and a freshwater mussel (Garton & Stoeckmann, 1992). The same has been shown for esterase in a freshwater fish (Koehn *et al.*, 1971) and a marine gastropod (Ushakov *et al.*, 1989), as well as for lactate dehydrogenase (LDH) in many different species of marine (for an example, see Powers *et al.*, 1986; Dahlhoff *et al.*, 1990) and freshwater (Serov *et al.*, 1988) fish. Functional implications of the different isozymes have not yet been resolved for esterase. PGI, however, occupies an important regulatory region between glycogen storage

and glycolysis, while LDH oxidises the (toxic) end-product of glycolysis. Furthermore, collective findings for PGI and LDH convincingly show how minor differences in habitat (body) temperature are sufficient for the selection of specific alleles with different kinetic properties.

3. Probably the most powerful evidence that natural selection acts on the thermal properties of different allozymes has been provided by breeding experiments. For example, in the crested blenny, the frequencies of LDH alleles in offspring from the same parents reared at different temperatures were similar to those in natural populations living at those same temperatures (Johnson, 1971). In addition, allele frequencies at three loci among killifish restricted within a warm basin shifted towards those typical of warmer habitats (Mitton & Koehn, 1975); and two strains of guppy selected over several generations for high temperature tolerance showed significant changes at each of the four polymorphic loci studied, with a predominant increase in the frequency of faster alleles (Shah, 1991).

Throughout this section, no attempt has been made to describe the specific nature of the associations between temperature and genotype at each locus. This is because these associations are highly variable, with no consistent trends either within or between species. In explanation, one simply cannot expect any single genotype to be most fit under all circumstances. Genetic components of fitness are affected by separate interactions with independent environmental variables such as both temperature and food availability. Furthermore, both experimental evidence and common sense dictate that the selective forces on each locus may vary according to species, sex, life-history stage and physiological condition (for a review, see van Delden, 1994). It should not be suprising that there is no general advantage for either heterozygosity or homozygosity. Hummel & Patarnello (1994) explained that differential consequences of heterozygosity may depend on the enzyme in question. For a monomeric enzyme, then the heterozygote is expected to show intermediate activity between the two homozygotes, but if non-monomeric, then heteroduplex chains may confer the heterozygote with higher (overdominant) activity than for both homozygotes (Hummel & Patarnello, 1994).

Multi-locus heterozygosity

In contrast with single-locus effects, ecogeographical associations with multi-locus heterozygosity appear to follow several common trends in the aquatic environment. Within species, heat resistance is known to: (i) decrease with depth of habitat in the water column (Schlieper *et al.*, 1967) or sediment (Wilson & Elkaim, 1991), where temperature variations are reduced, and (ii) increase with height on the shore (Wilson & Elkaim, 1991), where temperature variations are greater. Multi-locus heterozygosity within animal species also decreases with depth of habitat in the water column or sediment (Levinton, 1973), and increases with intertidal exposure (Koehn *et al.*, 1973; Lavie & Nevo, 1986).

Also within species, there are two especially convincing studies showing that genetic diversity is higher in habitats that are subject to large temperature fluctuations. In one example, average heterozygosity computed as Nei's index of gene diversity correlates in positive relation with the scale of temperature fluctuations experienced by separate populations of *Fundulus heteroclitus* along the east coast of North America (Powers & Place, 1978; Powers *et al.*, 1986). The other example stems from work studying the effects of depressed and unpredictable thermal regimens resulting from power generation at a hypolimnion-discharge dam on the Brazos river in Texas. A comprehensive series of investigations in the red shiner fish has established that: (i) the number of structural gene loci that are polymorphic and multi-locus heterozygosity at those loci both correlate in positive relations with the greatest daily temperature changes observed at different distances downstream of the dam; (ii) fully one-third of the gene variation may be attributed to variation between sites, rather than to differences among samples within sites; and (iii) there were no associated correlations with water flow rate, oxygen levels and 15 water quality variables (Zimmerman, 1984; King *et al.*, 1985).

These associations within species establish that multi-locus heterozygosity is greater under circumstances where temperature is likely to show the largest fluctuations. Such findings are consistent with the long-standing realisation that allozyme polymorphism may vary with ecological heterogeneity in space and time (Hedrick *et al.*, 1976), associated with the early theory that genetic diversity is adaptive in the face of environmental variability (Lerner, 1954).

Understanding evolutionary responses to temperature change

Influences of biochemical genetic polymorphism on metabolism and physiology

Allozyme frequency

Numerous studies have analysed the functional properties of allelic isozymes, with some excellent reviews on biochemical adaptation to temperature and other variables (Somero & Hochachka, 1976; Clarke & Koehn, 1992).

Among aquatic poikilotherms, there are several examples of biochemical differences between allozymes that help to explain how allozyme polymorphism may be adaptive given environmental temperature change. Of these examples, work on the latitudinal cline in an esterase locus in the freshwater fish *Catostomus clarkii* (Fig. 1) provided one of the first explanations for genotype–environment associations (Koehn, 1969). *In vitro* comparisons of thermal responses of esterases extracted from each of the three genotypes showed that: (i) the $Es\text{-}I^{a/a}$ homozygote most common at warmer latitudes had the greatest activity at the highest temperature; (ii) the $Es\text{-}I^{b/b}$ homozygote most common at the colder latitudes showed greatest activity at the lowest temperature; and (iii) the $Es\text{-}I^{a/b}$ heterozygote showed a broad range of high activity in the middle range of latitudes (Fig. 1); (Koehn, 1969). These findings are consistent with theoretical expectations that enzyme activities in heterozygous individuals should be intermediate between enzyme activities in the associated homozygotes (for an example, see Kacser & Burns, 1981). Similar adaptive differences have been observed within north–south clines at the LDH locus both in the fathead minnow

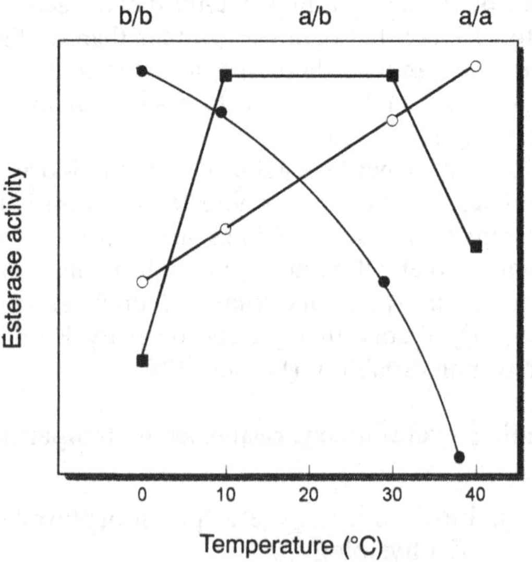

Fig. 1. Temperature responses of the three esterase genotypes $Es\text{-}I^{a/a}$, $Es\text{-}I^{a/b}$ and $Es\text{-}I^{b/b}$ in the freshwater fish *Catostomus clarkii* (After Koehn, 1969.)

Pimephales promelas (Merritt, 1972) and the killifish *Fundulus heteroclitus* (Place & Powers, 1979; Crawford *et al.*, 1990).

There is general agreement on the biological significance of maximising enzyme activity/flux; however, other characteristics of enzyme function that influence the balance between the productivity of a metabolic pathway and the energy costs of maintaining the constituent enzymes of that pathway may also be adaptive (Clarke & Koehn, 1992). Interactive effects of other loci on biochemical properties of allozymes further complicate matters. While biochemical associations give additional support for natural selection, we are a long way from fully understanding the metabolic and physiological consequences of genetic polymorphism.

Multi-locus heterozygosity

A more complete understanding of how biochemical genetic polymorphism may influence metabolism and physiology has emerged from studies of multi-locus heterozygosity, that varies in positive relation with growth and/or fecundity among many plant and animal species (Mitton & Grant, 1984; Zouros & Foltz, 1987). The strength of heterozygosity associations may change with age, reproductive behaviour, environmental influences, and background genetic effects (Gaffney *et al.*, 1990; Pecon Slattery *et al.*, 1993). Nevertheless, these associations are widespread, and confirm the general influence of biochemical genetic polymorphism on metabolism and performance.

Among animals, findings from aquatic poikilotherms have provided most information on the mechanistic basis of faster heterozygous production. Greater heterozygosity is associated with lower mass-specific energy requirements for maintenance metabolism in a variety of aquatic poikilotherms that include shellfish, finfish and salamanders, in which lower energy costs and/or higher efficiencies of growth among faster-growing individuals result at least in part from decreased intensities of whole-body protein turnover (for a review, see Hawkins & Day, 1996). Protein turnover is defined as the renewal and replacement of intracellular proteins, and functions primarily in maintenance, metabolic regulation and acclimation. These observations are complemented by others showing slower protein turnover and associated reductions in maintenance energy expenditure within many species of mammal selected for high growth rate, as well as the genetic regulation of specific proteolytic pathways (Hawkins, 1991). Collective findings therefore indicate energy requirements are affected by genetic controls on protein turnover, that those genetic controls are in some way associated with multi-locus heterozygosity, and that metabolic consequences affect performance among animals generally (Hawkins, 1991, 1995; Hawkins & Day, 1996).

Considering the genetic basis of heterozygosity associations, no study has yet unambiguously established whether the studied loci are causative agents of the correlation, or neutral markers of other loci that are more directly responsible for those associations (for a review, see Zouros & Pogson, 1994). In addition, work in marine shellfish has shown that heterozygosity associations may be restricted to loci that show deficiencies of heterozygotes within the sample population (Zouros & Foltz, 1987; Gaffney, 1994). More heterozygous individuals are certainly advantaged, displaying enhanced viability by living to a greater age (for examples see Zouros *et al.*, 1983; Pecon Slattery *et al.*, 1993). These natural patterns indicate that background genetic effects influencing genetic polymorphism may include subpopulation mixing, aneuploidy, null alleles and/or molecular imprinting (Gaffney *et al.*, 1990; Gaffney, 1994).

Metabolic influences on the thermal sensitivity of energy expenditure

Among poikilotherms, active metabolic adjustments that compensate for the immediate effects of temperature change differ according to species, tissue, animal size, physiological status, nutrition and thermal history, including both the rate and amplitude of temperature change (see Introduction, and for reviews, see Newell, 1979; Johnston & Dunn, 1987). These interrelations are complex, within few apparent generalisations.

The remainder of this subsection will describe how comparisons of physiological performance between individual mussels (*Mytilus edulis*) have helped to resolve metabolic influences on the thermal sensitivity of energy expenditure (Hawkins *et al.*, 1987; Hawkins, 1995). This work has established that temperature-dependent shifts in oxygen consumption may primarily reflect costs associated with changes in the rate of whole-body protein synthesis. Furthermore, greater metabolic sensitivities to either increased or decreased temperature were consistently associated with higher intensities of metabolism when previously acclimated to a constant temperature, and more specifically, metabolic sensitivity varied in positive relation with the proportion of protein synthesis that effected turnover (Fig. 2) (Hawkins *et al.*, 1987; Hawkins, 1995).

All of the above responses in *M. edulis* were observed among animals that had initially been acclimated to low rations that were barely sufficient to maintain positive net energy balance (Hawkins, 1995). Most

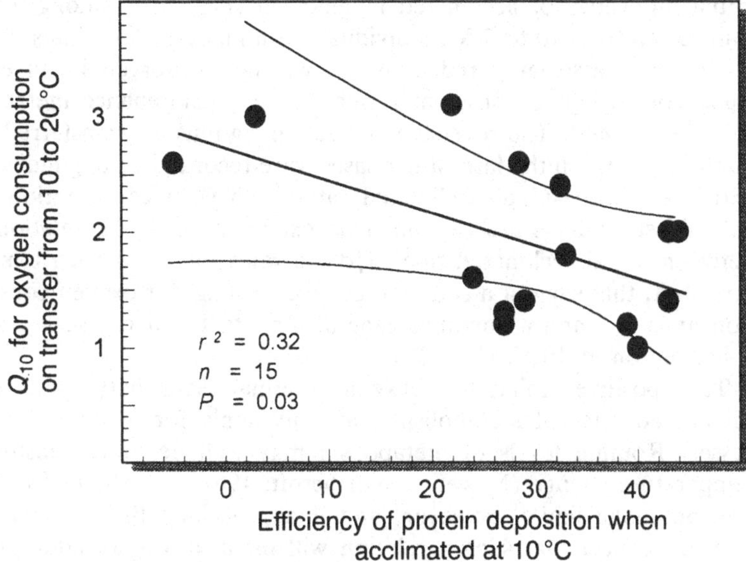

Fig. 2. Q_{10} values for oxygen consumption immediately following transfer of individual *Mytilus edulis* from 10 to 20 °C in relation to the proportions of total whole-body protein synthesis that were effecting protein turnover among those same individuals when previously acclimated at 10 °C. $r^2 = 0.32$; $n = 15$; $P = 0.03$. (After Hawkins *et al.* (1987), and the line ±95% CL fitted by least-squares).

metabolisable nutrients were therefore used to support maintenance processes, and only a small proportion utilised for growth. This is significant, for it has been suggested that maintenance metabolism in marine invertebrates is relatively insensitive to temperature change, compared with higher 'routine' rates of oxygen consumption in fed or active animals (Newell & Northcroft, 1967; Newell, 1979). To test whether relations between the initial intensity and thermal sensitivity of metabolism are similar at high routine rates, we have recently compared individual thermal responses in *M. edulis* (51 ± 11 mg dry tissue weight) that were first acclimated to laboratory seawater at 10 ± 1 °C while fed the unicellular alga *Isochrysis galbana* at rations that resulted in an average initial growth rate of 2.6 ± 0.6% per day (A.J.S. Hawkins *et al.*, unpublished results). Oxygen consumption was measured both before and immediately after transfer to either 20 or 5 °C. Figure 3 illustrates how absolute changes and the initial rates at each new temperature varied with initial acclimated rates. It is clear that the absolute changes varied in consistent negative relations with initial intensities of acclimated metabolism (Fig. 3*a*). Among mussels transferred from 10 to 5 °C, individuals with highest initial rates showed the greatest associated reduction, as we have observed in an earlier study comparing responses at lower rates of maintenance metabolism (Hawkins, 1995). The reverse was true among mussels transferred from 10 to 20 °C, when the largest increases were recorded among individuals with the lowest initial acclimated rates of oxygen consumption (Fig. 3*a*). These findings are explained on examination of the relationships between initial acclimated rates and the rates following each temperature shift, that suggest a common general 'ceiling' for oxygen consumption at 20 °C, and a common general limit to the decrease in oxygen consumption at 10 °C (Fig. 3*b*).

The positive relation between thermal sensitivity and initial acclimated rates of metabolism may only apply for maintenance processes. Routine levels of metabolism may well be more sensitive to temperature change (Newell & Northcroft, 1967; Newell, 1979). There are metabolic limits, however, and if approaching those limits, then one can expect that such a relation will break down. In addition, we have shown that temperature-dependent shifts in oxygen consumption may primarily reflect changes in the costs associated with different rates of protein synthesis, and that metabolic sensitivity varied in positive relation with the proportion of protein synthesis that effected turnover (Hawkins *et al.*, 1987; Hawkins, 1995). Our findings therefore reflect close links between protein metabolism and energy metabolism, as are consistent with high energy requirements both for protein synthesis and protein breakdown (Hawkins & Day, 1996). It is also logical that the influence of protein turnover on thermal sensitivity of energy expenditure should be less evident at faster rates of metabolism, becoming obscured by the increasing ancillary energy costs of feeding and growth.

A primary function of protein turnover is to mobilise amino acids for selective redistribution or catabolism (Hawkins, 1991). Thus, it seems likely that protein turnover may influence the temperature sensitivity of metabolism through its effect on cellular substrate concentration. When high substrate levels exceed the K_m value for a particular enzyme, then reaction (synthesis) rates are temperature-dependent (Somero & Hochachka, 1976). Alternatively, if substrate levels fall below that K_m value, then reductions in enzyme-substrate affinity may maintain reaction rate essentially independent of environmental temperature increase (Somero, 1969). Other factors affecting the thermal sensitivity of metabolic processes are known to involve mechanisms for transporting amino acids, membrane permeability, phospholipid

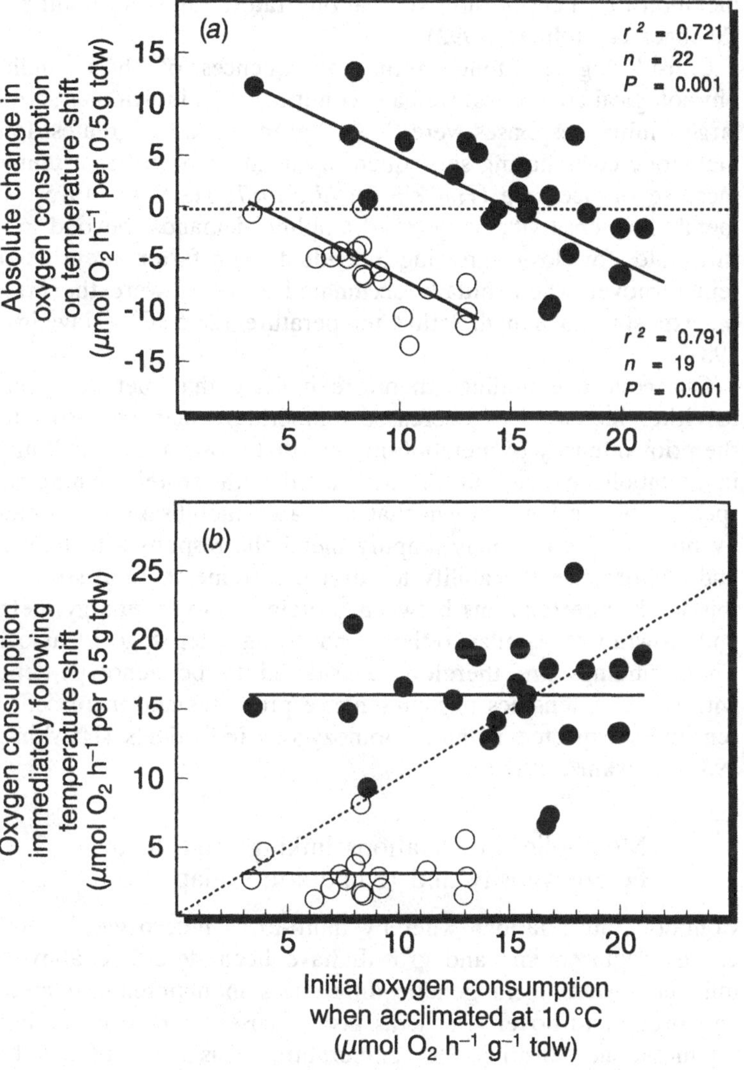

Fig. 3. Initial rates of oxygen consumption in individual mussels *Mytilus edulis* acclimated to high ration at 10 °C in relation to both (*a*) absolute changes in oxygen consumption and (*b*) initial rates of oxygen consumption following transfer of the same individuals to either 5 or 20 °C. tdw: total dry weight. All data are standardised per g total dry tissue weight, and lines ±95% CL fitted by linear least-squares regression (Hawkins, Bayne, Beuzet, Jones & Smith, unpublished data).

metabolism, heart and ventilation rates, and swimming speeds (Saenger & Holmes, 1992).

Considering the longer-term consequences of these findings for physiological energetics, we have confirmed that individuals showing the largest initial responses were those that incurred the greatest long-term metabolic costs during subsequent acclimation to either a temperature increase or decrease (Hawkins *et al.*, 1987; Hawkins, 1995). Greater metabolic sensitivity may also amplify demands beyond a critical threshold, for slower-growing individuals with faster intensities of protein turnover when initially acclimated at 10 °C were the first to die on exposure to a high lethal temperature (28.5 °C) (Hawkins *et al.*, 1987).

Our collective findings therefore indicate that metabolic sensitivity to either increased or decreased temperature may be associated with the prior intensity of metabolism, and that most of the resulting change in metabolic expenditure is associated with protein synthesis. More specifically, we have shown that increased mobilisation of amino acids by protein turnover may amplify metabolic responses to temperature, and compromise the ability to survive extreme thermal stress. During this work, interrelations between protein turnover, energy metabolism and growth were similar to those underlying heterozygosity associations. These findings are therefore considered to be genotype-dependent, with higher intensities of maintenance processes and increased thermal sensitivity expected in more homozygous individuals (Hawkins *et al.*, 1987; Hawkins, 1995).

Metabolic interrelations linking multi-locus heterozygosity and temperature adaptation

Common interrelations whereby multi-locus heterozygosity influences energy requirements and growth have been described above. Those influences stem from genetic differences in maintenance metabolism and protein turnover which, as also discussed, are primary influences on metabolic sensitivity to temperature. This subsection will discuss how these interrelations may represent the basis for ecological and evolutionary consequences of environmental temperature change.

Reduced energy costs of maintenance and growth in more heterozygous individuals leave more energy available for other requirements, with additional consequences for physiological performance (for a review, see Hawkins, 1991). A number of studies have shown how heterozygotes appear to be generally advantaged under stressful conditions (Mork & Sundnes, 1985; Nevo *et al.*, 1986, 1989; Gentili &

Beaumont, 1988; Blot & Thiriot-Quievreux, 1989; Newman *et al.*, 1989; Borsa, *et al.*, 1992), including elevated temperature in marine oysters *Crassostrea virginica* (Rodhouse & Gaffney, 1984) and clams *Mulinia lateralis* (Scott & Koehn, 1990). And in each case, it seems likely that heterozygotes performed better by virtue of lower energy requirements for maintenance processes, which helped to minimise net energy losses under conditions when metabolisable energy was either limited by anoxia, the inhibition of feeding processes (temperature and pollutants), or reduced food availability (starvation and competition).

Addressing the specific metabolic interrelations that link multi-locus heterozygosity and temperature adaptation, then findings reviewed in the above sections indicate that it is essential to consider the interactive effects of nutrition (energy intake). Under conditions of food abundance and growth, then three traits may enhance the average performance of multi-locus heterozygotes in the face of temperature change.

1. Lower energy requirements for maintenance processes mean that net production is maintained across a wider span of temperatures, that tissue loss only occurs under more extreme variations of temperature, and that phenotypic variability is reduced (Koehn & Bayne, 1989). Figure 4*a* illustrates these consequences for two individuals with different rates of respiration when initially acclimated under conditions of abundant food and growth at 15 °C. Assuming the same metabolic sensitivity to temperature ($Q_{10} = 2$), then the individual with lower initial energy expenditure will maintain net production over a greater range of temperature increase (range 1 versus range 2 in Fig. 4*a*).
2. Lower maintenance requirements in more heterozygous individuals may allow larger 'scope for activity' between standard and active levels of metabolism, as has been described for the tiger salamander *Ambystoma tigrinum* (Mitton *et al.*, 1986). Results presented in Fig. 3 suggest a common metabolic maximum between individual *Mytilus edulis*. Differential scope for activity may therefore be of general significance, but there is no related information from other animals.
3. Higher rates of feeding and absorption in faster-growing and/or more heterozygous individuals with reduced maintenance requirements have been reported in four different species of marine shellfish (Garton, 1984; Garton, *et al.*, 1984; Hawkins *et al.*, 1986, 1989; Holley & Foltz, 1987).

Assuming that the relationship between absorption and temperature remains independent of genotype, then reduced energy absorption by less heterozygous individuals with higher initial rates of acclimated energy expenditure will further reduce the range of temperature increase over which net production can be maintained (range 3 versus range 2 in Fig. 4*a*) (Hawkins, 1995).

Alternatively, under conditions when food intake is limited, and all energy is being utilised for maintenance processes, then we need to consider the influence of metabolic intensity upon thermal sensitivity of energy expenditure. Increasingly amplified metabolic responses to temperature, with maximal Q_{10} values of at least 3.5 (Hawkins *et al.*, 1987; Hawkins, 1995), represent a further disadvantage associated with the faster intensities of maintenance metabolism in less heterozygous individuals. The result is that energy losses and associated tissue wasting will increase more in response to the same temperature increment (range 2 versus range 3 in Fig. 4*b*).

This distinction between metabolic interrelations that link multi-locus heterozygosity and temperature adaptation under separate conditions of maintenance and growth is conceptually useful, but oversimplifies the situation in nature. Maintenance processes may not remain independent of growth rate. For example, protein turnover has been reported to show an anabolic increase with growth in the cod *Gadus morhua* (Houlihan *et al.*, 1988) and various mammals (Waterlow, 1984); therefore, there can be no clear distinction of consequences for metabolic

---

Fig. 4. Responses representing the immediate effects of temperature change on rates of energy absorption (A1 and A2) and energy expenditure (represented by respiration, R1 to R4) in the mussel *Mytilus edulis* previously acclimated at 15 °C. Responses are illustrated separately for animals (*a*) under conditions of growth, and (*b*) under conditions of tissue wasting/maintenance. A1 and R1 are based on measured responses described by Bayne (1973), assuming a Q_{10} value of 2 for energy expenditure. A2 and R2 illustrate the consequences of a 10% reduction in energy absorption and a 20% increase in acclimated energy expenditure, as have been observed in slower-growing and relatively homozygous individuals (see text). R3 and R4 show the effects of higher Q_{10} values of 2.75 and 3.5 that have been observed for individuals with higher initial energy expenditure under conditions of growth and maintenance, respectively (see text).

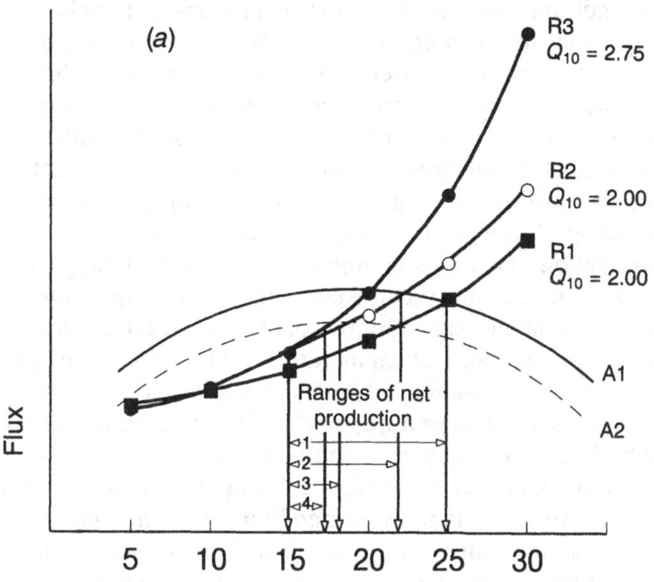
(a)
R3
Q_{10} = 2.75
R2
Q_{10} = 2.00
R1
Q_{10} = 2.00
A1
A2
Ranges of net
production
1
2
3
4
Flux
5
10
15
20
25
30

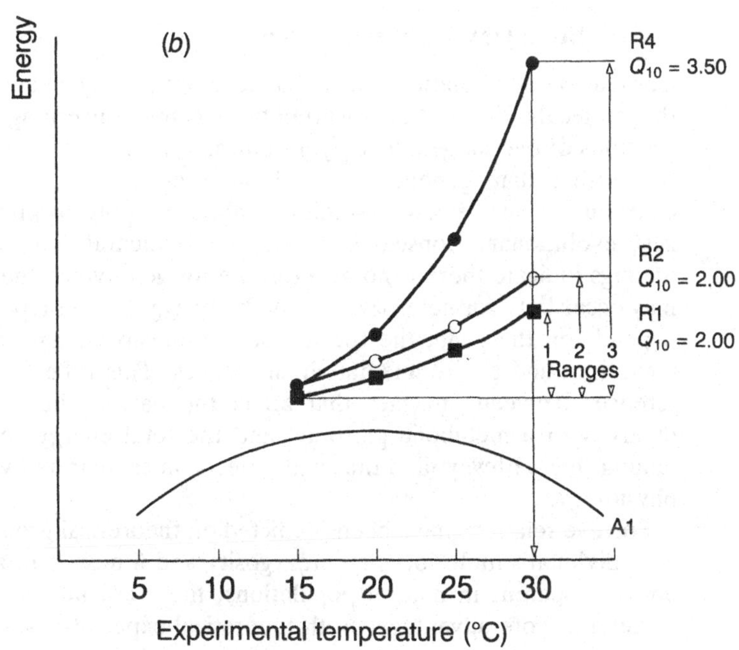
Energy
(b)
R4
Q_{10} = 3.50
R2
Q_{10} = 2.00
R1
Q_{10} = 2.00
1 2 3
Ranges
A1
5
10
15
20
25
30
Experimental temperature (°C)

expenditure associated with maintenance metabolism or growth. Especially in the range of metabolism that exceeds maintenance levels alone, but which falls well short of upper metabolic limits, then increasing thermal amplification of metabolic responses associated with higher rates of protein turnover in less heterozygous individuals may still result in significant but proportionally reduced enhancement of total energy expenditure (R3 in Fig. 4*a*). Under conditions of growth, this means a further decrease in the range of temperature increase over which net production can be maintained (range 3 versus range 4 in Fig. 4*a*).

One of the most consistent advantages highlighted above is that lower metabolic expenditure enables net production to be maintained over a wider range of temperatures. This must be of particular significance within aquatic poikilotherms subjected to temporal and spatial variations in temperature, and helps to explain widespread ecogeographical associations with multi-locus heterozygosity (reviewed earlier). It is also clear that advantages accruing from reduced thermal sensitivity with lower maintenance expenditure become increasingly significant under circumstances of limited energy acquisition. This or other similar interrelations may help to explain seasonal variation in heterozygosity-assocations observed within the same population (for an example, see Diehl & Koehn, 1985).

Summary and conclusions

Consequences of natural selection have frequently been observed at the molecular level; but uncertainty remains concerning the relative fractions of overall genetic polymorphism that result from natural selection rather than genetic drift and other neutral forces. The present contribution has reviewed studies of allozyme polymorphism that suggest evolutionary consequences of environmental temperature. The findings indicate there is no general rule for genotypic fitness. Selection may occur both for heterozygous and homozygous genotypes, and which depends on the gene, the species, sex, life-history stage, physiological condition, and environmental circumstances. This reflects adaptive differences between enzymes that affect the balance between the productivity of a metabolic pathway, and the total energy costs of maintaining that pathway, and thus with effects on cellular and whole-animal physiology.

Positive relations have been predicted on theoretical grounds between an individual's multi-locus heterozygosity and fitness, and although not always apparent in natural populations, there are numerous examples of heterozygote advantage. Both theoretical expectations and empirical findings indicate that enzyme activities in heterozygous individuals are intermediate to those in the corresponding homozygotes, and with functional intermediacy along latitudinal temperature clines. Given fluctuations of temperature and other factors, multi-locus heterozygotes show an average phenotypic advantage, with net heterozygote superiority. Greater ability to resist environmental fluctuations is reflected by reduced phenotypic variance in multi-locus heterozygotes (for examples, see King *et al.*, 1985; Parsons, 1990; Hawkins *et al.*, 1994).

Mechanistic links between allozyme polymorphism and the intensity of maintenance processes represent at least part of the metabolic and physiological basis of heterozygote advantage. Lower maintenance energy expenditure and faster growth in multi-locus heterozygotes are associated with reduced intensities of whole-body protein turnover. The specific genetic basis of these associations remains unclear; however, under circumstances of fluctuating environmental temperature, reduced metabolic intensity and slower mobilisation of amino acids by protein turnover may result in greater potential scope for activity over the normal temperature range, decreased metabolic sensitivity to temperature, lower energy costs incurred during acclimation to temperature change, greater independence of net production from temperature changes, and enhanced viability on exposure to extreme temperatures (Hawkins, 1995). These advantages will be most evident in the face of reduced energy acquisition resulting from limited food availability, physiological condition, reproductive status or other environmental stressors. Both the intensity of whole-body protein turnover (Hawkins, 1985) and metabolic sensitivity to temperature (for an example, see Worrall & Widdows, 1984) are known to change with season, so that an immediate challenge is to resolve how these environmental and endogenous variables interact to affect the integrated response to temperature.

Environmental fluctuation is not a prerequisite for heterozygote advantage, which remains evident under stable conditions. The above interrelations, however, appear especially adaptive among aquatic poikilotherms, that are frequently subject to fluctuations of temperature and other factors, and in which temperature acts directly on metabolism. For reasons described in the previous paragraph, the association whereby intensities of maintenance metabolism are on average lower in multi-locus heterozygotes appears to underly ecogeographical associations whereby both heterozygosity and heat resistance increase in habitats that are subject to greater thermal fluctuations. There are many models that help to account for genetic variation (for a recent review, see van Delden, 1994), and no one model can explain all of that variation.

However, the genetic and physiological interrelations reviewed above provide an adaptive mechanistic explanation for heterozygote advantage, and support the long-standing theory that high protein polymorphism results from a balance of selective forces stemming from ecological and/or environmental heterogeneity.

Given combined evidence that selection may occur both for heterozygous and homozygous genotypes, and that fitness associations stem from locus-specific effects, then future preservation should be of overall genetic polymorphism/diversity, rather than of multi-locus heterozygosity *per se*. Any reduction in genetic polymorphism following the elimination of temperature-sensitive genotypes may compromise the ability of impacted populations either for phenotypic acclimation or for genetic adaptation to further change in the thermal regimen.

References

Bayne, B.L. (1973). Physiological changes in *Mytilus edulis* L. induced by temperature and nutritive stress. *Journal of the Marine Biological Association, UK* **53**, 39–58.

Blot, M. & Thiriot-Quievreux, C. (1989). Multiple-locus fitness in a transfer of adult *Mytilus desolationis* (Mollusca: Bivalvia). In *Reproduction, Genetics and Distributions of Marine Organisms* (ed. J.S. Ryland & P.A. Tyler), pp. 259–64. 23rd European Marine Biology Symposium, Fredensborg: Olsen and Olsen.

Borsa, P., Jousselin, Y. & Delay, B. (1992). Relationships between allozyme heterozygosity, body size, and survival to natural anoxic stress in the palourde *Ruditapes decussatus* L. (Bivalvia: Veneridae). *Journal of Experimental Marine Biology and Ecology* **155**, 169–81.

Charlesworth, D. (1991). The apparent selection on neutral marker loci in partially inbreeding populations. *Genetical Research* **57**, 159–75.

Clark, A.G. & Koehn, R.K. (1992). Enzymes and adaptation. In *Genes in Ecology* (ed. R.J. Berry, T.J. Crawford & G.M. Hewitt), pp. 192–228. Oxford: Blackwell Scientific Publications.

Crawford, D.L., Place, A.R. & Powers, D.A. (1990). Clinal variation in the specific activity of lactate dehydrogenase-B. *Journal of Experimental Zoology* **255**, 110–13.

Dahlhoff, E., Schneidemann, S. & Somero, G.N. (1990). Pressure-temperature interactions on M sub(4)-lactate dehydrogenases from hydrothermal vent fishes: evidence for adaptation to elevated temperatures by the zoarcid *Thermarces andersoni*, but not by the bythitid, *Bythites hollisi*. *Biological Bulletin of the Marine Biological Laboratory at Woods Hole* **179**, 134–9.

Diehl, W.J. & Koehn, R.K. (1985). Multiple-locus heterozygosity, mortality, and growth in a cohort of *Mytilus edulis*. *Marine Biology* **88**, 265–71.

Gaffney, P.M. (1994). Heterosis and heterozygote deficiencies in marine bivalves: more light? In *Genetics and Evolution of Aquatic Organisms* (ed. A.R. Beaumont), pp. 146–53. London: Chapman and Hall.

Gaffney, P.M., Scott, T.M., Koehn, R.K. & Diehl, W.J. (1990). Interrelationships of heterozygosity, growth rate and heterozygote deficiencies in the coot clam, *Mulinia lateralis*. *Genetics* **124**, 687–99.

Garton, D.W. (1984). Relationship between multiple-locus heterozygosity and physiological energetics of growth in the estuarine gastropod *Thais haemastoma*. *Physiological Zoology* **57**, 530–43.

Garton, D.W., Koehn, R.K. & Scott, T.M. (1984). Multiple-locus heterozygosity and physiological energetics of growth in the coot clam, *Mulinia lateralis*, from a natural population. *Genetics* **108**, 445–5.

Garton, D.W. & Stoeckmann, A.M. (1992). Genotype-dependent metabolism at the phosphoglucose isomerase locus at ambient and elevated temperatures. *Journal of Shellfish Research* **11**, 226.

Gentili, M.R. & Beaumont, A.R. (1988). Environmental stress, heterozygosity and growth rate in *Mytilus edulis*. *Journal of Experimental Marine Biology and Ecology* **120**, 145–53.

Hawkins, A.J.S. (1985). Relationships between the synthesis and breakdown of protein, dietary absorption and turnovers of nitrogen and carbon in the blue mussel, *Mytilus edulis* L. *Oecologia* **66**, 42–9.

Hawkins, A.J.S. (1991). Protein turnover: a functional appraisal. *Functional Ecology* **5**, 222–33.

Hawkins, A.J.S. (1995). Effects of temperature change on ectotherm metabolism and evolution: metabolic and physiological interrelations underlying the superiority of multi-locus heterozygotes in heterogeneous environments. *Journal of Thermal Biology* **20**, 23–33.

Hawkins, A.J.S. & Day, A.J. (1996). Interrelations between genotype, protein metabolism and growth efficiency in marine animals. *Journal of Experimental Marine Biology and Ecology* (in press).

Hawkins, A.J.S., Bayne, B.L. & Day, A.J. (1986). Protein turnover, physiological energetics and heterozygosity in the blue mussel, *Mytilus edulis*: the basis of variable age specific growth. *Proceedings of the Royal Society of London* **229**, 161–76.

Hawkins, A.J.S., Bayne, B.L., Day, A.J., Rusin, J. & Worrall, C.M. (1989). Genotype-dependent interrelations between energy metabolism, protein metabolism and fitness. In *Reproduction, Genetics and Distributions of Marine Organisms* (ed. J.S. Ryland & P.A. Tyler, P.A.), pp. 283–92. 23rd European Marine Biology Symposium. Fredensborg: Olson and Olson.

Hawkins, A.J.S., Day, A.J., Gerard, A., Naciri, Y., Ledu, C., Bayne, B.L. & Heral, M. (1994). A genetic and metabolic basis for faster growth among triploids induced by blocking meiosis I but not meiosis II in the larviparous European flat oyster, *Ostrea edulis* L. *Journal of Experimental Marine Biology and Ecology* **184**, 21–40.

Hawkins, A.J.S., Wilson, I.A. & Bayne, B.L. (1987). Thermal responses reflect protein turnover in *Mytilus edulis*. *Functional Ecology* **1**, 339–51.

Hedrick, P.W., Ginevan, M.E. & Ewing, E.P. (1976). Genetic polymorphism in heterogeneous environments. *Annual Review of Ecology and Systematics* **7**, 1–32.

Hoffmann, K.H. (1983). Metabolic and enzyme adaptation to temperature and pressure. In *The Mollusca*, vol. 2 (ed. P.W. Hochachka), pp. 220–56. New York: Academic Press.

Holley, M.E. & Foltz, D.W. (1987). Effects of multiple-locus heterozygosity and salinity on clearance rate in a brackish water clam, *Rangia cuneata* (Sowerby). *Journal of Experimental Marine Biology and Ecology* **111**, 121–31.

Houlihan, D.F., Hall, S.J., Gray, C. & Noble, B.S. (1988). Growth rates and protein turnover in Atlantic cod, *Gadus morhua*. *Canadian Journal of Fisheries and Aquatic Sciences* **43**, 951–64.

Hummel, H. & Patarnello, T. (1994). Genetic effects of pollutants on marine and estuarine invertebrates. In *Genetics and Evolution of Aquatic Organisms* (ed. A.R. Beaumont), pp. 425–34. London: Chapman and Hall.

Johnson, M.S. (1971). Adaptive lactate dehydrogenase variation in the crested blenny, *Anoplarchus*. *Heridity* **27**, 205–26.

Johnston, I.A. & Dunn, J. (1987). Temperature acclimation and metabolism in ectotherms with particular reference to teleost fish. In *Temperature and Animal Cells* (ed. K. Bowler & B.J. Fuller), pp. 67–94. Cambridge: The Company of Biologists.

Kacser, H. & Burns, J.A. (1981). The molecular basis of dominance. *Genetics* **97**, 639–66.

King, T.L., Zimmerman, E.G. & Beitinger, T.L. (1985). Concordant variation in thermal tolerance and allozymes of the red shiner, *Notropis lutrensis*, inhabiting tailwater sections of the Brazos River, Texas. *Environmental Biology of Fishes* **13**, 49–57.

Koehn, R.K. (1969). Esterase heterogeneity: dynamics of a polymorphism. *Science* **163**, 943–4.

Koehn, R.K. & Bayne, B.L., (1989). Towards a physiological and energetic understanding of the energetics of the stress response. *Biological Journal of the Linnean Society* **37**, 157–71.

Koehn, R.K., Perez J.E. & Merritt, R.B. (1971). Esterase enzyme function and genetical structure of populations of the freshwater fish, *Notropis stramineus*. *American Naturalist* **105**, 51–69.

Koehn, R.K., Turano, F.J. & Mitton, J.B. (1973). Population genetics of marine pelecypods. II. Genetic differences in microhabitats of *Modiolus demissus*. *Evolution* **27**, 100–5.

Lavie, B. & Nevo, E. (1986). Genetic diversity of marine gastropods: contrasting strategies of *Cerithium rupestre* and *C. scabridum* in the Mediterranean Sea. *Marine Ecology Progress Series* **28**, 99–103.

Lerner, I.M. (1954). *Genetic homeostasis*, p. 134. Edinburgh: Oliver and Boyd.

Levinton, J.S. (1973). Genetic variation in a gradient of environmental variability. *Science* **180**, 75–6.

Levinton, J.S. & Lassen, H.H. (1978). Selection, ecology and evolutionary adjustment within bivalve mollusc populations. *Philosophical Transactions of the Royal Society of London* **284**, 403–15.

Merritt, R.B. (1972). Geographic distribution and enzymatic properties of lactate dehydrogenase allozymes in the fathead minnow, *Pinaphales spromelas*. *American Naturalist* **106**, 173–84.

Mitton, J.B., Carey, C. & Kocher, J.D. (1986). The relation of enzyme heterozygosity to standard and active oxygen consumption and body size of tiger salamanders, *Ambystoma tigrinum*. *Physiological Zoology* **59**, 574–82.

Mitton, J.B. & Grant, M.C. (1984). Associations among protein heterozygosity, growth rate and developmental homeostasis. *Annual Review of Ecology and Systematics* **15**, 479–99.

Mitton, J.B. & Koehn, R.K. (1975). Genetic organisation and adaptive response of allozymes to ecological variables in *Fundulus heteroclitus*. *Genetics* **79**, 97–111.

Mork, J. & Sundnes, G. (1985). O-group cod (*Gadus morhua*) in captivity: differential survival of certain genotypes. *Helgolander Meeresunters* **39**, 63–70.

Nevo, E. (1983). Adaptive significance of protein variation. In *Protein Polymorphism: Adaptive and Taxonomic Significance* (ed. G.S. Oxford & G. Rollinson), pp. 239–82. London: Academic Press.

Nevo, E., Roy, R., Lavie, B., Beiles, A. & Muchtar, S. (1986). Genetic diversity and resistance to marine pollution. *Biological Journal of the Linnean Society* **29**, 139–44.

Nevo, E., Shimony, T. & Libni, M. (1977). Thermal selection of allozyme polymorphism in barnacles. *Experientia* **43**, 1562–4.

Newell, R.C. (1979). *Biology of intertidal animals*, p. 781. Faversham: Marine Ecological Surveys.

Newell, R.C. & Northcroft, H.R. (1967). A re-interpretation of the effect of temperature on the metabolism of certain marine invertebrates. *Journal of Zoology* **151**, 277–98.

Newman, M.C., Diamond, S.A., Mulvey, M. & Dixon, P. (1989). Allozyme genotype and time to death of mosquitofish, *Gambusia affinis* (Baird and Girard) during acute toxicant exposure: a comparison of arsenate and inorganic mercury. *Aquatic Toxicology* **15**, 141–56.

Nyman, L. (1975). Allelic selection in a fish (*Gymbocephalus ceruna* L.) subjected to hotwater effluents. *Report of the Institute of Freshwater Research* **54**, 75–82.

Parsons, P.A. (1990). Fluctuating asymmetry: an epigenetic measure of stress. *Biological Reviews* **65**, 131–45.

Pecon Slattery, J., Lutz, R.A. & Vrijenhoek, R.C. (1993). Repeatability of correlations between heterozygosity, growth and survival in a natural population of the hard clam *Mercenaria mercenaria* L. *Journal of Experimental Marine Biology and Ecology* **165**, 209–24.

Place, A.R. & Powers, D.A. (1979). Genetic variation and relative catalytic efficiencies: lactate dehydrogenase B allozymes of *Fundulus heteroclitus. Proceedings of the National Academy of Sciences of the USA* **76**, 2354–8.

Powers, D.A. & Place, A.R. (1978). Biochemical genetics of *Fundulus heteroclitus* (L.). I. Temporal and spatial variation in gene frequencies of *Ldh-B, Mdh-A, Gpi-B, and Pgm-A. Biochemical Genetics* **16**, 593–607.

Powers, D.A., Ropson, I., Brown, D.C., van Beneden, R., Cashon, R., Gonzalez-Villasenor, L.I. & Dimichele, L. (1986). Genetic variation in *Fundulus heteroclitus*: geographical distribution. *American Naturalist* **26**, 131–44.

Prosser, C.L. (1973). *Comparative animal physiology*, p. 966. Philadelphia: W.B. Saunders.

Rodhouse, P.G. & Gaffney, P.M. (1984). Effect of heterozygosity on metabolism during starvation in the American oyster *Crassostrea virginica. Marine Biology* **80**, 179–87.

Saenger, P. & Holmes, N. (1992). Physiological, temperature tolerance, and behavioural differences between tropical and temperate organisms. In *Pollution in Tropical Aquatic Systems* (ed. D. W. Connell & D.W. Hawker), pp. 69–95. London: CRC Press.

Schlieper, C., Flugel, H. & Theede, H. (1967). Experimental investigations of the cellular resistance ranges of marine temperate and tropical bivalves: results of the Indian Ocean expedition of the German Research Association. *Physiological Zoology* **40**, 345–60.

Scott, T.M. & Koehn, R.K. (1990). The effect of environmental stress on the relationship of heterozygosity to growth rate in the coot clam *Mulinia lateralis* (Say). *Journal of Experimental Marine Biology and Ecology* **135**, 109–16.

Serov, D.V., Nikonorov, S.I. & Asadullaeva, Eh.S. (1988). Genetic selection in young stellate sturgeon. *Rybnoe Khoziaistvo* **4**, 64–5.

Shah, M.S. (1991). Thermal selection of allozyme polymorphism in the guppy *Poecilia reticulata. Asian Fisheries Science* **4**, 279–94.

Somero, G.N. (1969). Enzymatic mechanisms of temperature compensation: immediate and evolutionary effects of temperature on enzymes of poikilotherms. *American Naturalist* **103**, 518–30.

Somero, G.N. (1992). Biochemical ecology of deep-sea animals. *Experientia* **48**, 537–43.

Somero, G.N. & Hochachka, P.W. (1976). Biochemical adaptation to temperature. In *Adaptation to Environment: Essays on the Physiology of Marine Animals* (ed. R.C. Newell), pp. 125–90. London: Butterworths.

Ushakov, B.P., Chernokozheva, I.S. & Nikiforov, S.M. (1989). Neutrality of biochemical polymorphism of the gastropod *Notoacmea concinna* detected in short-term exposure to elevated temperature. *Biologiya Morya* **3**, 44–50.

Van Delden, W. (1994). Genetic diversity and its role in the survival of species. In *Biodiversity and Global Change* (ed. O.T. Solbrig, H.M. van Emden & P.G.W.J. Oordt), pp. 41–56. New York: CAB International.

Waterlow, J.C. (1984). Protein turnover with special reference to man. *Quarterly Journal of Experimental Physiology* **69**, 409–38.

Wilson, J.G. & Elkaim, B. (1991). Tolerances to high temperature of infaunal bivalves and the effect of geographical distribution, position on the shore and season. *Journal of the Marine Biological Association of the United Kingdom* **71**, 169–77.

Worrall, C.M. & Widdows, J. (1984). Investigation of factors influencing mortality in *Mytilus edulis* L. *Marine Biology Letters* **5**, 85–97.

Zimmerman, E.G. (1984). Genetic and physiological correlates in fish adapted to regulated streams. In *Regulated Rivers* (ed. A. Lillehammer & S.V. Saltveit), pp. 273–92. Oslo: Universitetsforlaget AS.

Zimmerman, E.G. & Richmond, M.C. (1981). Increased heterozygosity at the *Mdh-B* locus in fish inhabiting a rapidly fluctuating thermal environment. *Transactions of the American Fisheries Society* **110**, 410–16.

Zouros, E. & Foltz, D.W. (1987). The use of allelic isozyme variation for the study of heterosis. In *Isozymes: Current Topics in Biological and Medical Research* (ed. M.C. Rattazzi, J.G. Scandalios & G.S. Whitt), pp. 255–70. New York: Alan R. Liss.

Zouros, E. & Pogson, G.H. (1994). Heterozygosity, heterosis and adaptation. In *Genetics and Evolution of Aquatic Organisms* (ed. A.R. Beaumont), pp. 135–46. London: Chapman and Hall.

Zouros, E., Singh, S.M., Foltz, D.W. & Mallet, A.L. (1983). Post-settlement viability in the American oyster (*Crassostrea virginica*): an overdominant phenotype. *Genetical Research* **41**, 259–70.

H.E. GUDERLEY and J. ST PIERRE

Phenotypic plasticity and evolutionary adaptations of mitochondria to temperature

Introduction

Temperature is a critical determinant of physiological performance in ecotherms. For the majority of fish and gill-breathing aquatic invertebrates, thermal fluctuations in the environment change body temperature as a result of the rapid thermal equilibration across gill surfaces. Thermal fluctuations affect physiological processes both through effects on reaction rates as well as on the equilibria determining the non-covalent interactions that stabilise macromolecules and membranes. In response to either short- or long-term thermal fluctuations, individual organisms often adjust their exact biochemical composition as well as the rates of physiological and metabolic processes. Adaptation to different thermal habitats on an evolutionary time-scale is likely to have favoured qualitative and quantitative modifications in biochemical and physiological properties.

As the centres of oxidative phosphorylation, mitochondria are critical sites of adenosine triphosphate (ATP) provision in aerobic tissues. Whereas mitochondria are implicated in anabolic and catabolic pathways in liver, kidney and other organs, the primary role of mitochondria in oxidative muscle fibres is the provision of ATP for muscle contraction. Given the importance of locomotion in foraging, migration, prey capture and predator avoidance, both sustained and burst locomotor capacity are likely to have been modified during evolutionary adaptation to new thermal environments. These considerations suggest that the thermal sensitivity of mitochondria in oxidative fibres is likely to have changed during thermal adaptation on an evolutionary time-scale. During cold acclimation of fish, the mitochondrial volume density of muscle increases, much as observed in response to endurance training in mammalian muscle. This suggests that the aerobic capacity of muscle becomes limiting at low temperatures. This chapter examines how the properties of mitochondria from oxidative muscle fibres of ectotherms have been modified during long-term (evolutionary) adaptation to different thermal habitats, as well as examining the plasticity of muscle mitochondria during short-term thermal acclimation. Data from non-muscle systems will be presented when pertinent to the question of mitochondrial adaptation to temperature.

The thermal sensitivity of muscle mitochondria should resemble that of the locomotor activity in which they are implicated. For any given species of fish, sustained swimming performance increases with temperature to an optimum and then decreases (Fry & Hart, 1948; Brett, 1967). The optimal temperature for sustained swimming varies with the thermal habitat and with acclimation status. Antarctic fish have their optimum for sustained swimming around 0 °C, temperatures at which most temperate fish are virtually inactive and tropical fish are dead. In contrast, maximum aerobic performance in temperate species occurs at higher temperatures (15–25 °C) and the fish can become torpid below 5 °C. Numerous species, such as salmon (*Oncorhynchus nerka*), goldfish (*Carassius auratus*) and striped bass (*Morone saxatilis*) (Fry & Hart, 1948; Brett, 1967; Sisson & Sidell, 1987) shift the thermal optimum of sustained swimming during thermal acclimation. Adjustments in the thermal sensitivity of the metabolic and contractile properties of muscle could well accompany these short- and long-term changes in sustained swimming capacity.

In fish, a common response to short-term cold acclimation is an increase in the aerobic capacity of the swimming musculature. This has been observed in the form of enhanced mitochondrial volume densities in red and white muscle or as increases in the activity of mitochondrial enzymes (for reviews, see Egginton & Sidell, 1989; Guderley, 1990; Johnston, 1993; Sänger, 1993). The proportion of aerobic fibres in the swimming musculature may also rise during cold acclimation (Johnston & Lucking, 1978; Sidell, 1980). Such responses typically occur when eurythermal fish are acclimated to temperatures at least 8 °C below their optimum for locomotion (Guderley, 1990). The overall response of the swimming muscles to thermal acclimation is particularly well documented for striped bass (*M. saxatilis*), goldfish (*C. auratus*), carp (*Cyprinus carpio*) and the sculpin (*Myoxocephalus scorpius*). Cold acclimation also enhances muscle aerobic capacity for other temperate zone fish, including chain pickerel *Esox niger* (Kleckner & Sidell, 1985), crucian carp *Carassius carassius* (Johnston & Maitland, 1980) and the sticklebacks, *Pungitius pungitius* and *Gasterosteus aculeatus* (Guderley & Foley, 1990; Vézina & Guderley, 1991; Guderley *et al.*, 1994). Modifications of the contractile apparatus seem less frequent than increases in aerobic capacity as certain species, such as striped bass and chain pickerel, in which muscle aerobic capacity rises with cold acclimation show few or no changes in their contractile properties or contractile proteins (Sidell & Johnston, 1985; Sisson & Sidell, 1987). In muscle, oxidative capacity and hence mitochondrial function seem the primary target of thermal compensation.

Do mitochondria show thermal compensation of maximal oxidative capacities?

An increase in mitochondrial volume density during cold acclimation suggests that mitochondrial function becomes limiting during activity at low temperatures. The functional limitations could reflect: (i) the maximal oxidative capacities of mitochondria; (ii) limitations in diffusive transfer between the myofibrillar and mitochondrial compartments (Egginton & Sidell, 1989); or (iii) a loss of regulatory sensitivity (Dudley *et al.*, 1987) with a decrease in temperature. The first two possibilities are supported by the following studies. Johnston (1987) reported higher adenosine diphosphate (ADP)-stimulated rates of oxygen uptake at 0 °C in skinned muscle fibre segments isolated from Antarctic than temperate zone fishes. The rates of oxygen consumption per cm^3 mitochondria are, however, more similar among species, reflecting the higher volume densities of mitochondria in the red muscles of Antarctic fish (Johnston, 1987). Thus, expressing oxygen uptake per unit mitochondria removes the evidence for significant cold-adaptation of rates of oxygen uptake by muscle in Antarctic fish and suggests only limited evolutionary adjustments of the thermal sensitivity of mitochondrial respiration. This suggests that mitochondrial numbers must be increased to overcome the effects of low temperature on mitochondrial respiration.

The hypothesis that muscle mitochondria of organisms inhabiting cold habitats have not significantly enhanced their maximum capacities at low temperatures can be evaluated by comparison of the thermal sensitivity of maximal rates of substrate oxidation by mitochondria isolated from red muscle of fish inhabiting a variety of thermal habitats (Blier & Guderley, 1993*a*; Johnston *et al.*, 1994; J. St Pierre & H. Guderley, unpublished results). These studies examine teleost species living at the piscean temperature extremes, including the Lake Magadi tilapia, *Oreochromis alcalcus grahami*, which has one of the highest temperature tolerances of any fish (40 °C), the tilapias, *Oreochromis niloticus* and *O. andersoni* which live at warm temperatures (20–30 °C), the sculpin *M. scorpius* and the rainbow trout *Oncorhynchus mykiss* which live in temperate waters (1–18 °C) and the Antarctic notothenid,

Notothenia coriiceps, which requires special adaptations to avoid freezing. These data can be compared with results obtained with similar methods for mitochondria from muscle of carp, *C. carpio*, tuna, *Katsuwonus pelamis* and mako shark, *Isurus oxyrinchus*, as well as from the swordfish (*Xiphias gladius*) heater organ (Moyes *et al.*, 1988, 1992*a*; Ballantyne *et al.*, 1992). As differences in isolation and assay conditions can modify the measured properties of mitochondria, interspecific comparisons of mitochondrial properties are facilitated by the use of similar isolation and assay methods.

An indication of mitochondrial quality is provided by the respiratory control ratio (RCR) which is the ratio of the maximal, ADP-stimulated (state 3) rate of oxygen uptake over the rate obtained once the ADP has been depleted. The RCR values indicate the extent to which oxygen uptake is coupled to the phosphorylation of ADP. Low RCR values (below 4) may reflect contamination by myofibrillar ATPases or simply poor coupling of damaged mitochondria. Whereas maximal rates of oxygen uptake can be obtained by chemically uncoupling mitochondria, preparations with intrinsically low RCR values may be contaminated by other cellular fractions. Oxygen uptake rates are expressed per mg protein in the mitochondrial fraction, so contamination by other cellular structures would decrease the reliability of maximal rate estimates. The RCR values for the mitochondrial preparations are 10.0±0.5 for *O. a. grahami*, 7.0±2.4 for *N. coriiceps*, 5.8±2.8 for *M. scorpius*, 5.9±1.2 for *O. andersoni*, 12.0±1.1 for *O. mykiss*, 9–17 for *C. carpio*, 11.2±1.4 for *L. oxyrinchus*, 14.9±2.2 for *K. pelamis* and 9.8±3.0 for *X. gladius* (J. St Pierre & H. Guderley, unpublished results; Moyes *et al.*, 1988, 1992*a*; Ballantyne *et al.*, 1992; Johnston *et al.*, 1994). In general, RCR values do not vary with assay temperature although for trout the values were higher at intermediate assay temperatures (12 °C) (J. St Pierre & H. Guderley, unpublished results) and for carp, RCR values were higher at 10 °C than at 30 °C (Moyes *et al.*, 1988).

To extrapolate from *in vitro* maximal capacities to *in vivo* rates, mitochondria must be provided with physiologically relevant substrates. Questions of mitochondrial substrate preferences are particularly critical in interspecific comparisons. In the teleosts examined to date, muscle mitochondria oxidise pyruvate at maximal rates at all temperatures (Moyes *et al.*, 1988, 1992*a*; Johnston *et al.*, 1994; J. St Pierre & H. Guderley, unpublished results). Pyruvate and palmitoyl carnitine generally support similar rates of oxygen uptake by mitochondria isolated from red muscle of teleosts (Moyes *et al.*, 1989, 1992*a*; Kiessling & Kiessling, 1993; Johnston *et al.*, 1994; J. St Pierre & H. Guderley, unpublished results). In this respect, mitochondria from sculpin red muscle differ as palmitoyl carnitine was oxidised at only 55% of the rate of pyruvate (Johnston *et al.*, 1994). Given that mitochondria can have marked preferences for specific fatty acids or acyl carnitines (Ballantyne *et al.*, 1989; Crockett & Sidell, 1993), the thermal dependence of palmitoyl carnitine oxidation may not entirely reflect that of maximal capacities of lipid oxidation. In elasmobranchs, the substrate preferences of muscle mitochondria are quite different, with glutamine and β-hydroxybutyrate leading to oxidation rates 1.5 to two-fold those of pyruvate (Ballantyne *et al.*, 1992; Chamberlin & Ballantyne, 1992) and palmitoyl carnitine oxidation being undetectable (Ballantyne *et al.*, 1992). Physiologically pertinent substrates must be tested for the thermal sensitivity to reflect potential *in vivo* responses of the mitochondria. Unfortunately, many studies of isolated mitochondria use succinate as the carbon substrate, making extrapolation to *in vivo* conditions difficult.

Temperature has similar effects on maximal rates of pyruvate and palmitoyl carnitine oxidation by isolated muscle mitochondria (Figs. 1 and 2). Essentially for each species, rates rise with temperature and tend to increase with habitat temperature. The rates obtained for mitochondria from the swordfish heater organ lie well above any values obtained at 20 °C. The rates for the Lake Magadi tilapia (*O. a. grahami*) at their habitat temperatures are markedly higher than those of the Antarctic species at their habitat temperatures. Mitochondria from *N. coriiceps* oxidise pyruvate at rates of 15.5 nmole O mg protein^{-1} min^{-1} at −1.5 °C whereas those from Lake Magadi tilapia show rates of 253 nmole O mg protein^{-1} min^{-1} at 40 °C. Some thermal compensation is apparent, however, in *N. coriiceps* mitochondria as extrapolation of the temperature rate curves suggests that mitochondria from Lake Magadi tilapia would stop oxidising palmitoyl carnitine at 5 °C. Rate curves for pyruvate oxidation are more exponential, but double logarithmic plots intersect the ordinate at 0.4 °C. Furthermore, for the species living at intermediate temperatures, thermal compensation is indicated by the similarity of maximal rates of pyruvate oxidation at habitat temperatures. Thus, red muscle mitochondria from *M. scorpius, O. mykiss, O. niloticus, O. andersoni, C. carpio, K. pelamis* and *I. oxyrinchus* show rates between 60 and 110 nmole O mg protein^{-1} min^{-1} at habitat temperatures ranging from 12 to 30 °C. Whereas mitochondria from the Antarctic notothenid, functioning at their habitat temperature, cannot attain the rates that characterise the Lake Magadi tilapia at 40 °C, oxygen uptake rates of mitochondria from cold-adapted

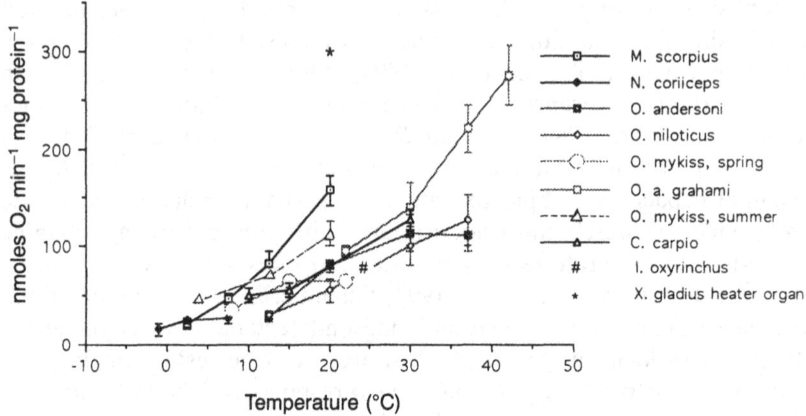

Fig. 1. Thermal sensitivity of maximal rates of pyruvate oxidation by fish red muscle mitochondria. The species used are indicated in the figure. Mitochondria were isolated in 140 mM KCl, 10 mM EDTA, 5 mM $MgCl_2$ and 20 mM Hepes, pH 7.3 at 20 °C (after Moyes *et al.*, 1988). Bovine serum albumin (BSA) is present at 2% (*O. mykiss*), 1% (*O. andersoni*) or 0.5% (other species). Oxygen consumption is measured polarographically. Oxygen concentrations are calculated using the relationship between temperature and the oxygen concentration of the medium obtained by couloximetry (Johnston *et al.*, 1994) or that obtained by chemical titration (Graham, 1987). The assay medium contains 140 mM KCl, 5 mM Na_2HPO_4, 20 mM Hepes, pH 7.3 at 20 °C, 0.5% BSA. Oxygen uptake is typically measured following the addition of 0.1 mM malate to spark the Krebs cycle and of a saturating concentration of substrate (2.5 mM pyruvate or 0.025 mM palmitoyl carnitine). The maximal oxidation rates are obtained by adding ADP to final concentrations of 0.25–0.5 mM. (After Moyes *et al.*, 1988, 1992; Blier & Guderley, 1993*a*; Johnston *et al.*, 1994 and J. St Pierre & H. Guderley, unpublished results.)

and temperate species are higher than those which would be predicted by extrapolation from the temperature-rate curve of Lake Magadi tilapia.

Muscle mitochondria from temperate zone species seem capable of more complete compensation of maximal capacities than those from the Antarctic species. In response to the constraints of functioning at continuously low temperatures, Antarctic fish have enhanced the levels of mitochondrial enzymes in the red swimming muscle (Crockett & Sidell, 1990). Citrate synthase and cytochrome oxidase activities per wet muscle mass are 1.5–5 times higher at 1 °C in Antarctic than in

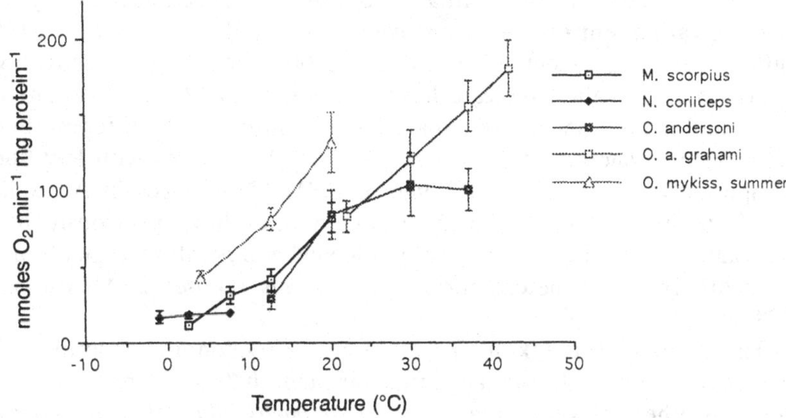

Fig. 2. Thermal sensitivity of maximal rates of palmitoyl carnitine oxidation by fish red muscle mitochondria. (After Johnston *et al.*, 1994; J. St Pierre & H. Guderley, unpublished results.)

temperate zone species. Similarly, the mitochondrial volume densities in the muscles of Antarctic species are markedly higher than those in temperate zone species, as in some Antarctic species the values reach 60% (Johnston *et al.*, 1988). Decreased mitochondrial clustering and higher intracellular lipid contents may minimise diffusive limitations and maximise rates of mitochondrial ATP generation in the more active Antarctic fish (Londraville & Sidell, 1990).

Thermal sensitivity of mitochondrial capacities: estimates from Arrhenius plots

The correlation between habitat temperatures and the thermal sensitivity of mitochondrial respiration has been examined with a somewhat different approach using hydrothermal vent and shallow-living invertebrates (Dahlhoff *et al.*, 1991; Dahlhoff & Somero, 1993). In these experiments, mitochondria were prepared from a variety of tissues and the maximal respiration rates of uncoupled mitochondria were measured at a range of temperatures, extending well above habitat values. The Arrhenius plot is constructed by plotting the natural logarithm of the maximal respiration rate against the reciprocal of the absolute temperature. The slope obtained at physiological temperatures can be used to calculate the apparent Arrhenius activation energy (E_a). The temperature at which the slope of the relationship changes sharply is the Arrhenius break temperature. The E_a is a function of the activation enthalpy, but as the activation enthalpy and the activation entropy are positively correlated (Low *et al.*, 1973), differences in E_a reflect changes in the Gibbs free energy of activation. Thus lower E_a values indicate lower barriers to catalysis. Comparisons of homologous enzyme systems from organisms with different body temperatures show that E_a values are lower for species with low body temperatures (Hochachka & Somero, 1984). Differences in Arrhenius break temperatures of mitochondrial enzymes have previously been correlated with differences in physiological temperature regulation of homeothermic and heterothermic endotherms (Geiser & McMurchie, 1984).

The Arrhenius break temperature for mitochondrial respiration is well correlated with the apparent maximal habitat temperatures of the hydrothermal vent organisms, *Riftia pachyptila, Alvinella pompejana, Bythograea thermydron, Calyptogena magnifica* and *Bathymodiolus thermophilus* as well as those of the shallow-living species, *Mytilus galloprovincialis, Cancer antennarius, Slemya reidi* and *Calyptogena elongata* (Dahlhoff *et al.*, 1991) and with the acclimation temperature of five species of abalones (genus *Haliotus*) (Dahlhoff & Somero, 1993). The combined data set shows a convincing positive correlation (Fig. 3). The correlations between habitat temperature and the Arrhenius break temperatures of cytochrome C oxidase and succinate dehydrogenase are as strong as those of mitochondrial respiration rates (Dahlhoff *et al.*, 1991; O'Brien *et al.*, 1991). Changes in fluidity are not detected at the temperatures at which respiration rates decrease, but disruption of membrane hydrophobic interactions reduces the temperatures at which loss of enzyme activity occurs. Membrane fluidity, as assessed by the fluorescence probe 1,6 diphenyl 1,3,5 hexatriene, is lower in the species living at warm temperatures than in those living at cooler temperatures (O'Brien *et al.*, 1991).

Apparent Arrhenius activation energies (E_a) of mitochondria isolated from the hydrothermal vent and shallow living invertebrates mentioned above are correlated with the habitat temperatures, but less well than the Arrhenius break temperatures (Dahlhoff *et al.*, 1991). In contrast, the E_a values derived from the temperature-rate curves in Figs. 1 and 2 are not correlated with habitat temperature ($r^2 = 0$). For these fish species, the E_a of mitochondrial respiration generally lies around 40 kJ mol^{-1} with mitochondria from *M. scorpius* showing a value of 72 kJ mol^{-1}. Nonetheless, within species, E_a can change with acclimatisation status (see section on phenotypic plasticity).

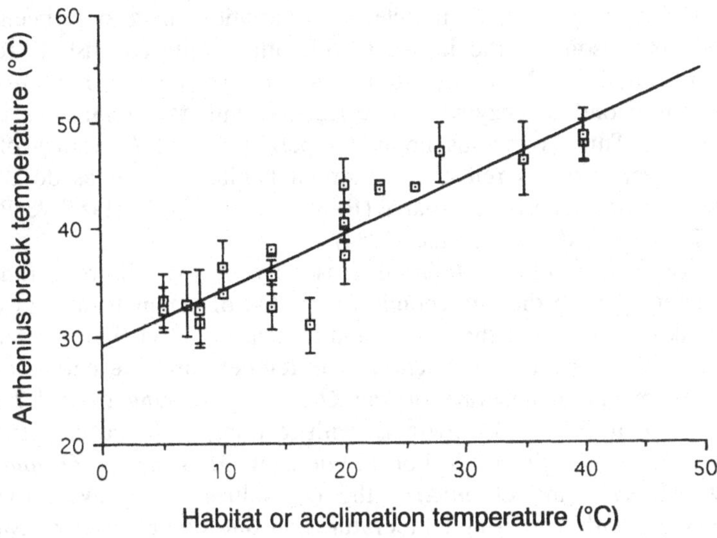

Fig. 3. Arrhenius break temperatures as a function of habitat or acclimation temperature for mitochondria isolated from hydrothermal vent organisms, shallow-living invertebrates and eastern Pacific abalones. The equation describing this relationship is $y = 29.31 + 0.51\ x$ ($R^2 = 0.82$) where y is the Arrhenius break temperature and x is the habitat or acclimation temperature. The data are from Dahlhoff *et al.* (1991) and Dahlhoff & Somero (1993).

Mechanisms explaining interspecific differences in mitochondrial temperature sensitivity and oxidative capacities

Compositional modifications such as changes in membrane properties, in cristae density, and in variants of mitochondrial enzymes and membrane proteins could be implicated in evolutionary adaptation of mitochondrial properties to distinct thermal habitats. Mitochondrial membrane fluidity changes with adaptation temperature as much or more than in other cellular membranes (Hazel & Williams, 1990). Maintenance of similar membrane fluidity at habitat temperatures may be a central mechanism in the thermal compensation of maximal oxidative capacities, even though no direct correlation between the temperatures at which fluidity changes and parameters such as the Arrhenius break temperatures of oxidative phosphorylation is apparent. Cristae densities of mitochondria from different tissues vary (i.e. Moyes *et al.*, 1992*a*), but, within fish aerobic muscle, such variations have only been found in comparisons of the icefishes with other Antarctic fish (Archer & Johnston, 1991). Numerous studies indicate that the thermal sensitivities of mitochondrial enzymes are correlated with the organism's thermal habitat. Thus Arrhenius break temperatures and E_a values of mitochondrial enzymes reflect the thermal habitats, much as do those of their cytoplasmic counterparts (Raison *et al.*, 1971; Hazel & Prosser, 1974; Geiser & McMurchie, 1984).

Comparison of the thermal sensitivity of the different functional elements within the mitochondria with that of maximal rates of oxygen uptake could reveal the components which establish the thermal sensitivity of mitochondrial function. The temperature-rate curves for substrate oxidation generate overall Q_{10} values ranging from 1.2 for *N. coriiceps* to 3.2 for *M. scorpius*, with sections of the curves generating Q_{10} values as high as 7.3. For *N. coriiceps, M. scorpius, O. andersoni, O. niloticus*, and *O. mykiss*, the Q_{10} values for pyruvate oxidation decrease with increasing temperatures, at times by as much as two-fold. Higher Q_{10} values also occur at low temperatures for palmitoyl carnitine oxidation by mitochondria from *M. scorpius* and *O. andersoni* and for pyruvate oxidation by liver mitochondria from brook trout (*Salvelinus namaycush*) (Ballantyne *et al.*, 1989). In contrast, the enzymatic components of pyruvate oxidation show a low and constant thermal sensitivity. Pyruvate dehydrogenase (PDH) isolated from *O. mykiss* red muscle has a Q_{10} of 1.33 between 8 and 15 °C and of 1.25 between 15 and 22 °C (Blier & Guderley, 1993*a*). The Q_{10} values of citrate synthase and cytochrome C oxidase from fish muscle lie between 1.35 and 1.7 (Blier & Guderley, 1988; Crockett & Sidell, 1990). This suggests that thermal limitations on pyruvate oxidation do not reside on the level of PDH, the Krebs cycle or the electron transport chain. The increase in Q_{10} values of pyruvate oxidation with falling temperatures may reflect the impact of localised changes in membrane fluidity (Raison *et al.*, 1971; Hazel & Williams, 1990) on the functioning of membrane transporters. Alternatively, this thermal sensitivity could reflect the properties of the pyruvate or adenine nucleotide transporters. The pyruvate transporter in rat liver mitochondria is highly sensitive to temperature with a Q_{10} value of 4.6 between 9 and 37 °C (Halestrap, 1975). The mammalian adenine nucleotide transporter has a higher Q_{10} at low than at high temperatures (Klingenberg *et al.*, 1982). Little is known about the properties of these transporters in ectotherms.

Shifts in the thermal sensitivity of mitochondrial substrate oxidation over physiological temperatures are not universal. Thus, in mitochondria from red muscle of Lake Magadi tilapia and carp, the Q_{10} value of pyruvate or palmitoyl carnitine oxidation changes little over the thermal range studied. Furthermore, for palmitoyl carnitine oxidation by mitochondria from *N. coriiceps* red muscle and *S. namaycush* liver (Ballantyne *et al.*, 1989) as well as pyruvate oxidation by mitochondria from summer trout, Q_{10} values are similar at low and high temperatures.

Thermal effects on mitochondrial regulatory properties: general considerations

Little is known about the thermal sensitivity of mitochondrial regulatory properties, particularly in the context of evolutionary adaptation to temperature. Oxidative phosphorylation involves numerous processes, including the translocation of adenine nucleotides and carbon substrates, the creation and maintenance of the proton gradient, proton leak, the oxidation of carbon substrates and ATP production by the H^+-ATP synthase. The contribution of these processes to the control of oxidative phosphorylation varies with respiratory rate and the exact experimental conditions (Groen *et al.*, 1982). In rat liver mitochondria oxidising succinate with an excess of hexokinase to maintain maximal rates, control is distributed almost equally among the adenine nucleotide translocator, the dicarboxylate carrier and cytochrome C oxidase. In state 4, proton leak is the prime determinant of respiration rate (Groen *et al.*, 1982; Brand *et al.*, 1994). Extrapolation of these results to *in vivo* conditions is difficult as succinate is generated intramitochondrially *in vivo*, suggesting that control by the dicarboxylate carrier would be less important. For mitochondria oxidising pyruvate or lipid, some control is likely to reside in the membrane transporters. Overall, it is clear that control of mitochondrial respiration is distributed among a number of processes and that this distribution changes with respiratory state (Brand & Murphy, 1987).

The availability of ADP for oxidative phosphorylation has often been implicated in the control of mitochondria respiration, particularly in muscle (Brand & Murphy, 1987). During muscle contraction, however, changes in the concentrations of ATP, ADP and AMP are minimised by tight metabolic regulation. Furthermore, estimates of adenylate concentrations under various physiological states suggest that mitochondria rarely encounter the high ADP and low ATP levels required for maximal state 3 rates. In the mammalian heart, creatine phosphate is thought to transfer ATP equivalents between mitochondria and the myofibrils. Heart mitochondria preferentially phosphorylate ADP generated by the mitochondrial creatine phosphokinase (CPK). The creatine phosphate produced would be used by the myofibrillar CPK to provide ATP to the myosin ATPase (Gellerich & Saks, 1982; Jacobus, 1985). Whereas such preferential phosphorylation is not found in heart mitochondria of the horseshoe crab, *Limulus polyphemus* (Doumen & Ellington, 1990) and mitochondrial phosphagen kinases are absent from molluscs and insect flight muscle (Schneider *et al.*, 1989), the occurrence of this mechanism in muscle mitochondria from ectothermal vertebrates has not been systematically examined. Meyer *et al.* (1984) suggest that the mere presence of a near-equilibrium phosphagen kinase in mitochondria allows the phosphagen to carry most high-energy phosphate equivalents to the cytoplasm. Mitochondria are frequently associated with substantial amounts of other ATP-using enzymes, primarily hexokinase and adenylate kinase. Adenylate kinase enhances the rate and efficiency of the transport of ATP equivalents from isolated rabbit heart mitochondria to hexokinase (Dzeja *et al.*, 1985). Adenylate kinase has been shown to preferentially provide ADP for oxidative phosphorylation in rat liver mitochondria (Gellerich *et al.*, 1993). These results suggest that ADP delivery to the mitochondria is likely to occur via channelling through ADP-generating enzymes rather than direct diffusion as used in experiments with isolated mitochondria. Ideally, considerations of the thermal sensitivity of mitochondrial regulation should take the properties of the ADP delivery mechanism into account.

Some information is available concerning the thermal sensitivity of regulatory processes in mammalian mitochondria, particularly for activation by Ca^{2+} adenine nucleotide exchange and pyruvate transport. The calcium activation of maximal rates of succinate oxidation by rat liver mitochondria is sharply temperature dependent with a much weaker activation occurring at low (25–28 °C) than at high (32–34 °C) temperatures (Moreno-Sanchez & Torres-Marquez, 1991). Decreasing temperature markedly slows the phosphorylation of exogenous ADP by rat liver mitochondria while the phosphorylation of endogenous ADP is little affected (Kemp *et al.*, 1969; Pfaff *et al.*, 1969; Klingenberg *et al.*, 1982). The Q_{10} of adenine nucleotide exchange by rat liver and beef heart mitochondria is 8–10 at low temperatures (<15 °C) while it lies around 2 at higher temperatures (Duée & Vignais, 1969; Klingenberg *et al.*, 1982). The pyruvate transporter of rat liver mitochondria is also highly sensitive to temperature (Halestrap, 1975). Membrane fluidity changes may explain the thermal sensitivity of mitochondrial transporters. The absence of lipid phase changes at the Arrhenius break temperatures and differences in the responses of specific transporters, however, suggest that the thermal sensitivities reflect those of localised lipid-transporter interactions or of the transporters themselves (Klingenberg *et al.*, 1982).

Thermal sensitivity of mitochondrial regulatory properties: ectotherms

The thermal adaptations of kinetic and regulatory properties of enzymes are a guide to possible adaptations of mitochondria from ectotherms. As has been particularly well demonstrated for lactate dehydrogenase and pyruvate kinase, enzyme substrate affinities tend to drop with increasing temperature, but are conserved within a narrow range of values at physiological temperatures (Hochachka & Somero, 1984). These patterns provide a null hypothesis for examining the thermal sensitivity of mitochondrial regulatory properties.

Despite the paucity of studies of the thermal sensitivity of mitochondrial regulatory properties from non-mammalian species, certain conclusions can be drawn. The oxygen affinity of isolated mitochondria ($K_{m_{app}}$ = 0.9 μM) from goldfish lateral line red muscle is not influenced by temperature (Bouwer & Van den Thillart, 1984). In these experiments, succinate was the substrate and the mitochondria had RCR values of 7.58±1.71. The mitochondrial affinity for oxygen depends on the respiratory state as it was decreased by accumulation of a respiratory inhibitor (presumably oxaloacetate). Although the significance of the mitochondrial oxygen affinities depends on the intracellular levels of oxygen, these data suggest that the mitochondrial affinity for oxygen would be less of a determinant of mitochondrial activity than thermally-induced changes in tissue oxygen saturation because of changes in oxygen solubility and delivery.

Physiological shifts in pH with temperature could modify mitochondrial properties. The creation and maintenance of the proton gradient is critical for oxidative phosphorylation. The entry of phosphate, pyruvate, ketone bodies, glutamate, di- and tri-carboxylic acids into the mitochondria depends directly or indirectly on proton symports. The pH gradient of isolated carp red muscle mitochondria is inversely related to external pH (Moyes *et al.*, 1988). Increases in the pH gradient should facilitate metabolite entry into the mitochondria. Accordingly, the affinity of scallop muscle mitochondria for pyruvate is increased by decreases in pH (Guderley *et al.* 1995). At any given external pH, increases in temperature decrease the pH gradient of carp muscle mitochondria. When external pH, however, covaries with temperature as *in vivo*, the mitochondrial pH gradient is maintained constant at approximately 0.4 U (Moyes *et al.*, 1988). Thus, during normal adaptation to thermal fluctuations, the mitochondrial pH gradient would not be compromised.

Two studies indicate that thermal influences on the affinity of mitochondria for their carbon substrates are minimised when pH is allowed to fluctuate with temperature in a physiological fashion (Yacoe, 1986; Blier & Guderley, 1993*a*). The pyruvate affinity of mitochondria isolated from *O. mykiss* red muscle remains approximately 45 μM between 8 and 22 °C, when pH covaries with temperature (Fig. 4) (Blier & Guderley, 1993*a*). At constant pH, the pyruvate affinity decreases with rising temperature (Fig. 4). PDH isolated from trout red muscle has a higher affinity for pyruvate than the mitochondria, which suggests that PDH creates a gradient facilitating pyruvate entry into the mitochondria. As the pyruvate affinities of PDH and of mitochondria respond similarly to temperature and pH (Fig. 4), the pyruvate affinity of the isolated mitochondria probably reflects the properties of pyruvate dehydrogenase. The $K_{m_{app}}$ of red muscle mitochondria is close to that of mitochondria from trout white muscle (35–40 μM; Moyes *et al.*, 1992*b*) and somewhat higher than that of scallop phasic muscle mitochondria (13–26 μM; Guderley *et al.*, 1995). The $K_{m_{app}}$ for succinate of mitochondria isolated from iguana liver only changes with temperature at high pH and high pCO_2 (Yacoe, 1986). Physiological shifts in pH and pCO_2 with temperature tend to minimise shifts in succinate affinity of mitochondria from iguana liver. In conclusion, physiological shifts in the pH and pCO_2 with temperature allow mitochondrial affinities for their carbon substrates to remain quite stable.

In contrast, the ADP affinity of *O. mykiss* red muscle mitochondria decreases markedly with a fall in temperature, even when pH is allowed to covary with temperature (Fig. 5) (Blier & Guderley, 1993*b*). This is reminiscent of the response of mammalian mitochondria to a fall in temperature where phosphorylation of exogenous ADP is diminished and the phosphorylation of endogenous ADP is little affected (Kemp *et al.*, 1969; Klingenberg *et al.*, 1982). Given the importance of ADP as a regulator of and a substrate for oxidative phosphorylation in muscle (Brand & Murphy, 1987), such a loss of sensitivity to ADP would require concommitant changes in other facets of energy metabolism. As trout readily live at temperatures at which their muscle mitochondria lose sensitivity to ADP, this surprising thermal sensitivity would require compensatory modifications, such as rapid changes in the composition of mitochondrial membranes (Hazel & Williams, 1990) during exposure of trout to low temperatures. Compensatory changes in the concentrations of adenylates would be unlikely given their widespread implication in energy metabolism.

The kinetic characteristics of the adenine nucleotide translocator are similar to those of the mitochondrial affinity for ADP, which suggests that the translocator is a major determinant of the mitochondrial affinity for ADP. ADP/ATP transport kinetics in mammalian mitochondria as

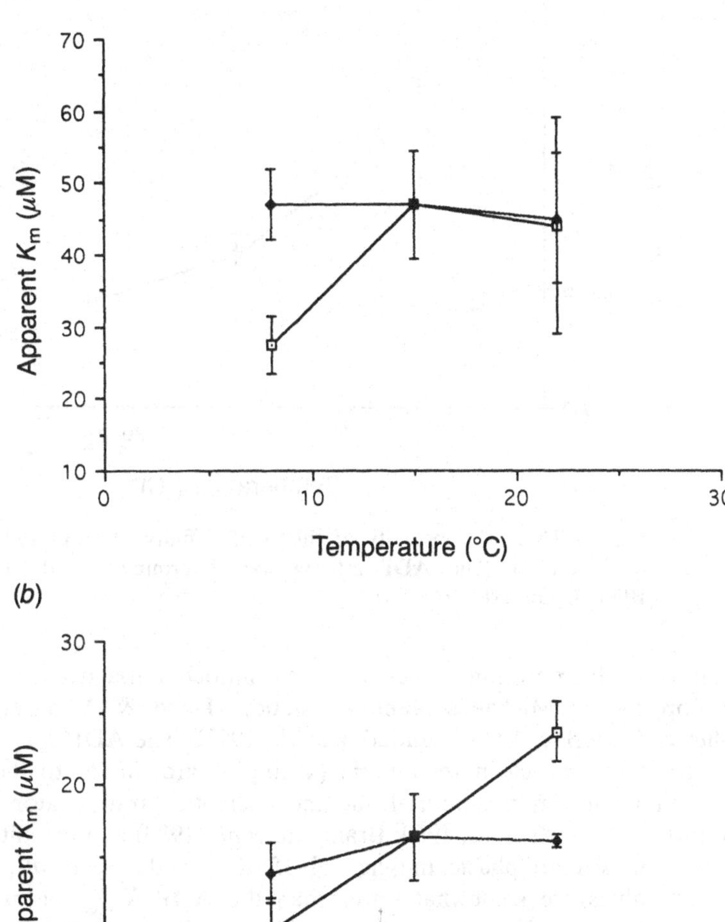

Fig. 4. Thermal sensitivity of the pyruvate affinity of isolated mitochondria (a) and of pyruvate dehydrogenase (b) purified from trout red muscle. (a) ⊡, constant pH (7.4), ◆, adjusted pH; (b) ⊡, constant pH (7.8), ◆, adjusted pH. (After Blier & Guderley, 1993*a*).

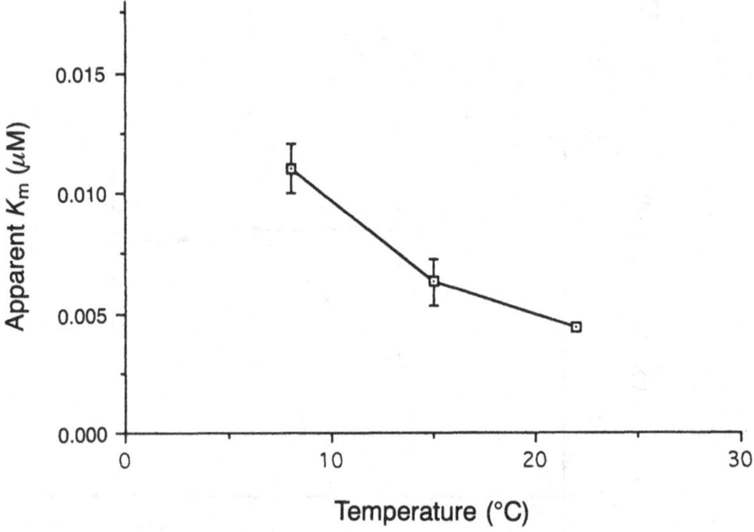

Fig. 5. Thermal sensitivity of the ADP affinity of trout red muscle mitochondria. The ADP affinity was determined as described in Blier & Guderley (1993*b*).

well as ADP saturation curves of muscle mitochondria from trout and scallop follow Michaelis–Menten kinetics (Duée & Vignais, 1969; Blier & Guderley, 1993*b*; Guderley *et al.*, 1995). The ADP $K_{m_{app}}$ values for trout red muscle mitochondria (4–10 μM) are similar to those for the mammalian mitochondrial adenine nucleotide translocator (1 and 10 μM ADP; Pfaff *et al.*, 1969; Brandolin *et al.*, 1980) and for mitochondria from scallop phasic muscle (12–15 μM; Guderley *et al.*, 1995). These values are somewhat lower than the ADP $K_{m_{app}}$ for oxidative phosphorylation (20–60 μM) of isolated mammalian heart mitochondria (Harris & Das, 1991). Nonetheless, considering the phylogenetic and methodological differences among these studies, the similarity of the ADP $K_{m_{app}}$ values is impressive.

Overall, these data suggest that the mitochondrial affinities for oxygen and carbon substrates are fairly constant over physiological temperatures. Definitive interpretation of these results requires knowledge of the impact of temperature upon intracellular concentrations of these substrates as well as upon the affinities of competing enzymatic systems. Nonetheless, oxygen and carbon substrate affinities seem unlikely to limit mitochondrial function at cold temperatures. Similarly, the mitochondrial pH gradient remains constant when external pH covaries with temperature. In contrast, the ADP sensitivity of trout muscle mitochondria may well limit function at low temperatures. The interactions between the adenine nucleotide translocator and ADP-generating enzymes, the activation of the H^+-ATPase, the regulation of cytochrome oxidase and proton leak may also change their importance with shifts in temperature.

Phenotypic plasticity of mitochondrial catalytic and regulatory properties

As mentioned earlier, cold acclimation markedly increases the aerobic capacity of fish muscle. This is manifested by rises in the mitochondrial volume density and in the specific activity of mitochondrial enzymes. Mitochondrial oxidative capacities and regulatory properties may be altered by cold acclimation. Several studies have examined this question, but few present data for sufficiently well-coupled mitochondria to give full credence to the results. Considerably more information is available concerning shifts in the phospholipid composition and fluidity of mitochondrial membranes (Van den Thillart & de Bruin, 1981; Wodtke, 1981*a*; Hazel & Williams, 1990) than concerning changes in capacities and regulatory properties. Homeoviscous adaptation of mitochondrial membranes during thermal acclimation is more complete than that of other subcellular fractions in green sunfish (*Lepomis cyanellus*) liver and goldfish brain. Virtually complete compensation of the permeability of liver mitochondria is attained after thermal acclimation of goldfish (for a review, see Hazel & Williams, 1990). The compensation of membrane fluidity during thermal acclimation seems to reflect changes in membrane composition more than physicochemical changes in the surrounding fluids (Hazel *et al.*, 1992).

Changes in the maximal capacities of mitochondria with thermal acclimation are suggested by the following studies. In muscle mitochondria isolated from goldfish acclimated to different temperatures, the Arrhenius break temperature shifts with acclimation temperature as does the nature of the Arrhenius plot (Van den Thillart & Modderkolk, 1978). Whereas the rate of miochondrial respiration per gram of muscle rises with cold acclimation, the specific activity of cytochrome C oxidase increases considerably more than the mitochondrial respiration rate (Van den Thillart & Modderkolk, 1978). The rise in oxidation rate per gram of tissue seems to reflect the increased mitochondrial abundance (Tyler & Sidell, 1984) rather than changes in mitochondrial properties. The increases in succinic dehydrogenase and cytochrome C oxidase activity, however, during cold acclimation of goldfish and carp reflect changes in the lipid composition of the mitochondrial membrane (Hazel, 1972; Wodtke, 1981*b*), suggesting changes in the intrinsic properties of mitochondria with cold acclimation. Long-term (>12 weeks) cold acclimation of the sea bass, *Dicentrarchus labrax*, decreases the maximal rates of glutamate oxidation at 20 °C by mitochondria isolated from liver and heart (Trigari *et al.*, 1992). No changes in membrane lipid unsaturation accompany cold acclimation in sea bass. Considering the strength of the response of mitochondrial abundance to cold acclimation, surprisingly few data are available concerning the effect of thermal acclimation on mitochondrial oxidative capacities.

To examine whether the maximal capacities of red muscle mitochondria change with environmental temperature, mitochondria isolated from fall-acclimatised *O. mykiss* were compared to those isolated from summer-acclimatised trout. The maximal capacities for pyruvate oxidation are considerably higher in mitochondria isolated from fall-acclimatised trout (Fig. 6) (J. St Pierre & H. Guderley, unpublished results). This shift leads to complete compensation of rates at the respective acclimatisation temperatures (12 versus 8 °C). Placing the data for fall-acclimatised trout on the temperature-rate curves accentuates the evidence for evolutionary temperature adaptation of the maximal

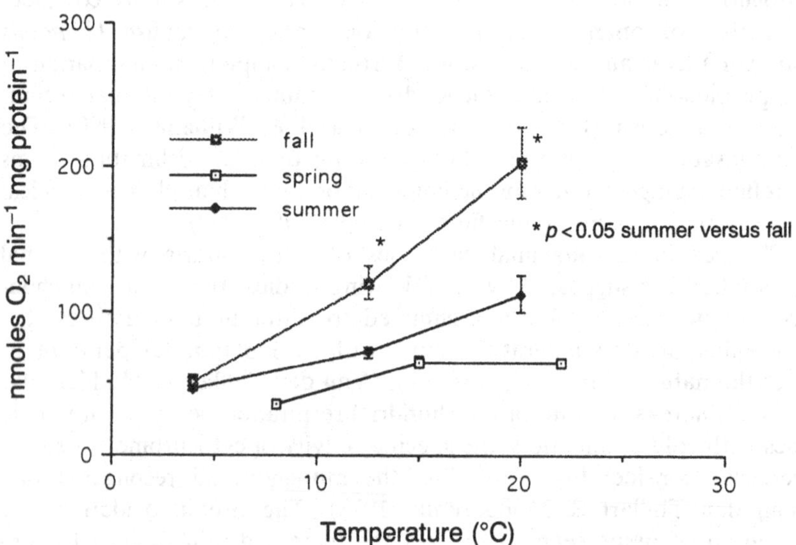

Fig. 6. Seasonal changes in the thermal sensitivity of pyruvate oxidation by mitochondria isolated from the red muscle of the rainbow trout. (After J. St Pierre & H. Guderley, unpublished results.)

capacities of mitochondrial pyruvate oxidation (compare Figs. 1 and 7).

The thermal sensitivity of maximal oxidation rates also changes with acclimatization to fall conditions. Maximal oxidation rates of mitochondria isolated from fall-acclimatised trout had higher Q_{10} values between 4 and 12 °C than those of summer-acclimatised trout (3.07 versus 1.72). At higher temperatures, Q_{10} values were similar (1.75 and 1.82 for summer and fall trout). Corresponding shifts in the E_a values for the temperature-rate curve were apparent. Mitochondria isolated from trout acclimatised to cooler conditions (spring and fall trout) show higher E_a values (57.2 and 58.3 kJ mol^{-1} than mitochondria isolated from summer trout (36.9 kJ mol^{-1}). This response contrasts with the positive correlation between E_a values and physiological temperatures found for isolated enzymes and for mitochondria from hydrothermal vent organisms (Hochachka & Somero, 1984; Dahlhoff *et al.*, 1991).

The catalytic capacities as well as the thermal sensitivity of fish muscle mitochondria can change during thermal acclimatisation. The increases in muscle aerobic capacity during cold acclimation may therefore reflect modifications in both the properties and the numbers of mitochondria present. The contrasting responses of rainbow trout and sea bass to thermal acclimation suggest that the specific responses

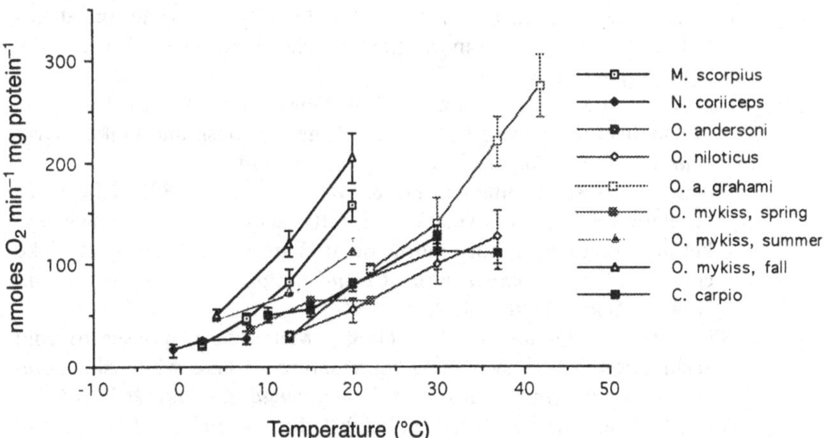

Fig. 7. Thermal sensitivity of maximal rates of mitochondrial pyruvate oxidation. Comparison of rate temperature curves obtained for a wide range of fish species, including summer- and fall-acclimatised rainbow trout. (After Moyes *et al.*, 1988, 1992; Blier & Guderley, 1993*a*; Johnston *et al.*, 1994 and J. St Pierre & H. Guderley, unpublished results.)

depend on the species' thermal ecology or phylogeny. During evolutionary adaptation to temperature, compensation of maximal mitochondrial capacities is apparent, particularly for species living at intermediate temperatures. A lack of adjustment, therefore, of mitochondrial capacities with cold acclimation would not reflect the impossibility of such changes but may indicate that they are too costly. In comparison with the mitochondria from fall-acclimatised trout, the mitochondria from the Antarctic fish have relatively lower capacities, suggesting that specialisation for life at continuously low temperatures has limited the expansion of the oxidative capacities of individual mitochondria.

Acknowledgements

The writing of this chapter was greatly facilitated by the excellent conditions for working, reflection and integration provided by the laboratory of Professor Osmar Nusetti at the Escuela de Ciencias, Universidad de Oriente, Cumana, Venezuela. The authors' research is supported by funds from NSERC (Canada) and has benefited considerably from our collaboration with Professor Ian Johnston.

References

Archer, S.D. & Johnston, I.A. (1991). Density of cristae and distribution of mitochondria in the slow muscles of Antarctic fish. *Physiological Zoology* **64**, 242–58.

Ballantyne, J.S., Chamberlin, M.E. & Singer, T.D. (1992). Oxidative metabolism in thermogenic tissues of the swordfish and mako shark. *Journal of Experimental Zoology* **261**, 110–14.

Ballantyne, J.S., Flannigan, D. & White, T.B. (1989). Effects of temperature on the oxidation of fatty acids, acyl carnitines and ketone bodies by mitochondria isolated from the liver of the lake charr, *Salvelinus namaycush*. *Canadian Journal of Fisheries and Aquatic Sciences* **46**, 950–4.

Blier, P.U. & Guderley, H.E. (1988). Metabolic responses to cold acclimation in the swimming musculature of lake whitefish, *Coregonus clupeaformis*. *Journal of Experimental Zoology* **264**, 244–52.

Blier, P.U. & Guderley, H.E. (1993*a*). Effects of pH and temperature on the kinetics of pyruvate oxidation by muscle mitochondria from rainbow trout (*Oncorhynchus mykiss*). *Physiological Zoology* **66**, 474–89.

Blier, P.U. & Guderley, H.E. (1993*b*). Mitochondrial activity in rainbow trout red muscle: the effect of temperature on the ADP-dependence of ATP synthesis. *Journal of Experimental Biology* **176**, 145–57.

Bouwer, S. & Van den Thillart, G. (1984). Oxygen affinity of mitochondrial state III respiration of goldfish red muscle: the influence of temperature and O_2 diffusion on K_m values. *Molecular Physiology* **6**, 291–306.

Brand, M.D., Chien, L.-F., Ainscow, E.K., Rolfe, D.F.S. & Porter, R. K. (1994). The causes and functions of mitochondrial proton leak. *Biochimica Biophysica Acta* **1187**, 132–9.

Brand, M. D. & Murphy, M. P. (1987). Control of electron flux through the respiratory chain in mitochondria and cells. *Biological Reviews* **62**, 141–93.

Brandolin, G., Doussiere, J., Gulik, A., Gulik-Krzywicki, T., Lauquin, G.J.M. & Vignais, P.V. (1980). Kinetic, binding and ultrastructure properties of the beef heart adenine nucleotide carrier protein after incorporation into phospholipid vesicles. *Biochimica Biophysica Acta* **592**, 592–614.

Brett, J.R. (1967). Swimming performance of sockeye salmon in relation to fatigue time and temperature. *Journal of Fisheries Research Board, Canada* **24**, 1731–41.

Chamberlin, M.E. & Ballantyne, J.S. (1992). Glutamine metabolism in elasmobranch and agnathan muscle. *Journal of Experimental Zoology* **264**, 267–72.

Crockett, E.L. & Sidell, B.D. (1990). Some pathways of energy metabolism are cold adapted in Antarctic fishes. *Physiological Zoology* **63**, 472–88.

Crockett, E.L. & Sidell, B.D. (1993). Substrate selectivities differ for hepatic mitochondrial and peroxisomal β-oxidation in an Antarctic fish, *Notothenia gibberifrons*. *Biochemical Journal* **289**, 427–33.

Dahlhoff, E., O'Brien, J., Somero, G.N. & Vetter, R.D. (1991). Temperature effects on mitochondria from hydrothermal vent invertebrates: evidence for adaptation to elevated and variable habitat temperature. *Physiological Zoology* **64**, 1490–508.

Dahlhoff, E. & Somero, G.N. (1993). Effects of temperature on mitochondria from abalone (Genus *Haliotis*): adaptive plasticity and its limits. *Journal of Experimental Biology* **185**, 151–68.

Doumen, C. & Ellington, W.R. (1990). Mitochondrial arginine kinase from the heart of the horseshoe crab, *Limulus polyphemus*. II. Catalytic properties and studies of potential functional coupling with oxidative phosphorylation. *Journal of Comparative Physiology* B **160**, 459–68.

Dudley, G.A., Tullson, P.C. & Terjung, R.L. (1987). Influence of mitochondrial content on the sensitivity of respiratory control. *Journal of Biological Chemistry* **262**, 9109–14.

Duée, E.D. & Vignais, P.V. (1969). Kinetics and specificity of the adenine nucleotide translocation in rat liver mitochondria. *Journal of Biological Chemistry* **244**, 3920–31.

Dzeja, P., Kalvenas, A., Toleikis, A. & Praskevicius, A. (1985). The effect of adenylate kinase activity on the rate and efficiency of energy transport from mitochondria to hexokinase. *Biochemistry International* **10**, 259–65.

Egginton, S. & Sidell, B.D. (1989). Thermal acclimation induces adaptive changes in subcellular structure of fish skeletal muscle. *American Journal of Physiology* **256**, R1–R9.

Fry, F.E.J. & Hart, J.S. (1948). Cruising speed of goldfish in relation to water temperature. *Journal of Fisheries Research Board, Canada* **7**, 169–75.

Geiser, F. & McMurchie, E.J. (1984). Differences in the thermotropic behaviour of mitochondrial membrane respiratory enzymes from homeothermic and heterothermic endotherms. *Journal of Comparative Physiology* **155**, 125–33.

Gellerich, F. & Saks, V.A. (1982). Control of heart mitochondrial oxygen consumption by creatine kinase: the importance of enzyme localization. *Biochemical and Biophysical Research Communications* **105**, 1473–81.

Gellerich, F.N., Khuchua, Z.A. & Kuznetsov, A.V. (1993). Influence of the mitochondrial outer membrane and the binding of creatine kinase to the mitochondrial inner membrane on the compartmentation of adenine nucleotides in the intermembrane space of rat heart mitochondria. *Biochimica Biophysica Acta* **1140**, 327–34.

Graham, M. (1987). The solubility of oxygen in physiological salines. *Fish Physiology and Biochemistry* **4**, 1–4.

Groen, A.K., Wanders, R.J.A., Westerhoff, H.S., Van Der Meer, R. & Tager, J.M. (1982). Quantification of the contribution of various steps to the control of mitochondrial respiration. *Journal of Biological Chemistry* **257**, 2754–7.

Guderley, H. (1990). Functional significance of metabolic responses to thermal acclimation in fish muscle. *American Journal of Physiology* **259** (*Regulatory Integrative Comparative Physiology* **28**), R245–R252.

Guderley, H. & Foley, L. (1990). Anatomic and metabolic responses to thermal acclimation in the ninespine stickleback, *Pungitius pungitius*. *Fish Physiology and Biochemistry* **8**, 465–74.

Guderley, H., Lavoie, B.A. & Dubois, N. (1994). The interaction among age, thermal acclimation and growth rate in determining muscle metabolic capacities and tissue masses in the threespine stickleback, *Gasterosteus aculeatus*. *Fish Physiology and Biochemistry* **13**, 419–31.

Guderley, H., Mundarain-Rojas, F. & Nusetti, O. (1995). Metabolic specialization of mitochondria from scallop phasic muscle. *Marine Biology* **122**, 409–16.

Halestrap, A.P. (1975). The mitochondrial pyruvate carrier. Kinetics and specificity for substrates and inhibitors. *Biochemical Journal* **148**, 85–96.

Harris, D.A. & Das, A.M. (1991). Control of mitochondrial ATP synthesis in the heart. *Biochemical Journal* **280**, 561–73.

Hazel, J.R. (1972). The effect of temperature acclimation upon succinic dehydrogenase activity from the epaxial muscle of the common goldfish (*Carassius auratus* L.). I. Properties of the enzyme and the effect of lipid extraction. *Comparative Biochemistry and Physiology* **43B**, 837–61.

Hazel, J.R., McKinley, S.J. & Williams, E.E. (1992). Thermal adaptation in biological membranes: interacting effects of temperature and pH. *Journal of Comparative Physiology B* **162**, 593–601.

Hazel, J.R. & Prosser, C.L. (1974). Molecular mechanisms of temperature compensation in poikilotherms. *Physiological Reviews* **54**, 620–77.

Hazel, J.R. & Williams, E.E. (1990). The role of alterations in membrane lipid composition in enabling physiological adaptation of organisms to their physical environment. *Progress in Lipid Research* **29**, 167–227.

Hochachka, P.W. & Somero, G.N. (1984). *Biochemical Adaptation*. New Jersey: Princeton.

Jacobus, W.E. (1985). Respiratory control and the integration of heart high-energy phosphate metabolism by mitochondrial creatine kinase. *Annual Review of Physiology* **47**, 707–25.

Johnston, I.A. (1987). Respiratory characteristics of muscle fibres in a fish (*Chaenocephalus aceratus*) that lacks haeme pigments. *Journal of Experimental Biology* **133**, 415–28.

Johnston, I.A. (1993). Phenotypic plasticity of fish muscle to temperature change. In *Fish Ecophysiology* (ed. J.C. Rankin & F.B. Jensen), pp. 322–340. London: Chapman and Hall.

Johnston, I.A., Camm, J.-P. & White, M. (1988). Specialisations of swimming muscles in the pelagic Antarctic fish *Pleuragramma antarcticum*. *Marine Biology* **100**, 3–12.

Johnston, I.A., Guderley, H.E., Franklin, C.E., Crockford, T. & Kamunde, C. (1994). Are mitochondria subject to evolutionary temperature adaptation? *Journal of Experimental Biology* **195**, 293–306.

Johnston, I.A. & Lucking, M. (1978). Temperature induced variation in the distribution of different types of muscle fibre in the goldfish (*Carassius auratus*). *Journal of Comparative Physiology* **124**, 111–16.

Johnston, I.A. & Maitland, B. (1980). Temperature acclimation in crucian carp (*Carassius carassius* L.): morphometric analysis of muscle fibre ultrastructure. *Journal of Fish Biology* **17**, 113–25.

Kemp, Jr., A., Groot, G.S.P. & Reitsma, H.J. (1969). Oxidative phosphorylation as a function of temperature. *Biochemica et Biophysica Acta* **180**, 24–34.

Kiessling, K.H. & Kiessling, A. (1993). Selective utilization of fatty acids in rainbow trout (*Oncorhynchus mykiss* Walbaum). *Canadian Journal of Zoology* **71**, 248–51.

Kleckner, N. & Sidell, B.D. (1985). Comparisons of maximal activities of enzymes from tissues of thermally-acclimated and naturally-acclimatized chain pickerel (*Esox niger*). *Physiological Zoology* **58**, 18–28.

Klingenberg, M., Grebe, K. & Appel, M. (1982). Temperature dependence of ADP/ATP translocation in mitochondria. *European Journal of Biochemistry* **126**, 263–9.

Londraville, R.L. & Sidell, B.D. (1990). Ultrastructure of aerobic muscle in Antarctic fishes may contribute to the maintenance of diffusive fluxes. *Journal of Experimental Biology* **150**, 205–20.

Low, P.S., Bada, J.L. & Somero, G.N. (1973). Temperature adaptation of enzymes: roles of the free energy, the enthalpy and the entropy of activation. *Proceedings of the National Academy of Sciences of the USA* **70**, 430–2.

Meyer, M.A., Sweeney, H.L. & Kushmerick, M.J. (1984). A simple analysis of the 'phosphocreatine' shuttle. *American Journal of Physiology* **246**, C265–C377.

Moreno-Sanchez, R. & Torres-Marquez, M.E. (1991). Control of oxidative phosphorylation in mitochondria, cells and tissues. *International Journal of Biochemistry* **23**, 1162–74.

Moyes, C.D., Buck, L.T. & Hochachka, P.W. (1988). Temperature effects on pH of mitochondria isolated from carp red muscle. *American Journal of Physiology* **254**, (*Regulatory, Integrative and Comparative Physiology* **23**), R611–R615.

Moyes, C.D., Buck, L.T., Hochachka, P.W. & Suarez, R.K. (1989). Oxidative properties of carp red and white muscle. *Journal of Experimental Biology* **143**, 321–31.

Moyes, C.D., Mathieu-Costello, O.A., Brill, R.W. & Hochachka, P.W. (1992*a*). Mitochondrial metabolism of cardiac and skeletal muscles from a fast (*Katsuwonus pelamis*) and a slow (*Cyprinus carpio*) fish. *Canadian Journal of Zoology* **70**, 1246–53.

Moyes, C.D., Schultes, P.M. & Hochachka, P.W. (1992*b*). Recovery metabolism of trout white muscle: role of mitochondria. *American Journal of Physiology* **262** (*Regulatory, Integrative and Comparative Physiology* **31**), R295–R304.

O'Brien, J., Dahlhoff, E. & Somero, G.N. (1991). Thermal resistance of mitochondrial respiration: hydrophobic interactions of membrane proteins may limit thermal resistance. *Physiological Zoology* **64**, 1509–26.

Pfaff, E., Heldt, H.W. & Klingenberg, M. (1969). Adenine nucleotide translocation in mitochondria. Kinetics of the adenine nucleotide exchange. *European Journal of Biochemistry* **10**, 484–93.

Raison, J.K., Lyons, J.M. & Thomson, W.W. (1971). The influence of membranes on the temperature-induced changes in the kinetics of some respiratory enzymes of mitochondria. *Archives of Biochemistry and Biophysics* **142**, 83–90.

Sänger, A.M. (1993). Limits to the acclimation of fish muscle. *Reviews in Fish Biology and Fisheries* **3**, 1–15.

Schneider, A., Weisner, R.J. & Grieshaber, M.K. (1989). On the role of arginine-kinase in insect flight muscle. *Insect Biochemistry* **19**, 471–80.

Sidell, B.D. (1980). Response of goldfish (*Carassius auratus* L.) muscle to acclimation temperature: alterations in biochemistry and proportions of different fibre types. *Physiological Zoology* **53**, 98–107.

Sidell, B.D. & Johnston, I.A. (1985). Thermal sensitivity of contractile function in chain pickerel, *Esox niger. Canadian Journal of Zoology* **63**, 811–16.

Sisson, III, J.E. & Sidell, B.D. (1987). Effect of thermal acclimation on muscle fibre recruitment of swimming striped bass (*Morone saxatilis*). *Physiological Zoology* **60**, 310–20.

Trigari, G., Pirini, M., Ventralla, V., Pagliarani, A., Trombetti, F. & Borgatti, A.R. (1992). Lipid composition and mitochondrial respiration in warm- and cold-adapted sea bass. *Lipids* **27**, 371–7.

Tyler, S. & Sidell, B.D. (1984). Changes in mitochondrial distribution and diffusion distances on muscle of goldfish upon acclimation to warm and cold temperatures. *Journal of Experimental Zoology* **232**, 1–9.

Van den Thillart, G. & De Bruin, G. (1981). Influence of environmental temperature on mitochondrial membranes. *Biochemica et Biophysica Acta* **640**, 439–47.

Van Den Thillart, G. & Modderkolk, J. (1978). The effect of acclimation temperature on the activation energy of state III respiration and on the unsaturation of membrane lipids of goldfish mitochondria. *Biochimica et Biophysica Acta* **510**, 38–51.

Vézina, D. & Guderley, H. (1991). Anatomic and enzymatic responses of the threespined stickleback, *Gasterosteus aculeatus* to thermal acclimation and acclimatization. *Journal of Experimental Zoology* **258**, 277–87.

Wodtke, E. (1981*a*). Temperature adaptation of biological membranes. The effects of acclimation temperature on the unsaturation of the main neutral and charged phospholipids in mitochondrial membranes of the carp (*Cyprinus carpio*). *Biochimica Biophysica Acta* **640**, 698–709.

Wodtke, E. (1981*b*). Temperature adaptation of biological membranes. Compensation of the molar activity of cytochrome c oxidase in the mitochondrial energy-transducing membrane during thermal acclimation of the carp (*Cyprinus carpio* L.). *Biochimica et Biophysica Acta* **640**, 710–20.

Yacoe, M.E. (1986). Effects of temperature, pH and CO_2 tension on the metabolism of isolated hepatic mitochondria of the desert iguana, *Dipsosaurus dorsalis*. *Physiological Zoology* **59**, 263–72.

I.A. JOHNSTON, V.L.A. VIEIRA and J. HILL

Temperature and ontogeny in ectotherms: muscle phenotype in fish

Introduction

The thermal environment experienced by an organism often changes dramatically during ontogeny. These changes may reflect a seasonal warming or cooling and/or a change in habitat as development progresses. For example, amphibians and numerous insects show a transition between an aquatic larval and a terrestrial adult stage. Aquatic and terrestrial environments differ markedly with respect to their thermal conductances, heat capacities and richness of microclimates. The transition between environments is frequently accompanied by changes in the function of tissues and by the development of new respiratory, circulatory, sensory and locomotor systems etc. (for an example, see Burggren, 1992). Thermal patterns during ontogeny are particularly complex in some parasitic animals that have successive ectothermic and endothermic hosts as well as free living stages. Temperature can have quite different effects on physiology, behaviour and survival at different stages of the life cycle. For example, in a wide range of ectotherms, temperature tolerance (Blaxter, 1988), and the temperature optimum for growth (Hovenkamp & Witte, 1991) vary during ontogeny in parallel with the changing thermal environment of the organism. Marden (1995) found that maturation in the dragonfly *Libellula pulchella* was accompanied by striking changes in the thermal physiology of the flight muscles. In the newly emergent adults, vertical force production during fixed flight attempts showed a broad plateau between 28 and 45 °C. In contrast, in fully mature adults, peak performance was only approached within a few degrees of the thermal optimum, which occurred at 38–48 °C. In spite of the importance of understanding how all stages of the life cycle respond to different thermal environments, the great majority of studies to date have focused exclusively on adults.

Developmental plasticity in ectotherms

The environment of an organism interacts with the developmental programme with major consequences for the phenotype. Temperature is one of a range of environmental factors, both abiotic (pH, humidity, salinity, daylength, etc.) and biotic (resource availability, population density, hormones, etc.) that can influence development, usually in a highly complex and interactive fashion. The range of phenotypes produced by a genotype in different environments is called the norm of reaction (Scheiner, 1993).

Environmental cues, including temperature, are commonly used to activate switches in the developmental programme leading to either the initiation of a developmental arrest or the production of alternative phenotypes differing in morphological features, behaviour and/or life-history characters. This type of plasticity has variously been called polyphenism, conditional choice or developmental conversion (Scheiner, 1993).

Facultative diapause is common in insects and provides a good example of a developmental arrest (Sternberg, 1995). It is under the control of specific cues of temperature and day length and represents a temporary interruption of the developmental programme to enable survival under unfavourable conditions. Pupal and adult diapause are mediated via a suite of sensory systems. Egg diapause is maternally controlled in species such as the bog-dwelling dragonfly, *Somatocholora alpestris*, providing an example of a cross-generational effect of temperature on phenotype (Sternberg, 1995).

A particularly striking example of how temperature can activate a developmental switch to produce alternate phenotypes is sex determination in some ectotherms. In many invertebrates, fish, amphibians and reptiles, sex is determined by environmental cues acting during a critical period in development rather than through distinct sex chromosomes. Species with temperature-dependent sex determination (TSD) are generally found in thermally patchy environments, allowing for production of both sexes. For example, in crocodilians, the sex of the embryo is determined by the accumulative effects of the nest temperature in the period from shortly after egg laying to the first half of embryonic development (Janzen & Paukstis, 1991). The molecular mechanisms advanced to explain reptilian TSD involve effects of temperature on the activity and/or expression of aromatase, which converts oestrogen to testosterone and on the transcription of genes coding for luteinising hormone and its receptor (Janzen & Paukstis, 1991).

Phenotypic variation in ectotherms also arises from the effects of temperature on the rates and degrees of expression of the developmental programme. In contrast to developmental switches which produce discontinuous phenotypes, environmental effects on the rate of development produce a continuum of phenotypes. This kind of phenotypic plasticity has been called continuous liability, dependent development or phenotypic modulation (Scheiner, 1993).

The effects of temperature on development have been shown to vary between traits and at different embryological stages (Smith-Gill, 1983). Temperatures close to the upper and lower thermal tolerance limits of a species produce a high probability for abnormal phenotypes, which could be viewed as a failure to buffer developmental processes under extreme conditions (Smith-Gill, 1983). Severe temperature shocks at critical phases in development can even result in teratogenesis and the formation of 'monsters' such as twin-headed embryos (Stockard, 1921). Morphogenetic processes which show little if any developmental plasticity are considered to be canalised, i.e. fixed at some optimal level by strong selection (Scharloo, 1991).

At a given point during embryogenesis, the organs and tissues will have reached different points in their developmental programmes and may therefore exhibit somewhat different temperature sensitivities. In some cases these differences manifest themselves as a change in the relative timing of growth and maturation of different parts of the body.

There is evidence that the phenotypic variation induced by variable temperatures in ectotherms, although not adaptive in itself, can be subject to strong selection. Perhaps the most compelling evidence comes from studies of amphibian life cycles. For example, field and laboratory experiments have shown that temperature can account for most of the observed variation in growth and differentiation in populations of the anuran *Rana clamitans* along an altitudinal gradient (Breven *et al.*, 1979). Growth in montane frog populations was shown to have a reduced temperature sensitivity relative to lowland populations, serving to compensate for the retarding effects of low temperature on growth. Lowland frogs that were transplanted to high altitude had their growth slowed by low temperature to such a extent that the larval stage was extended by a whole season relative to the larvae from montane populations. Such transplant experiments indicate that the phenotypic modulation of this particular trait is under genetic control (Breven *et al.*, 1979).

Scheiner (1993) has suggested that a single evolutionary model can be used to explain the plasticity of continuously and discontinuously variable phenotypes and that only the shape of the reaction norm need differ. For example, discontinuous plasticity might be explained by a continuous underlying response to the environment at the cellular or genetic level that has a distinct threshold for phenotypic expression. In this case the reaction norm would be S-shaped with a very steep slope at the inflection point as observed for arrested development. The phenotypic plasticity to temperature observed during ontogeny may have pathological, non-adaptive or adaptive explanations (pp. 127–152). In all cases there will presumably be some cost to the organism, for example, in terms of the maintenance of regulatory genes and enzymes (Moran, 1992).

Rationale for our studies on fish muscle

Most of the previous studies on developmental plasticity and temperature have been concerned with complex traits such as the timing of metamorphosis, growth rate and body size. The traits in question were treated essentially as 'back boxes' without trying to understand either their component features or the underlying mechanisms. Our studies have focused on the maturation of the myotomal muscles in fish, a tissue which constitutes 60–70% of the body. This chapter describes our experiments on the effects of temperature on muscle development in the Atlantic herring (*Clupea harengus* L.).

The Atlantic herring is a pelagic fish with a widespread distribution in the North Atlantic, North Sea and Baltic. Different stocks of herring can be recognised showing variation in body size at maturity and in numerous morphological characters, such as the number of vertebrae (Greer-Walker *et al.*, 1972). There is a considerable gene flow within coastal or oceanic stocks (King *et al.*, 1987), although evidence for the genetic structuring of stocks has been obtained for discrete populations in Norwegian fjords (Jørstad *et al.*, 1991). Herring stocks spawn in almost every month of the year and thus experience different temperature regimes during development. The eggs are deposited on the sea bed, and on hatching the yolk-sac larvae must swim continuously to avoid sinking (Batty, 1994). The larval stages have a relatively long life in the plankton, undergoing metamorphosis to the juvenile stage after 2–5 or more months, depending on the time of year and sea temperature (Doyle, 1977).

Ecological context

In the Firth of Clyde, Atlantic herring spawn from March to early April at depths of 15–25 m. Long-term temperature data are available from the North Channel between Scotland and Ireland. For March the near surface temperature has varied between 4.8 and 9.8 °C over the last 40 years, with an average of 7.2 °C (Jones & Jeffs, 1991). During the embryonic and early larval stages, the temperature remains relatively constant, but increases rapidly in mid-April to a seasonal high of around 15–16 °C in September. To facilitate comparison between groups, we have incubated eggs at constant temperatures of 5, 8, 12 and 15 °C. After yolk-sac absorption, however, the temperature was allowed to rise gradually to reflect the natural situation. To investigate whether the temperature experienced by the embryo can have a lasting impact on adult phenotype, fish reared at a series of constant temperatures to first feeding were subsequently grown at a common temperature for the remainder of the larval stage.

Some comparative studies have also been carried out on a summer spawning stock of Atlantic herring from the North Sea (Buchan stock). Buchan herring spawn at a depth of around 60 m in late August and the eggs develop at 11.5 °C. On hatching the larvae experience around 13 °C in the surface waters; however, temperature falls with the onset of winter, and metamorphosis is delayed until the following spring.

Myogenesis in fish

Recent advances in the understanding of development in zebrafish (*Brachydanio rerio*) provide important background information for comparative studies. By labelling cells of the gastrulating zebrafish embryo with a fluorescent dye, a fate map has been produced, and the cell movements resulting in organogenesis studied (Kimmel *et al.*, 1990). The antecedents of the lateral mesoderm from which the myotomal muscle is formed are present at the lateral and marginal region of the embryo at the onset of gastrulation, when anterior/posterior (A/P) polarity is established. These cells involute at the margin during gastrulation and move towards the A/P axis. At the same time the axial mesoderm (which forms the notochord and prechordal plate) involutes rapidly at the embryonic shield region (called the dorsal lip in *Xenopus*) (Moav, 1994).

By the end of gastrulation, a layer of ectodermal tissue overlies a rod of axial mesoderm, which is bounded by the lateral mesoderm. The tail bud also forms during gastrulation, extending posteriorly as development progresses. After gastrulation the neural tube begins to differentiate and the somites begin to form in a rostral to caudal direction. As in other vertebrates the newly formed somites consist of radially arranged epithelial balls of mesoderm (epithelial somites). The epithelial somites undergo a series of morphological changes that eventually result in their dissolution and differentiation into bone, cartilage, dermis and muscle (Keynes & Stern, 1988). During somite maturation, cells in the ventromedial part of the somite form a loosely arranged mesenchyme that gives rise to the sclerotome cells from which the axial skeleton: vertebrae, cartilage and ligaments are constructed. The cells of the lateral part of the somite form the precursors of the fin and trunk muscles. The somite cells closest to the neural tube migrate ventrally down the remaining epithelial portion of the somite to form the dermomyotome. Gap junctions are observed between adjacent somites in herring (Johnston *et al.*, 1995), which suggests that they are initially electrically coupled.

In all vertebrates, myogenesis is under the control of a family of basic helix-loop-helix DNA binding proteins such as myogenin, Myo-D and myf-5 (for a review, see Buckingham, 1994). Knockout mutations in mouse have shown that myogenin plus either Myo-D or myf-5 are required for the formation of an apparently normal muscle phenotype (Metzger *et al.*, 1995). Homologous muscle transcription factors have been identified in rainbow trout (Rescan *et al.*, 1995) and zebrafish (E.S. Weinberg, GenBank Accession No. Z36945), although their functional roles have still to be established.

The first cells in the somite that can be identified as muscle precursors lie as a column of large cuboidal cells adjacent and lateral to the notocord. They are the first cells in the zebrafish that stain positive for *MyoD* transcripts (M. Westerfield, unpublished results). These cells give rise to mononucleate myotubes which initially develop adjacent to the notocord and then migrate laterally through the somite to the lateral midline superficial to the rest of the somite (Raamsdonk *et al.*, 1974; Johnston *et al.*, 1995). These cells are fated to become at least a subset of the superficial red muscle cells (M. Westerfield, unpublished results) and a further subset of the mononucleate myotubes express the homeobox gene *engrailed* (Hatta *et al.*, 1991). These muscle pioneer cells are the first fibres to be innervated by the pioneer motor neurons and are also thought to have a role in patterning the characteristic chevron shape of the myotome (Kimmel *et al.*, 1991). Myotubes in the mid-somite are formed from another population of myoblasts by the fusion of two to six cells to give multi-nucleated myotubes, the majority of which gradually mature into the white muscle fibres. Again this process is essentially similar in zebrafish (Hanneman, 1992) and herring (Johnston *et al.* 1995).

Temperature affects the relative timing of development

In zebrafish, staging by somite number has been shown to predict the development of specific structures more accurately than staging the embryo by elapsed time after fertilisation (Westerfield, 1994). Somites form much faster as incubation temperature is increased. For example, in Clyde herring a new somite segregates from the paraxial mesoderm once every 176 min at 5 °C compared with once every 52 min at 15 °C (Q_{10} = 3.4 from the 5 to 55-somite stage) (Johnston *et al.*, 1995). Expressed as a fraction of the time from fertilisation to hatching (percentage development time, DT), somitogenesis occurs relatively earlier at 5 °C that at 15 °C. For example, after 32% DT there are around 42 somites at 5 °C but only 25 somites at 15 °C (Johnston *et al.*, 1995). Myotube formation occurs at the same somite stage at all temperatures in herring (Fig. 1). The maturation of the somites and initiation of myotube formation is thought to involve inducer molecules from other axial structures such as the notochord and neural tube (for a review, see Buckingham, 1994).

Synthesis of the myofibrillar proteins begins at about the same time that the somites become innervated (J. Hill & I.A. Johnston, unpub-

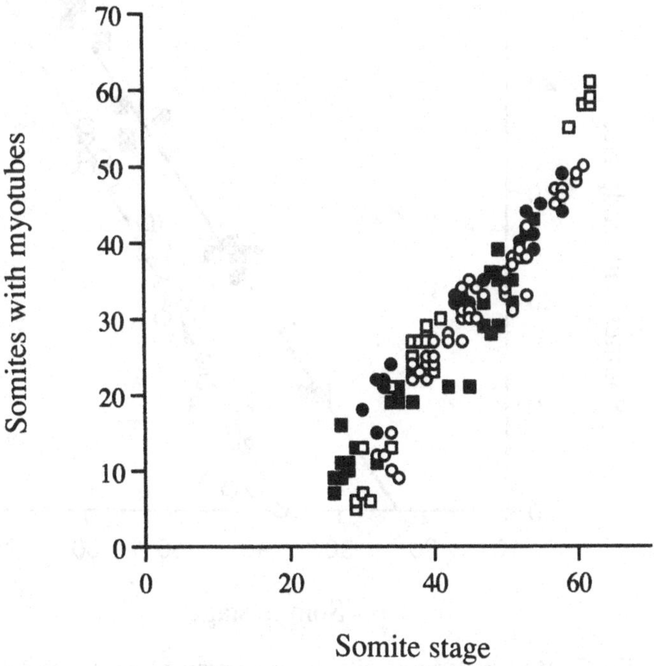

Fig. 1. Most posterior somite containing myotubes in Clyde herring embryos as a function of somite stage. The most posterior somite containing myotubes was identified in living embryos using Nomarski interference optics. Fish were reared at 5 °C (○), 8°C (●), 12 °C (□) and 15 °C (■). (After Johnston and Vieira, unpublished results.)

lished results). In Clyde herring, staining of embryos for acetyl cholinesterase reveals that functional muscle endplates form at progressively later somite stages up to the 61-somite stage as temperature is raised from 5 to 15 °C (Fig. 2).

Antibody staining with the antibodies HNK-1, a carbohydrate moiety involved in the NCAM/L2 cell adhesion reaction, and α-acetylated tubulin, which stains all neural processes, has been used to examine the timing of innervation of the trunk in embryos of Buchan herring (Hill & Johnston, 1996). The appearance of motor neuron cell bodies in the spinal cord was found to be relatively independent of temperature from 5 to 15 °C. Outgrowth of the motor axons, however, was retarded at low temperatures, occurring at the 30-somite stage at 12 °C, but not until after 40-somite stage at 5 °C (Fig. 3). The motor axons exit the spinal cord ventrally and are tightly fasiculated, projecting along the somitic cleft (Hill & Johnston, 1996). No effect of temperature was

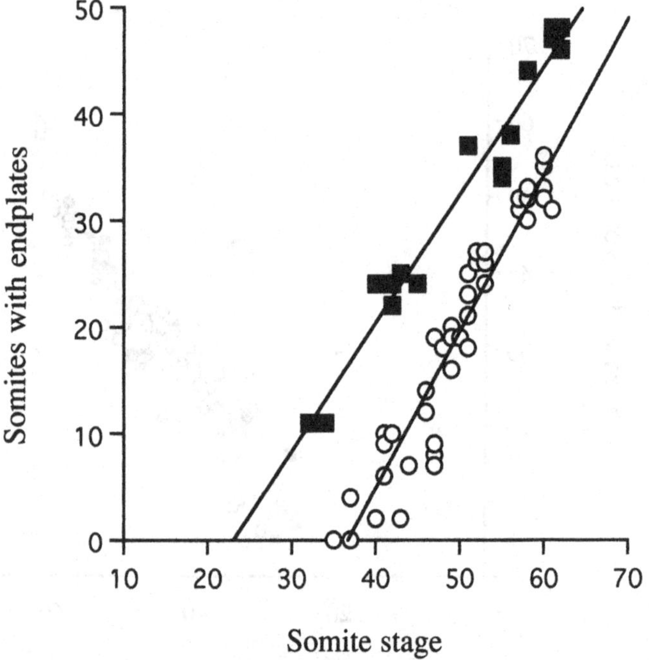

Fig. 2. Most posterior somite showing staining for acetylcholinesterase activity in Clyde herring embryos as a function of somite stage. Only fish reared at 5 °C (○) and 15 °C (■) are illustrated for clarity, data from fish reared at 8 °C and 12 °C being intermediate (After Johnston and Vieira, unpublished results.)

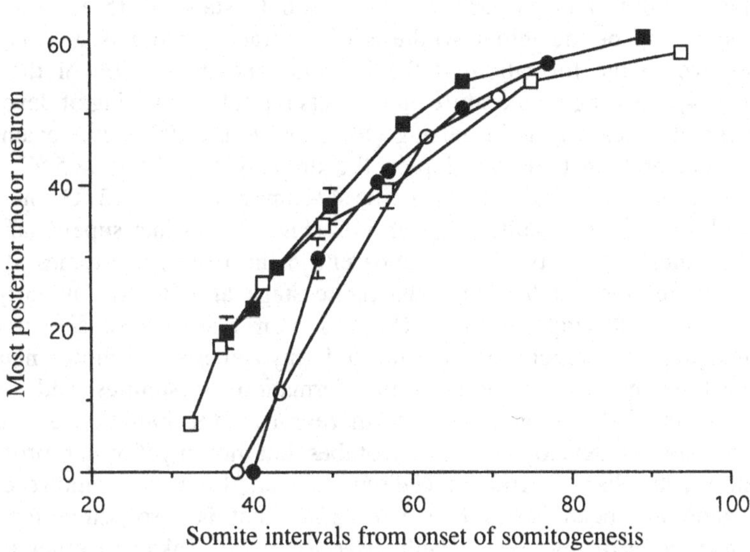

Fig. 3. Motor neuron axonogenesis with respect to somite stage in herring embryos. The embryos at 8 °C and 5 °C develop motor neurons at later somite stages than the embryos raised at higher temperatures. The x-axis is the somite interval (si) from development of the first somite so that the first 62si correspond exactly to somite stage of the embryo. Fish were reared at 5 °C (○), 8 °C (●), 12 °C (□) and 15 °C (■). (After Hill & Johnston, 1996).

seen on the timing of development of other identifiable trunk neurons or on the timing of myotube formation in the somite (as was observed for Clyde herring). Studies in zebrafish have shown that muscle synapse formation is dependent on a complex pattern of signals between motor neurons and muscle fibres (Sepich *et al.*, 1994). At low temperatures the mature motor neuron cell bodies in herring may be awaiting a signal, most probably from the somite, before initiating axogenesis (Hill & Johnston, 1996). By the 60-somite stage there had been some 'catch-up' in motor neuron development, such that at this point there was no difference in the position of the most posterior motor neuron relative to somite stage at 5 and 15 °C.

We have investigated the initial formation of myofilaments in the multi-nucleated myotubes of Clyde herring using electron microscopy (Johnston *et al.*, 1995). In rostral somites (6 to 10 counting from the head), actin and myosin filaments and Z-lines were observed around the periphery of myotubes at the 42-somite stage at 5 °C, the 38-somite stage at 8 °C and as early as the 27-somite stage at 15 °C (Fig. 4). The timing of the initial synthesis of contractile proteins was altered relative to the formation of the fin folds and maturation of the eye (Fig. 4). Rearing temperature also affects the relative timing of development of other organs in herring (Johnston *et al.*, 1995). For example, the pectoral fin buds develop in the order 8 °C > 12 °C > 5 °C, and the pronephric tubules form at the 40-somite stage at 12 °C, but not until after the 61-somite stage at 5 °C (Fig. 5). Distinct superficial and inner muscle fibre types were apparent in the rostral myotomes at the 62-somite stage at 5 °C, the 48-somite stage at 8 °C and as early as the 40-somite stage at 12 °C (Fig. 5). Our results have shown that changing the temperature can uncouple myogenesis and motor neuron development with respect to the formation of somites and other elements of the nervous system. In zebrafish, the mutation genotype *fub-1* allows development of myotubes but not myofibrillar proteins and myofibrillar construction (Felsenfeld *et al.*, 1990, 1991) and recently a gene has been isolated in *Drosophila* that is responsible for the construction of myotubes (Paululat *et al.*, 1995). Taken together these results suggest that myotube formation is not causally linked to myofibrillargenesis; in effect there are at least two separate developmental processes involved in constructing a fully functional muscle unit.

In herring, however, myogenesis and innervation of the myotomes are complete well before hatching, such that yolk-sac larvae probably have functionally equivalent locomotor systems, except perhaps at temperatures near their upper thermal tolerance level of 17 °C. For example, Batty *et al.* (1993) found that the maximum escape swimming speed of herring larvae reared at 17 °C was impaired relative to fish reared at 5–15 °C.

The larvae of many warm-water fish hatch at a much earlier stage of development than herring. For example, tambaqui (*Collosoma macropomum*) (Vieira & Johnston, 1996) and curimatã-pacú (*Prochilodus maggravi*) (Brooks *et al.*, 1995) hatch before all the somites have formed, and the larvae initially lack eye pigment, jaws, a gut and pectoral fins. At hatching the myotomes of the free-swimming larvae consist of a single layer of muscle fibres overlying a core of myoblasts and myotubes still in the earliest stages of myofibrillargenesis. In such cases, any changes in the relative timing of muscle fibre and motor neuron development may well produce different functional phenotypes over a wide range of temperatures.

Similar changes in the relative timing of organogenesis have been observed during the larval stages of other fish. For example, in the Japanese flounder (*Paralichthys olivaceus*), the relative timing of appearance of eye pigment, mouth opening and pectoral fin formation varies with rearing temperature over the range 15 to 21 °C (Fukuhara, 1990). In turbot reared at 12 °C, the appearance of the swim bladder and the development of a loop in the gut varies with respect to the formation of the caudal fin and first feeding compared with fish reared at 16 °C. Spines form on the head in the region of the optic capsule at first feeding in turbot larvae reared at 16 °C, but are absent in 12 °C larvae (Gibson & Johnston, 1995). Although the function of larval cephalic spines is uncertain they have been implicated as anti-predator devices (Blaxter, 1988). Some of the effects of temperature on the timing of development may influence mortality rates, and rigorous experimental testing of this idea is now required. Such studies are important because very few of the larvae of marine fish eventually give rise to adults, typically less than 0.1% (Ferron & Legget, 1994). Even relatively small effects of development temperature on mortality rates might therefore have a significant effect on population size.

Myofibrillar protein expression

During ontogeny there is a sequential expression of different isoforms of the myofibrillar proteins (Scapolo *et al.*, 1988; Veggetti *et al.*, 1993). In herring and many other fish, two muscle fibre types can be distinguished in the myotomes of the larvae (Vieira & Johnston, 1992). Immediately beneath the skin there is a single layer of muscle fibres with significantly smaller cross-sectional areas, lower volume densities of myofibrils, and higher volume densities of mitochondria than the inner core of muscle fibres. These superficial and inner muscle fibre types are the precursors of the red and white muscle fibres in adults, although they contain a number of different myofibrillar protein isoforms. In herring, two-dimensional polyacrylamide gel electrophoresis and peptide mapping have shown that the inner muscle of larvae, adult red muscle and adult white muscle contain distinct isoforms of myosin heavy chains (Crockford & Johnston, 1993). Although the inner fibres of larvae share myosin light chain 1 (MLC1), MLC3, tropomyosin and troponin C isoforms with adult white muscle, they contain unique isoforms of troponin T (TNT), troponin I and MLC2 (Crockford & Johnston, 1993).

Multiple TNT isoforms arise from the alternate splicing of a single gene (Breitbart *et al.*, 1985). Crockford & Johnston (1993) found that the number of TNT isoforms in the presumptive white muscle of herring larvae decreased over the course of the yolk-sac stage, consistent with a transition from embryonic to larval isoforms. One-day-old

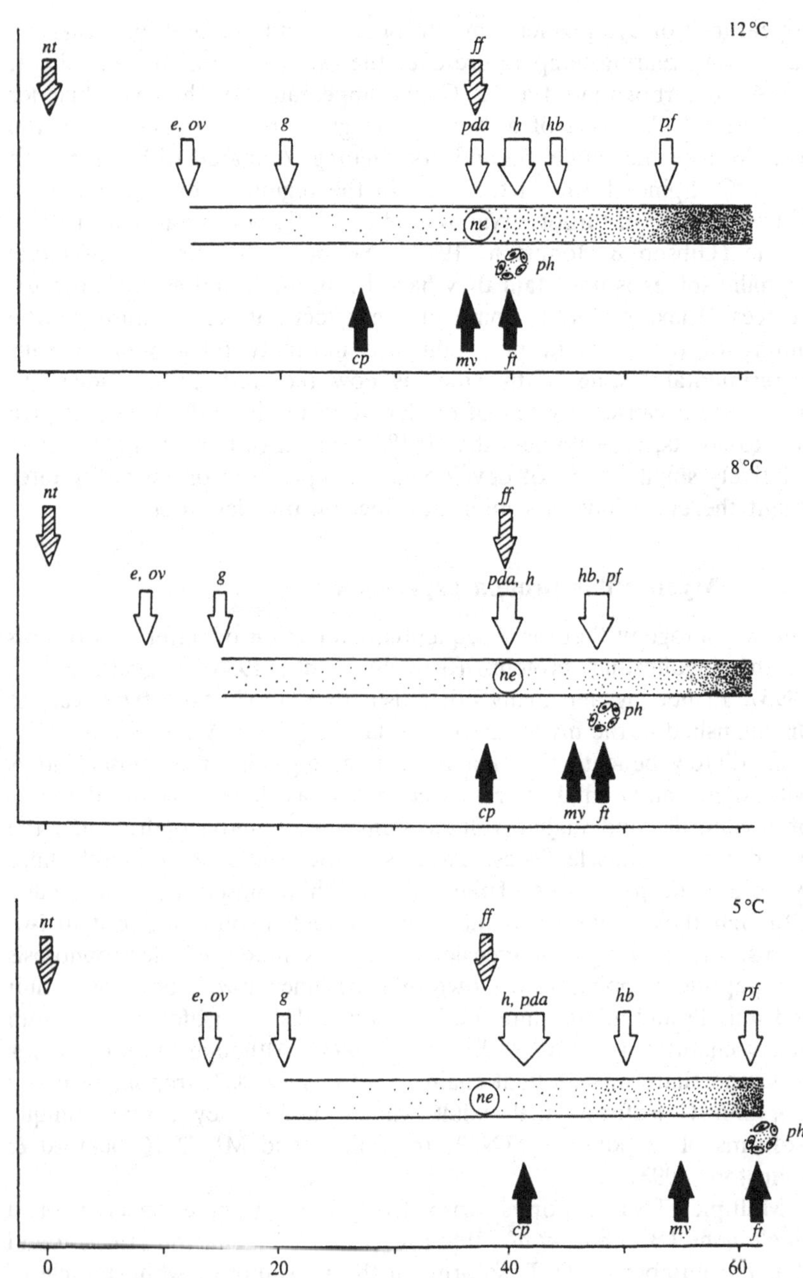
12 °C
nt
ff
e, ov
g
pda
h
hb
pf
ne
ph
cp
my
ft
8 °C
nt
ff
e, ov
g
pda, h
hb, pf
ne
ph
cp
my
ft
5 °C
nt
ff
e, ov
g
h, pda
hb
pf
ne
ph
cp
my
ft
0
20
40
60
Number of somites

larvae reared at 5 °C expressed a higher proportion of embryonic isoforms than larvae reared at 10 °C. Over a period of 7 days, there was a gradual reduction in the number of TNT isoforms, but at no stage did the patterns found in larvae reared at 5, 10 and 15 °C resemble each other (Crockford & Johnston, 1993). Myofibrillar genes are under independent regulation (Pette & Staron, 1990) and therefore unique composition of isoforms are possible at different temperatures during early larval life. Whether these differences are of any functional significance is unknown. Later in development, following absorption of the yolk sac, the myofibrillar isoform compositions of the white myotomal muscle in herring is similar at all temperatures (N. Cole & I. A. Johnston, unpublished results).

Muscle cellularity and ultrastructure

It has long been recognised that meristic characters in herring and other fish vary with both the incubation temperature and salinity at which the eggs develop (Tåning, 1952; Hempel & Blaxter, 1961). Recent research has shown that the temperature of embryonic development also influences the number and cross-sectional areas of presumptive red and white muscle fibres at hatching in a number of species, including herring (Vieira & Johnston, 1992), Atlantic salmon (Stickland *et al.*, 1988) and plaice (*Pleuronectes platessa*) (Brooks & Johnston, 1993). For example, in experiments with Clyde herring in 1993 the number of white muscle fibres increased with rearing temperature whereas the average cross-sectional area of the fibres at 5 °C was almost double that at either 8 or 12 °C (Fig. 6). As a result the total cross-sectional area of myotomal muscle increased in the order

---

Fig. 4. Relative timing of myogenesis in the rostral somites (6–10) of Clyde herring embryos reared at different temperatures. Serial sections of at least six embryos were examined by light microscopy at each somite stage, and three to five embryos by electron microscopy. The horizontal bar shows the maturation of the spinal cord and the stippling provides a qualitative indication of the increase in the number of neuronal cell bodies. Arrows indicate the formation/appearance of the following structures/characters: cp: contractile filaments; e: eye; ff: dorsal and ventral fin folds; ft: distinct muscle fibre types; g: gut; h: heart; hb: first heart beat; my: myofibrils; ne: neurocoel; nt: notochord; ov: otic vesicle; ph: pronephros; pda: presumptive dorsal aorta; pf: pectoral fins. (After Johnston *et al.*, 1995.)

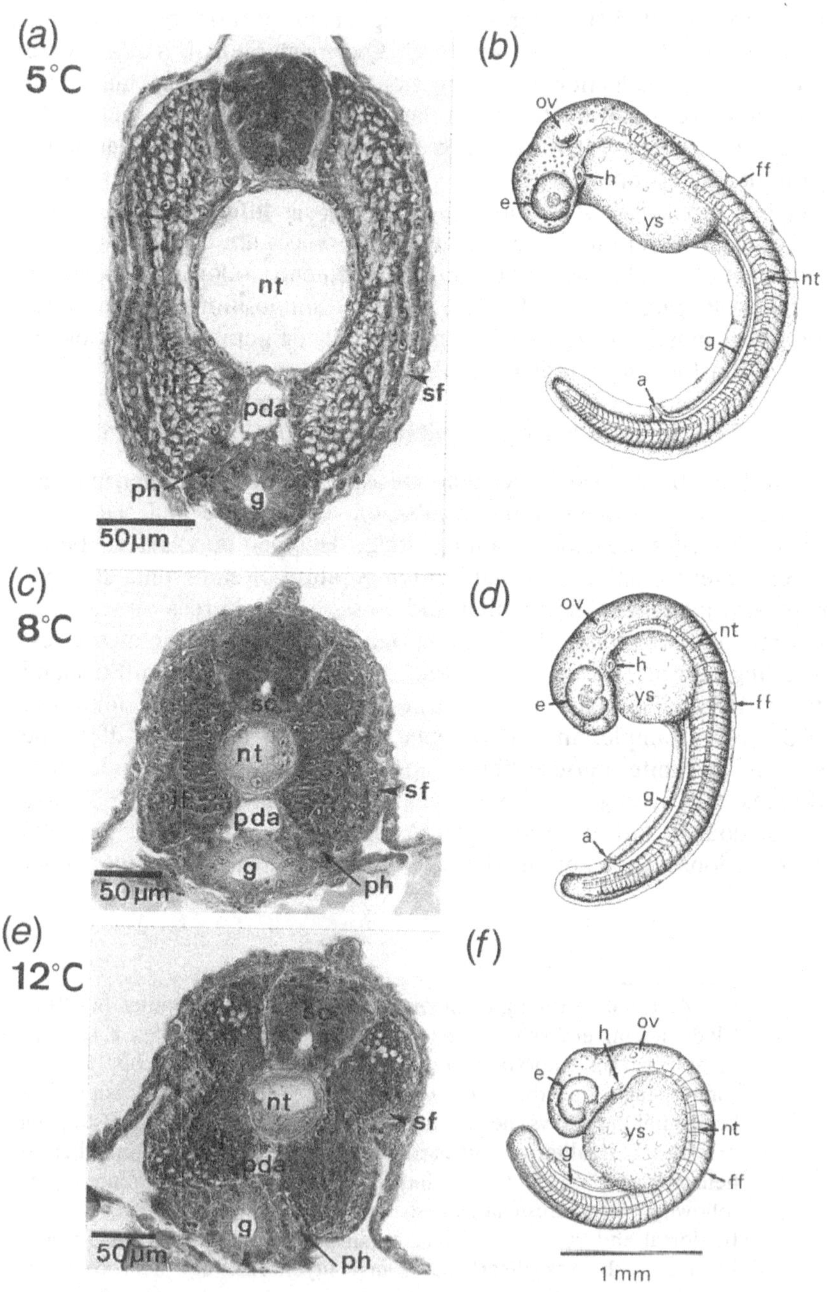
(a)
5°C
sc
nt
sf
pda
ph
g
50μm
(b)
ov
ff
h
e
ys
nt
g
a
(c)
8°C
sc
nt
sf
pda
g
ph
50 μm
(d)
ov
nt
h
ys
e
ff
g
a
(e)
12°C
sc
nt
sf
pda
g
ph
50μm
(f)
ov
h
e
ys
nt
g
ff
1 mm

5 °C > 12 °C > 8 °C fish (Johnston *et al.*, 1995). In herring, there is considerable interannual variation in the effects of temperature on muscle cellularity (Vieira & Johnston, 1992; Johnston, 1993; Johnston *et al.*, 1995) which may reflect genetic variation and/or differences in egg quality, including the concentrations of maternally-derived mRNAs and growth factors.

At hatching herring larvae lack gills and oxygen diffuses to the muscle fibres via the skin and intervening tissues. Most of the repertoire of swimming behaviour of larvae is supported by aerobic metabolism. Both the red and white muscle fibres of larvae contain higher volume densities of mitochondria than the red and white muscle fibres in adult stages (Vieira & Johnston, 1992). The volume density of mitochondria has been shown to increase with rearing temperature (Table 1). In contrast, the volume density of myofibrils in the red fibres was significantly lower at 5 than 15 °C, but was independent of temperature in the white muscle fibres (Table 1). Higher volume densities of muscle mitochondria with increased rearing temperature have also been reported for plaice larvae (Brooks & Johnston, 1993). It should be noted that the differences in mitochondrial volume density observed between different rearing temperatures are opposite to those observed with temperature acclimation in the adult stages of several fish species. Cold acclimation is associated with an increase in mitochondrial density in crucian carp (*Carassius carassius*) (Johnston & Maitland, 1980) and striped bass (*Morone saxitilis*) (Eggington & Sidell, 1989). In this case higher values of mitochondrial volume density at low temperatures are thought to compensate for reduced rates of oxidative phosphorylation and diffusion rates.

Fig. 5. Somite stage at which distinct muscle fibre types were recognisable in the rostral myotomes (6–10) of Clyde herring embryos. Note the single layer of small diameter superficial muscle fibres surrounding an inner mass of larger diameter inner fibres (presumptive white muscle). On the right-hand side of each section is a *camera lucida* drawing of the corresponding somite stage of the embryo: (*a*, *b*) 62-somite embryo at 5 °C; (*c*, *d*) 48-somite embryo at 8 °C; (*e*, *f*) 40-somite embryo at 12 °C. Abbreviations: a: anus; e: eye; ff: fin folds; g: gut; h: heart; if: inner muscle fibres; pda: presumptive dorsal aorta; ph: pronephric tubules; sf:superficial muscle fibres; ov: otic vesicle; nt: notochord; ys: yolk-sac. Scale bars a, c, e: 50 μm; b, d, f: 1 mm. (After Johnston *et al.*, 1995.)

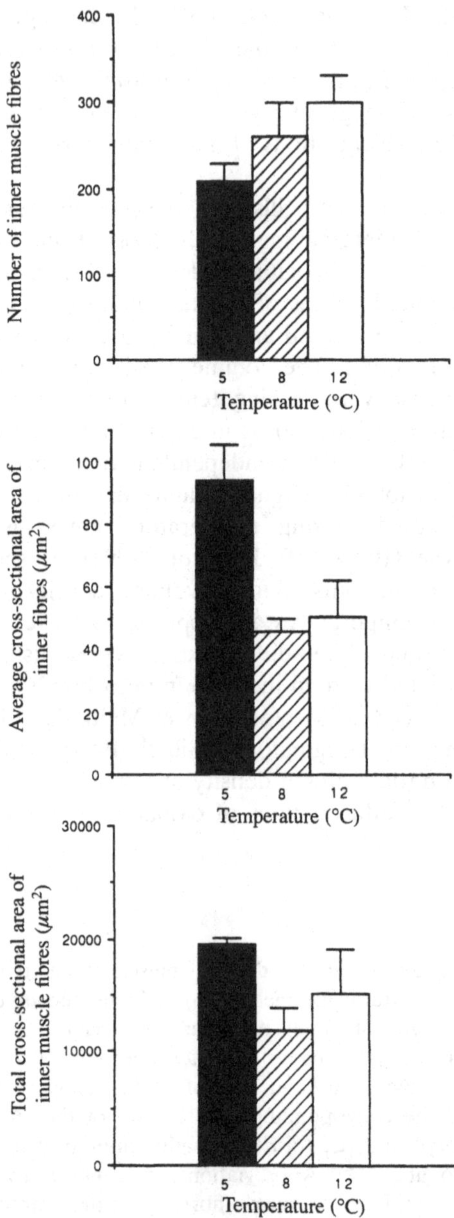

Fig. 6. Number and cross-sectional area of inner muscle fibres in 1-day-old Clyde herring larvae reared at different temperatures. Values represent mean ± SD of six fish per temperature. (After Johnston *et al.*, 1995.)

Table 1. *Ultrastructural characteristics of muscle fibres in herring* (Clupea harengus) *larvae reared at three temperatures*

Parameter	Temperature (°C)		
	5	10	15
Superficial muscle fibres			
V_v(mt,f)	37.6 ± 5.5	38.8 ±10.3	46.0 ± 6.4
S_v(cr,mt)	8.0 ± 2.4	10.2 ± 2.7	8.7 ± 2.8
V_v(my,f)	25.6 ± 4.2	21.0 ± 2.7	18.8 ± 4.6
Inner muscle fibres			
V_v(mt,f)	15.9 ± 5.0	20.5 ± 7.3	26.1 ± 6.6
S_v(cr,mt)	10.5 ± 1.2	9.9 ± 2.8	7.3 ± 1.3
V_v(my,f)	38.0 ± 7.4	35.3 ± 6.2	32.5 ± 6.6

Values are mean ± SD of 20 fibres from five fish.
V_v(mt,f): total volume density of mitochondria (%); S_v(cr,mt): surface density of mitochondrial cristae ($\mu m^2 \mu m^{-3}$); V_v(my,f): total volume density of myofibrils (%).
Source: After Vieira & Johnston (1992).

Metamorphosis

Many teleosts undergo a distinct metamorphosis from a larval to a juvenile form. In some cases, for example in flatfish, metamorphosis involves marked changes in morphology, habitat, lifestyle and diet over a relatively short period of time. Metamorphosis in the herring is associated with the development of functional gill filaments, pigmentation of the blood, the formation of adult slow muscle fibres in the region of the lateral line nerve, the silvering of the body and many changes in sensory systems and behaviour (Batty, 1984; Blaxter, 1988; Johnston & Horne, 1994). At 8–12 °C, metamorphosis starts at around 28 mm total length (TL) and is complete by around 40 mm TL. Prior to 28 mm TL the superficial muscle layer stains intensely against an antibody to myosin light chain 3. The expression of this protein is switched off in larger larvae. Red muscle fibres, that are also unstained with anti-MLC3, form externally to the larval superficial fibres which invaginate in the region of the lateral line nerve (Johnston & Horne, 1994). Increased rearing temperature affects the time taken to complete metamorphosis and to reach a given body length, but does not change the proportion of different fibre types (I. A. Johnston *et al.*, unpublished results) or the numbers and size classes of the white myotomal muscle fibres (Fig. 7).

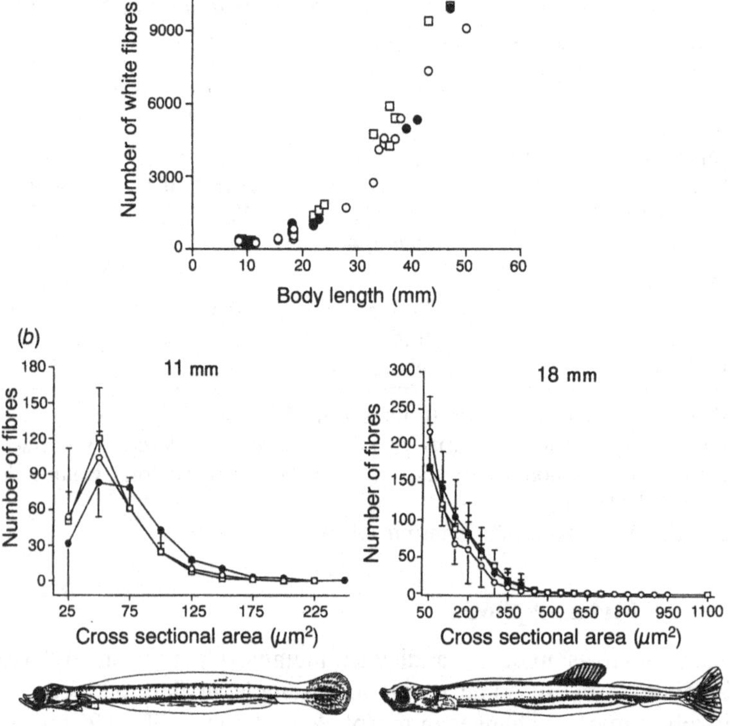

Fig. 7. Effects of temperature on the dynamics of muscle growth during the larval stages of Clyde herring. (*a*) Increase in number of white muscle fibres at the level of the dorsal fin. (*b*) Frequency distribution of cross-sectional area of white muscle fibres in 11 and 18 mm larvae which are illustrated below. At around 18 mm the dorsal and caudal fins develop and the larvae show a transition from an anguilliform to a sub-carangiform mode of swimming. Fish were reared at 5 °C (○), 8 °C (●) and 12 °C (□). (After I.A. Johnston *et al.*, unpublished results.)

In contrast, Calvo & Johnston (1992) found that turbot (*Scophthalmus maximus*) reared through metamorphosis at 17 °C had higher numbers of tonic muscle fibres/myotome and a greater cross-sectional area of red muscle than fish reared at 23 °C. Increases in the volume of red muscle with cold acclimation have also been reported in the adult stages of several teleost fish including goldfish (Johnston & Lucking, 1978), striped bass (Moerland & Sidell, 1986) and common carp (Rome *et al.*, 1992). In all these cases, however, much larger differences in acclimation temperature of at least 15 °C are required for a change in muscle fibre distribution (Guderley, 1990).

Post-larval muscle growth

In birds and mammals, muscle fibre number is fixed at birth and muscle growth is via an increase in the diameter of the existing fibres (hypertrophy) (Goldspink, 1972). Hypertrophy growth and muscle regeneration following injury are achieved by the division of a population of muscle stem cells (satellite cells) that are enclosed within the basal lamina of muscle fibres (Campion, 1984). Division of the satellite cells regenerates a stem cell and also produces a daughter cell committed to terminal differentiation following a limited number of subsequent divisions. Tissue culture experiments have shown that there are distinct classes of satellite cells for fast and slow muscles and at the embryonic, neonatal and adult stages of the life cycle (for a review, see Stockdale, 1992).

In contrast to mammals many fish continue to produce new muscle fibres (hyperplasia) throughout adult life. It is not known how many classes of muscle satellite cell occur in fish. In Clyde herring, labelling experiments with bromo-deoxyuridine (BrdU), which is incorporated into replicating DNA, have shown that just prior to hatching undifferentiated myoblasts present on the surface of the muscle fibres start to divide (Johnston *et al.*, 1995). Clyde herring hatch at around 8 mm, and muscle growth is entirely by hypertrophy until the larvae reach 20 mm (Fig. 7). During hypertrophic growth, muscle fibres absorb additional nuclei in order to maintain a relatively constant nuclear to cytoplasmic ratio (Koumans *et al.*, 1991). In many fish including Atlantic salmon (I.A. Johnston & A. McLay, unpublished results), plaice (Brooks & Johnston, 1993) and sea bass (Veggetti *et al.*, 1990) discrete germinal zones of myoblasts are present at hatching. By first feeding the germinal zones become exhausted and satellite cells are observed enclosed within the basal lamina of muscle fibres (I.A. Johnston & A. McLay, unpublished results). At this stage muscle growth is via hyperplasia and hypertrophy.

Weatherley & Gill (1985) found that for teleosts from temperate latitudes, small diameter fibres, consistent with new fibre production, continued to be produced until fish reached about 70% of their ultimate length. In general, a large body size was positively correlated with the ability to produce new muscle fibres for a longer fraction of the lifespan. Conversely, species or strains with a small ultimate body size were found to grow largely by fibre hypertrophy (Weatherley & Gill, 1987).

Meyer-Rochow & Ingram (1993) studied populations of southern smelt (*Retropina retropina*) from New Zealand locally adapted to riverine and lacustrine habitats. There is genetic evidence that the lacustrine populations in North Island lakes were introduced from the lower Waikato river, and hence are derived from the diadromous riverine stock. The lake stock reached a smaller ultimate body length than the riverine stock and this difference has been related to the earlier cessation of new muscle fibre recruitment (Meyer-Rochow & Ingram, 1993). If the conjecture of recent common ancestry is indeed correct, then these stock differences may reflect the cooler temperature of the riverine environment at the time the embryos and larvae were developing.

We have recently obtained more direct experimental support for the suggestion that the temperature experienced by the embryo can modify growth characteristics. In a quantitative electron microscopy study of a 1-day-old Clyde herring, the number of undifferentiated larval myoblasts per myotome or per mm muscle cross-sectional area was found to be two-fold higher at 8 °C than 5 °C (Johnston, 1993). In subsequent experiments we reared herring at 5 and 12 °C until first feeding, and then transferred them to ambient seawater temperature (9–13 °C). Although muscle fibre number was initially similar in the two groups after 60 days at a common temperature, the fish initially exposed to 5 °C had significantly more white muscle fibres/myotome than those at 12 °C, consistent with enhanced hyperplastic growth (Fig. 8). These results indicate that post-larval growth characteristics are determined at least in part by the thermal experience of the embryo, perhaps via altered numbers of muscle stem cells and/or the neuroendocrine secreting cells involved in growth regulation.

Summary and conclusions

Gastrulation results in three major tissues in fish embryos. Firstly, the axial mesoderm which develops into the notochord and which is now thought to be the structure which controls the development of many other organ systems. Secondly, the neural ectoderm which develops into the neural tube, and thirdly, the paraxial mesoderm which segments to form the somites (myomeres). Herring eggs can successfully hatch between 4 °C and 15–17 °C, although survival is poorer at the highest temperatures. Our studies have shown that most early events in embryogenesis including gastrulation, and the development of the notochord, neural tube and somites are highly canalised over this entire temperature range (Fig. 9). This implies strong selection of regulatory mechanisms to ensure the coordinated development of these structures.

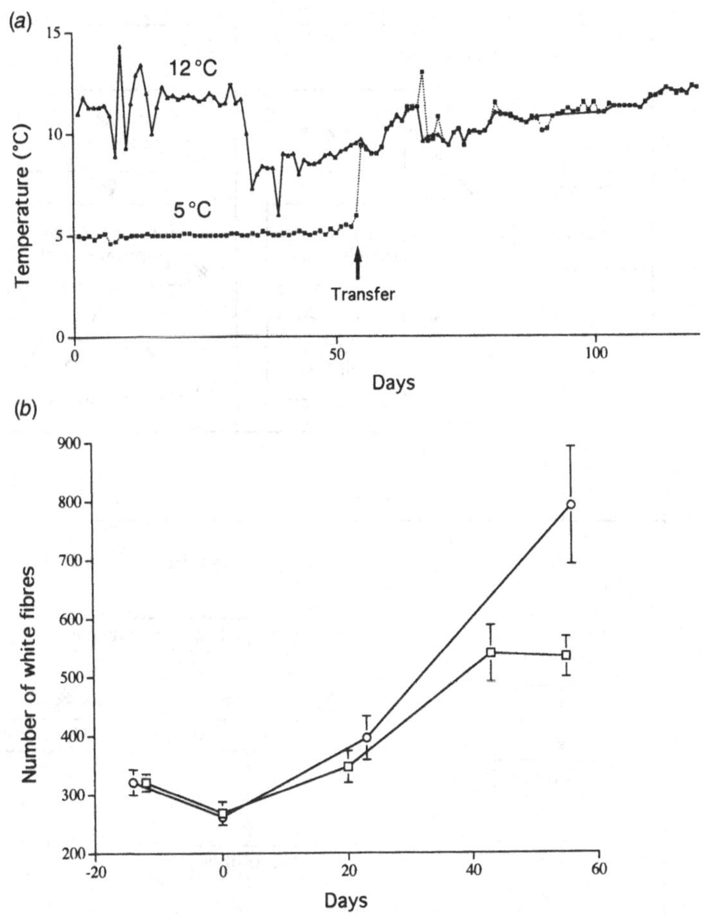

Fig. 8. Effect of embryonic temperature on subsequent growth characteristics in Clyde herring larvae transferred to a common temperature. (*a*) Herring were reared at temperatures of 5 °C (○) and 12 °C (□) until first feeding and then transferred to variable ambient temperaures. (*b*) Number of white muscle fibres at the level of the dorsal fin. (After Johnston & Palmer, unpublished results.)

The evolutionary cost of canalisation may be evident in the formation of the neural tube which involves a massive over-production of neuro-ectodermal cells, the excess of which are removed by apoptosis during larval development (Oppenheim, 1991). Other important processes are happening during the early part of embryogenesis. The neural tube is differentiating to form the neurons that will interact with the notochord

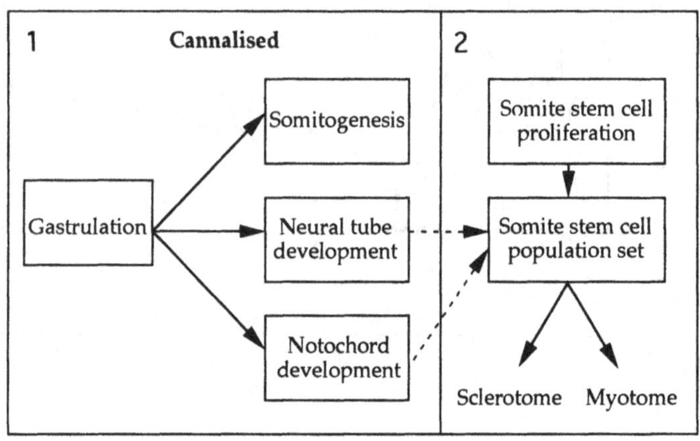
1
Cannalised
Gastrulation
Somitogenesis
Neural tube development
Notochord development
2
Somite stem cell proliferation
Somite stem cell population set
Sclerotome
Myotome

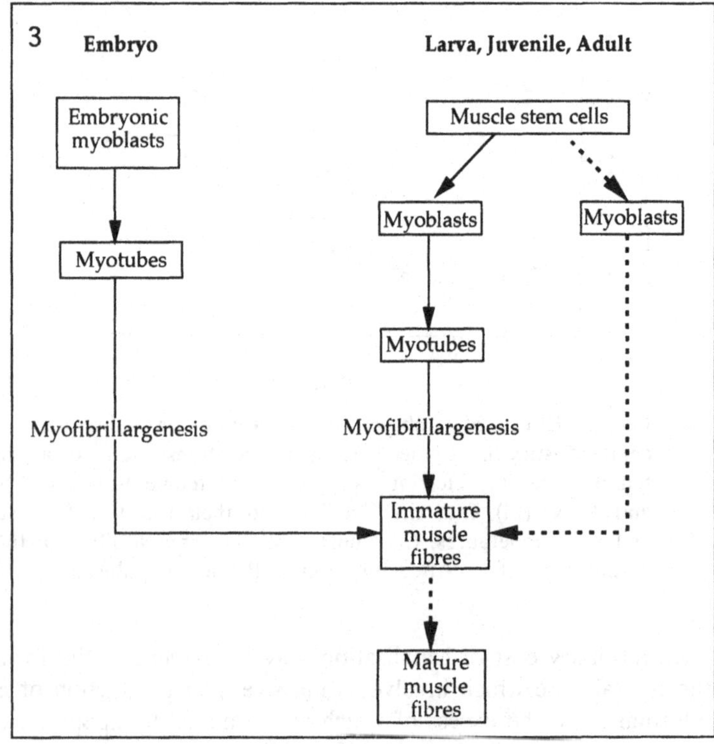
3
Embryo
Larva, Juvenile, Adult
Embryonic myoblasts
Myotubes
Myofibrillargenesis
Muscle stem cells
Myoblasts
Myoblasts
Myotubes
Myofibrillargenesis
Immature muscle fibres
Mature muscle fibres

to signal differentiation in the somites. At the same time cell proliferation in the somite builds up the population of stem cells that will differentiate to become myoblasts, myotubes and scleretomal derivatives (Fig. 9).

We have shown that a number of characters vary with rearing temperature, including the relative timing of myogenesis (Fig. 4), the number of myotubes/muscle fibres (Fig. 6), and the numbers of cells with the ultrastructural characteristics of myoblasts (Johnston, 1993). Our hypothesis to explain these observations involves differential effects of temperature on the rate of division of the somite stem cell population on the one hand, and the production of the signal(s) required for muscle differentiation and the fating of myoblast lineages on the other (Fig. 9). In herring, the phenotypic variation produced by changes in the timing of myogenesis is probably of little long-term importance as the motor system has reached a similar stage of development at all temperatures by hatching. In contrast, laboratory experiments have shown (Fig. 8) that development temperature fixes some component(s) of the growth potential of muscle tissue, resulting in different larval phenotypes. Field studies have revealed major differences in the total number of white muscle fibres in different herring stocks. For example, Blackwater herring, an estuarine stock which develop at lower water temperatures than the nearby offshore Bank herring, have around 1.9-times as many fibres at maturity (Greer-Walker *et al.*, 1972). The

---

Fig. 9. Summary of muscle development in fish. Box 1 shows early events in embryogenesis are highly canalised. For example, no relative difference is observed between the somite stage of the embryo and identifiable structures within the neural tube at different temperatures. The notochord has been implicated as the limiting factor on the overall rate of development because development cannot proceed faster than the posterior extension of the axial mesoderm. Box 2 demonstrates that as somitogenesis continues, cell division proceeds in each somite and is arrested on receipt of signals from the notochord and neural tube (Buckingham, 1994). The relative differences in cell division rate in the somite at different temperatures affects the final stem cell population for the somite. Box 3 shows a putative myogenic pathway for fish. The cellular equivalence of embryonic myoblasts and muscle stem cells is not known; here they are shown as separate cell types. The solid line indicates new fibre formation by hyperplasia and the dashed line indicates the hypertrophic pathway of fibre growth by fusion of myoblasts with new muscle fibres.

hypothesis that development temperature alters muscle stem cell populations in herring and thereby muscle fibre number rests on a number of critical assumptions that require testing. Firstly, have muscle stem cells been accurately identified by electron microscopy? Cells with the ultrastructural characteristics of myoblasts undoubtedly include muscle stem cells, myoblasts committed to terminal differentiation and non-muscle fibre stem cells. Furthermore, it is known that treatment of fibroblasts with myoD results in the synthesis of myofibrillar proteins although not myotube formation. Research is needed to characterise muscle transcription factors in fish and determine their expression patterns in the different classes of myogenic precursor cell(s). Our hypothesis also assumes that populations of somitic stem cells have a fixed clonal life, an idea that might be testable in cell culture. Research is also required on the regulatory mechanisms responsible for muscle stem cell proliferation. It is known that the growth of muscle and other tissues varies enormously with food availability, temperature, day length, exercise and with behavioural factors such as social dominance. An improved understanding of muscle growth regulation during ontogeny in fish is likely to find numerous practical applications, particularly in the management of wild populations and in the acquaculture industry.

Acknowledgements

Our work on environmental influences on fish development is supported by the Natural Environment Research Council of the UK.

References

Batty, R.S. (1984). Development of swimming movements and musculature of larval herring (*Clupea harengus*). *Journal of Experimental Biology* **110**, 217–29.

Batty, R.S. (1994). The effect of temperature on the vertical distribution of larval herring (*Clupea harengus* L.). *Journal of Experimental Marine Biology and Ecology* **177**, 269–76.

Batty, R.S., Blaxter, J.H.S. & Fretwell, K. (1993). Effect of temperature on escape responses of larval herring, *Clupea harengus*. *Marine Biology* **115**, 523–8.

Blaxter, J.H.S. (1988). Pattern and variety in development. In *Fish Physiology* (ed. W.S. Hoar & D. J. Randall), pp. 1–58. New York: Academic Press.

Breitbart, R.W., Nguyen, H.T., Medford, R.M., Destree, A.T., Mahdavi, V. & Nadal-Ginard, B. (1985). Intricate combinational patterns of exon splicing generate multiple troponin T isoforms from a single gene. *Cell* **41**, 67–82.

Breven, K.A., Gill, D.E., & Smith-Gill, J.S. (1979). Countergradient selection in the green frog, *Rana clamitans. Evolution* **33**, 609–23.

Brooks, S. & Johnston, I.A. (1993). Influence of development and rearing temperature on the distribution, ultrastructure and myosin subunit composition of myotomal muscle-fibre types in the plaice, *Pleuronectes platessa. Marine Biology* **117**, 501–13.

Brooks, S., Vieira, V.L.A., Johnston, I.A. & Macheru, P. (1995). Muscle development in larvae of a fast growing tropical freshwater fish, the Curimatã-pacú (*Prochilodus margravii*). *Journal of Fish Biology* **47**, 1026–37.

Buckingham, M. (1994). Molecular biology of muscle development. *Cell* **78**, 15–21.

Burggren, W.W. (1992). The importance of an ontogenetic perspective in physiological studies: amphibian cardiology as a case study. In *Physiological Adaptation in Vertebrates* (ed. S.C. Wood, R.E. Weber, A.R. Hargens & R.W. Millard), pp. 235–52. New York: Dekker.

Calvo, J. & Johnston, I.A. (1992). Influence of rearing temperature on the distribution of muscle fibre type in the turbot *Scopthalmus maximus* at metamorphosis. *Journal of Experimental Marine Biology and Ecology* **161**, 45–55.

Campion, D.R. (1984). The muscle satellite cell: a review. *International Review of Cytology* **87**, 225–51.

Crockford, T. & Johnston, I.A. (1993). Developmental changes in the composition of myofibrillar proteins in the swimming muscles of Atlantic herring, *Clupea herengus. Marine Biology* **115**, 15–22.

Doyle, M.J. (1977). A morphological staging system for the larval development of the herring, *Clupea harengus* L. *Journal of the Marine Biological Association of the United Kingdom* **57**, 859–67.

Egginton, S. & Sidell, B.D. (1989). Thermal acclimation induces adaptive changes in subcellular structure of fish skeletal muscle. *American Journal of Physiology* **256**, R1–9.

Felsenfeld, A.L., Curry, M. & Kimmel, C.B. (1991). The *fub-1* mutation blocks initial myofibril formation in zebrafish muscle pioneer cells. *Developmental Biology* **148**, 23–30.

Felsenfeld, A.L., Walker, C., Westerfield, M., Kimmel, C.B. & Streisinger, G. (1990). Mutations affecting skeletal muscle myofibril structure in the zebrafish. *Development* **108**, 443–59.

Ferron, A. & Leggett, W.C. (1994). An appraisal of condition measures for marine fish larvae. In *Advances in Marine Biology* (ed. J.H.S. Blaxter & A.J. Southward), pp. 217–86. London: Academic Press.

Fukuhara, O. (1990). Effect of temperature on yolk utilization, initial growth, and behaviour of unfed marine fish larvae. *Marine Biology* **106**, 169–74.

Gibson, S. & Johnston, I.A. (1995). Temperature and development in larvae of the turbot, *Scophthalmus maximus*. *Marine Biology* **124**, 17–25.

Goldspink, G. (1972). Postembryonic growth and differentiation of striated muscle. In *The Structure and Function of Muscle* (ed. G.H. Bourne), vol. 1, 2nd edn, p. 179. New York: Academic Press.

Greer-Walker, M.G., Bird, A.C. & Pull, G.A. (1972). The total number of white skeletal muscle fibres in cross section as a character for stock separation in North Sea herring (*Clupea harengus*). *Journal du Conseil International pour l'exploration de la Mer* **34**, 238–43.

Guderley, H. (1990). Functional significance of metabolic responses to thermal acclimation in fish muscle. *American Journal of Physiology* **259**, R245–R252.

Hanneman, E.H. (1992). Diisopropylfluorophosphate inhibits acetylcholinesterase activity and disrupts somitogenesis in the zebrafish. *Journal of Experimental Zoology* **263**, 41–53.

Hatta, K., BreMiller, R., Westerfield, M. & Kimmel, C.B. (1991). Diversity of expression of *engrailed*-like antigens in zebrafish. *Development* **12**, 821–32.

Hempel, G. & Blaxter, J.H.S. (1961). The experimental modification of meristic characters in herring (*Clupea harengus* L.). *Journal du Conseil International pour l'exploration de la Mer* **XXVI**, 336–46.

Hill, J. & Johnston, I.A. (1996). Early motor innervation of somites is delayed by low temperature in Atlantic herring (*Clupea harengus*) embryos. *Journal of Fish Biology* (in press).

Hovenkamp, F. & Witte, J.I.J. (1991). Growth, otolith growth and RNA/DNA ratios of larval plaice, *Pleuronectes platessa*, in the North Sea 1987 to 1989. *Marine Ecology Progress Series* **70**, 105–16.

Janzen, F.J. & Paukstis, G.L. (1991). Environmental sex determination in reptiles: ecology, evolution and experimental design. *The Quarterly Review of Biology* **66**, 149–79.

Johnston, I.A. (1993). Temperature influences muscle differentiation and the relative timing of organogenesis in herring (*Clupea harengus*) larvae. *Marine Biology* **116**, 363–79.

Johnston, I.A. & Horne, Z. (1994). Immunocytochemical investigations of muscle differentiation in the Atlantic herring (*Clupea harengus*: Teleostei). *Journal of the Marine Biological Association of the UK* **74**, 79–91.

Johnston, I.A. & Lucking, M. (1978). Temperature induced variation in the distribution of different muscle fibre types in the goldfish (*Carassius auratus*). *Journal of Comparative Physiology B* **124**, 111–16.

Johnston, I.A. & Maitland, B. (1980). Temperature acclimation in crucian carp, *Carcassius carassius* L., morphometric analysis of muscle fibre ultrastructure. *Journal of Fish Biology* **17**, 113–25.

Johnston, I.A., Vieira, V.L.A. & Abercromby, M. (1995). Temperature and myogenesis in embryos of the Atlantic herring *Clupea harengus. Journal of Experimental Biology* **198**, 1389–403.

Jones, S.R. & Jeffs, T.M. (1991). *Near-surface sea temperatures in coastal waters of the North Sea, English Channel and Irish Sea.* Data Report, No. 24. Ministry of Agriculture, Fisheries and Food Research, 24.

Jørstad, K.E., King, D.P.F. & Neavdal, G. (1991). Population structure of Atlantic herring, *Clupea harengus* L. *Journal of Fish Biology* **39(sA)**, 43–52.

Keynes, R. & Stern, C.D. (1988). Mechanisms of vertebrate segmentation. *Development* **103**, 413–29.

Kimmel, C.B., Schilling, T.F. & Hatta, K. (1991). Patterning of body segments of the zebrafish embryo. *Current Topics in Developmental Biology* **25**, 77–110.

Kimmel, C.B., Warga, R.M. & Schilling, T.F. (1990). Origin and organisation of the zebrafish fate map. *Development* **108**, 581–94.

King, D.P.F., Ferguson, A. & Moffett, I.J.J. (1987). Aspects of the population genetics of herring, *Clupea harengus*, around the British Isles and the Baltic Sea. *Fisheries Research* **6**, 35–52.

Koumans, J.T.M., Akster, H.A., Booms, G.H.R., Lemmens, C.J.J. & Osse, J.W.M. (1991). Numbers of myosatellite cells in white axial muscle of growing fish: *Cyprinus carpio* L. (Teleostei). *American Journal of Anatomy* **192**, 418–24.

Marden, J.H. (1995). Large-scale changes in thermal sensitivity of flight performance during adult maturation in a dragonfly. *Journal of Experimental Biology* **198**, 2095–102.

Metzger, J.M., Rudnicki, M.A. & Westfall, M.V. (1995). Altered Ca^{2+} sensitivity of tension in single-muscle fibers from MyoD gene-inactivated mice. *Journal of Physiology* **485**, 447–53.

Meyer-Rochow, V.B. & Ingram, J.R. (1993). Red white muscle distribution and fiber growth dynamics: a comparison between lacustrine and riverine populations of the southern smelt *Retropinna retropinna* (Richardson). *Proceedings of the Royal Society of London Series B* **252**, 85–92.

Moav, B. (1994). Development and molecular genetics in fish. *Israeli Journal of Zoology* **40**, 441–66.

Moerland, T.S. & Sidell, B.D. (1986). Contractile responses to temperature in the locomotory musculature of striped bass *Morone saxatilis. Journal of Experimental Biology* **240**, 25–33.

Moran, N.A. (1992). The evolutionary maintenance of alternative phenotypes. *American Naturalist* **139**, 971–89.

Oppenheim, R.W. (1991). Cell death during development of the nervous system. *Annual Review of Neuroscience* **14**, 453–501.

Paululat, L., Burchard, S. & Renkawitzpohl, R. (1995). Fusion from myoblasts to myotubes is dependent on the rolling stone gene (*rost*) of *Drosophila. Development* **121**, 2611–20.

Pette, D. & Staron, R.S. (1990). The molecular diversity of mammalian muscle fibres. *News in Physiological Sciences* **8**, 153–7.

Raamsdonk, van W., van der Stelt, A., Diegenbach, P.C., van de Berg, W., de Bruyn, H., van Dijk, J. & Mijzen, P. (1974). Differentiation of the musculature of the teleost *Brachydanio rerio. Z. Anat. Entwickl.-Gesch.* **145**, 321–42.

Rescan, P.Y., Gauvry, L., & Paboeuf, G. (1995). A gene with homology to myogenin is expressed in developing myotomal musculature of the rainbow trout and *in vitro* during the conversion of myosatellite cells to myotubes. *FEBS Letters* **362**, 89–92.

Rome, L.C., Sosnicki, A. & Choi, I.H. (1992). The influence of temperature on muscle function in the fast swimming scup. 2. The mechanics of red muscle. *Journal of Experimental Biology* **163**, 281–95.

Scapolo, P.A., Veggitti, A., Mascarallo, F. & Romanello, M.G. (1988). Developmental transitions of myosin isoforms and organisation of the lateral muscle in the teleost *Dicentrarchus labrax* (L.). *Anatomy and Embryology* **178**, 287–96.

Scharloo, W. (1991). Canalisation: genetic and developmental aspects. *Annual Review of Ecology and Systematics* **22**, 65–93.

Scheiner, S.M. (1993). Genetics and evolution of phenotypic plasticity. *Annual Review of Ecology and Systematics* **24**, 35–68.

Sepich, D.S., Ho, R.K. & Westerfield, M. (1994). Autonomous expression of the *nic1* acetylcholine receptor mutation in zebrafish muscle cells. *Developmental Biology* **161**, 84–90.

Smith-Gill, S.J. (1983). Developmental plasticity: Developmental conversion *versus* phenotypic modulation. *American Zoologist* **23**, 47–55.

Sternberg, K. (1995). Influence of oviposition date and temperature upon embryonic development in *Somatochlora alpestris* and *S. artica* (Odonata, Corduliidae). *Journal of Zoology* **235**, 163–74.

Stickland, N.C., White, R.N., Mascall, P.E., Crook, A.R. & Thorpe, J.E. (1988). The effect of temperature on myogenesis in embryonic development of the salmon (*Salmo salar* L.). *Anatomy and Embryology* **178**, 253–7.

Stockard, C.R. (1921). Developmental rate and structural expression: an experimental study of twins, 'double monsters' and single deformities, and the interaction among embryonic organs during their origin and development. *American Journal of Anatomy* **28**, 115–266.

Stockdale, F. (1992). Myogenic cell lineages. *Developmental Biology* **154**, 284–98.

Tåning, A.V. (1952). Experimental study of meristic characters in fish. *Biological Reviews* **27**, 169–93.

Veggetti, A., Mascerello, F., Scapolo, P.A. & Rowlerson, A. (1990). Hyperplastic and hypertrophic growth of lateral muscle in *Dicentrarchus labrax* (L.). *Anatomy and Embryology* **182**, 1–10.

Veggetti, A., Mascarello, F., Scapolo, P.A., Rowlerson, A. & Candia-Carnevali, M.D. (1993). Muscle growth and myosin isoform transitions during development of a small teleost fish, *Poecilia reticulata* (Peters) (Atheriniformes, Poeciliidae): a histochemical, immunohistochemical, ultrastructural and morphometric study. *Anatomy and Embryology* **187**, 353–61.

Vieira, V.L.A. & Johnston, I.A. (1992). Influence of temperature on muscle-fibre development in larvae of the herring *Clupea harengus*. *Marine Biology* **112**, 333–41.

Vieira, V.L.A. & Johnston, I.A. (1996). Muscle development in Tambaqui (*Colossoma macropomum*) (Characidae): an important Amazonian food fish. *Journal of Fish Biology* (in press).

Weatherley, A.H. & Gill, H.S. (1985). Dynamics of increase in muscle fibres in fishes in relation to size and growth. *Experientia* **41**, 353–4.

Weatherley, A.H. & Gill, H.S. (1987). *The Biology of Fish Growth*. London: Academic Press.

Westerfield, M. (1994). *The Zebrafish Book*: A guide for the laboratory use of zebrafish (*Brachydanio rerio*). 2.1. Eugene, OR, USA: University of Oregon Press.

D. ATKINSON

Ectotherm life-history responses to developmental temperature

Introduction

An organism's life history is its lifetime pattern of growth, differentiation, storage of reserves and reproduction (Begon *et al.*, 1990: 473). Life-history traits therefore include size at birth and maturity and the investment in reproduction at different ages and sizes (Roff, 1992; Stearns, 1992).

The temperatures under which ectothermic organisms evolve can sometimes have the same qualitative effects on life-history traits as temperatures experienced during development. The evolutionary effect comprises differences between genotypes arising from selection for several generations at different temperatures, and are observed when the different genetic lines are then reared under identical environmental conditions. By contrast, the developmental effect is the result of phenotypic plasticity and can even be observed among members of a clone reared at different temperatures. An example of such similar responses is the common decrease in adult size in *Drosophila* following evolution or development at increased temperatures (pp. 265–92; Partridge *et al.*, 1994). Whether this similarity arises because the same selection pressures cause evolution of both size and the developmental response (or reaction norm; Stearns, 1992) of size to different temperatures, is not known. Developmental responses of body size to temperature are reviewed and discussed in the present chapter, while both developmental and evolutionary responses are discussed by Partridge & French (pp. 265–92) with special reference to studies on *Drosophila*.

Temperature can initiate responses or affect rates of differentiation, growth and reproduction. An example of initiation is the requirement of a period of low temperature before rapid further development can proceed: this is seen in vernalisation in plants and diapause in insects which are generally interpreted as adaptations that match the timing of development to favourable times of year (Tauber *et al.*, 1986;

Roberts & Summerfield, 1987). The effect of temperature on rates depends on both mean temperature and the frequency and magnitude of any fluctuations (Ratte, 1985; Cossins & Bowler, 1987; Liu *et al.*, 1995). The present chapter focuses only on the effects of mean temperature.

Specifically, I describe those responses of life histories to mean developmental temperature that are widespread among ectothermic species. Of these, effects on body size at a given stage of development are shown to be particularly difficult to explain. A combination of theory and analysis of the scientific literature is then used to generate ideas and filter out those particularly worthy of experimental tests. The chapter concludes with a discussion of how the remaining ideas might be tested further.

Development, growth and reproduction

General form of the response to temperature

Development

To adequately compare and understand the effects of temperature on life histories, care needs to be taken over defining and choosing the appropriate stages of development. The continuous change of many characters, for instance, can make the definition of developmental stages rather arbitrary. Particular care needs to be taken when temperature affects the rate of change of some characters more than others or affects the number of moults before reproductive maturity (for examples, see Bellinger & Pienkowski, 1987; Atkinson, 1994). In addition, as development involves both tissue differentiation and the growth of the organism (except where growth ceases as during pupation in insects), it cannot be completely separated from the growth of the organism. Yet despite these difficulties it is commonly (albeit implicitly) assumed that insight can come from the examination of relationships between the timing of necessary physiological events which have considerable ecological importance and temperature. Such events – which typically include completion of hatching, completion of last juvenile moult and first release of gametes – are generally used by life-history biologists to define ‘developmental stages’. Moreover, it has been widely observed that the relationship between the rate of development, expressed as the reciprocal of the time taken for completion of a given developmental step, and rearing temperature follows a consistent form: it approximates to the shape of an asymmetrical wigwam (Fig. 1). Below a base temperature or null-point (T_b) no development occurs. With an increase in temperature the rate increases until a maximum beyond which it decreases sharply to zero.

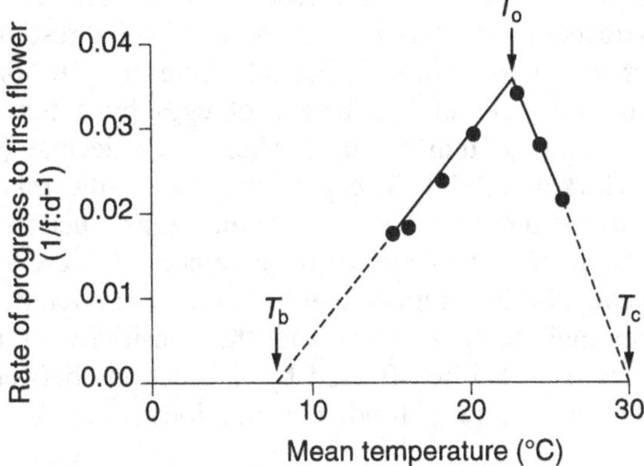

Fig. 1. Relation between mean temperature and rate of progression towards first flower (1/f) in common bean *Phaseolus vulgaris* cultivar Negro Patrizia. The values of base temperature, 'optimum' temperature and ceiling temperature are designated T_b, T_o and T_c, respectively. Temperatures below T_o are generally referred to as 'sub-optimal' while those above T_o are 'supraoptimal'. (After Roberts & Summerfield, 1987.)

The temperature at which maximum developmental rate is observed is usually referred to, rather misleadingly, as the temperature optimum (T_o). Whether it is optimal in the sense of maximising Darwinian fitness (which itself is often hard to define) is at least questionable because rates of survival often decrease considerably at this temperature (Bursell, 1974; Cossins & Bowler, 1987; Wagner *et al.*, 1987). Increases in developmental rates with temperature sometimes become less steep near T_o (Wagner *et al.*, 1987).

Base temperatures are often estimated by extrapolation from developmental rates at higher temperatures because experiments very near to the base temperature take a long time to complete (Trudgill, 1996). Values obtained from actual measurements of T_b are often slightly less than estimates based on linear extrapolation (Trudgill, 1996). Thus the relationship between developmental rate and temperature in the 'suboptimal' range can sometimes be sigmoidal rather than linear. Nonetheless, the assumption of a linear relationship is commonly used with considerable success for predicting the flowering times of crops and in pest control (Summerfield *et al.*, 1991; Trudgill, 1996).

Growth and reproduction

An ecological measure of post embryonic growth of an organism is its increase in dry weight or some close correlate of this (e.g. wet weight).

Growth can therefore result from increases in cell size, cell number and mass of extracellular material such as fat. When resources are abundant, an increase in temperature generally increases rates of growth and reproduction (e.g. rate of production of eggs by a female of a given size) except at high temperatures when rates decline (Bursell, 1974; Shrode & Gerking, 1977). The general form of the relationship between temperature and rates of growth and reproduction is thus superficially similar to that with rate of development. Differences have been found, however, in the width of band of permissible temperatures, the base and 'optimal' temperatures, and the sensitivity of rates to temperature. Rates can also be affected by interactions between temperature and other factors (e.g. food, photoperiod).

Variations in response to temperature

Width of permissible temperature band

The width of temperature band that permits reproduction has been found to be narrower than those permitting development in the fish *Cyprinodon nevadensis* (Shrode & Gerking, 1977). For insect species such as *Chaoborus crystallinus*, which however, produces eggs prior to metamorphosis the favourable range for egg production and larval development is identical (Ratte, 1985). The range of temperatures that permits development is generally narrowest for embryonic fish, wider for larvae and wider still for juveniles (Rombough, 1988).

Base and 'optimal' temperatures

Base temperatures for development increase with each successive developmental phase in species as different as the moth, *Lymantria monacha* (Zwolfer, 1933, cited by Ratte, 1985) and wheat, *Triticum aestivum* (Slafer & Savin, 1991). These increases correspond with warming normally experienced during the period of juvenile development. One advantage of this response would be to prevent individuals at late stages of development from maturing too soon before the favourable warm season (Atkinson, 1996). Moreover, in the fish *C. nevadensis*, the bottom end of the permissible range for growth and the temperature producing fastest growth are both about 7°C lower than the corresponding values for reproduction (Shrode & Gerking, 1977).

Interactions with other environmental factors

The effect of temperature on rates of development, growth and reproduction can depend on interactions with factors such as photoperiod (e.g. Summerfield *et al.*, 1991; Wayne & Block, 1992). The adaptive significance of these interactions presumably lies in the advantage of maturing at the optimum time of year (Roberts & Summerfield, 1987). Interactions between temperature and resource quality and availability have important effects on life histories (e.g. Vidal, 1980; Sweeney *et al.*, 1986) and, as will be discussed later, may explain why development is usually more sensitive to temperature than is growth.

Ratio of growth to development: size-at-stage

If rearing temperature alters the size of an organism at a given stage of development, then the ratio of growth to developmental rate has been altered. A reduced size at a given stage can therefore result despite faster average growth if development is accelerated more than growth. The shorter time period between developmental stages would thus reduce the total amount of growth.

It has long been recognised that increased developmental temperatures often result in reduced adult or final size (reviews by Bělehrádek, 1935; Ray, 1960; von Bertalanffy, 1960; Precht *et al.*, 1973). But some authors have drawn particular attention to the existence of increases in body size, as well as reductions, in response to increased developmental temperature (Laudien, 1973: 389; Ratte, 1985). A recent review (Atkinson, 1994) attempted to evaluate the evidence for a general rule by examining only those studies in which effects on size at specified developmental stages, usually near reproductive maturity, were shown to be statistically significant ($p < 0.05$), and in which effects of temperature were isolated from other variables. It therefore excluded studies in which conditions that can interact with temperature to affect rates of growth and development (e.g. food supply, photoperiod) were obviously not controlled. This review also attempted to exclude some obvious causes of reduced size, such as temperatures that were stressfully high for growth or development (i.e. those not normally encountered in nature, or which produced slower average rates of growth or development than did lower temperatures). Experiments were also excluded if the amount of energy available appeared to be limiting in any of the treatments such as when all food was eaten between feeds or when light levels produced slower plant growth than at the higher levels examined.

Of 109 studies examined, which included taxa from nine phyla in four kingdoms, and included species with determinate and indeterminate growth, 91 (83.5%) showed only size reductions (Table 1). This smaller size is attained at increased temperatures despite an expected or observed increase in average growth rate during the developmental

Table 1. *Effects of increased developmental temperature on body size in ectotherms*

Kingdom	Phylum	Numbers of size reductions	Numbers with some size increase[a]
Bacteria (Monera)		1	0
Protista	Chlorophyta	4	0
	Bacillariophyta	0	1
	Ciliophora	2	0
Plantae	Spermatophyta	6	0
Animalia	Aschelminthes	3	0
	Mollusca	4	0
	Arthropoda	63	17
	Chordata	8	0
Total		91	18

[a]Significant size increases with increased temperature were recorded in at least some part of the range. These also include cases in which size was reduced in certain parts of the temperature range and increased in others.
Source: After Atkinson (1994).

phase or prior to growth cessation (Fig. 2). Despite attempts to exclude examples in which experimental protocol was weak or temperatures were inappropriate, data were often incomplete and so inadmissible studies may have been included inadvertently. Nonetheless, the review included enough detailed and meticulous studies to support the general rule that increased temperatures lead to reduced size at a given developmental stage; however, apparently genuine exceptions to this rule (e.g. Fig. 3) deny it the status of a biological law applicable to all ectotherms.

Approaches to understanding temperature–size relationships

Simplicity and generality

The discovery of a widespread relationship such as a temperature–size rule tends to direct research towards general explanations which apply throughout the ectotherms, and away from those specific to particular populations, species or groups of species. Simple, rather than complex explanations are also usually sought. This does not deny the likelihood that many factors will affect the relationship between temperature and body size, and that only future empirical tests using particular genotypes

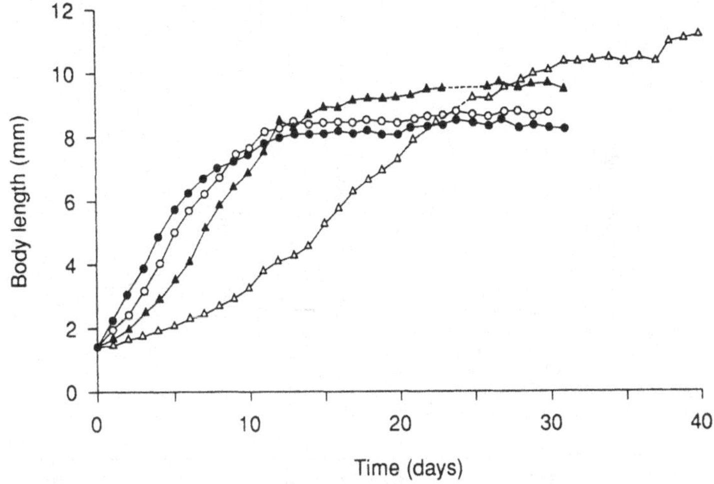

Fig. 2. Growth in mean body length of larvae of the midge *Chaoborus* at different temperatures: ●, 30 °C; ○, 25 °C; ▲, 20 °C; △, 15 °C. This shows the typical effect of temperature when resources are abundant; mean growth rate increases but final size is reduced with increased temperature. (After Hanazato & Yasuno, 1989.)

or populations will reveal the full complexity of these effects. Rather, the search for simple explanations which are capable of explaining much of the observed variation in size is a convenient first step to prioritise areas for detailed investigation. This approach is adopted in the present chapter.

Von Bertalanffy's explanation

Until recently, the explanation usually offered for the general reduction in ectotherm size with increased rearing temperature was that of von Bertalanffy (1960) (Precht *et al.*, 1973; Perrin, 1988). Von Bertalanffy suggested that as catabolic processes were mainly of a chemical nature they would have a high temperature coefficient, whereas anabolic processes ultimately depend on physical processes such as permeation and diffusion, which would have a low temperature coefficient. He also argued that anabolism was limited potentially by the rate of intake of substances such as respiratory gases and hence by the size of the areas through which they were absorbed. His growth equation expresses the rate of growth as the difference between anabolism and catabolism:

$$dw/dt = aw^m - bw^n \tag{1}$$

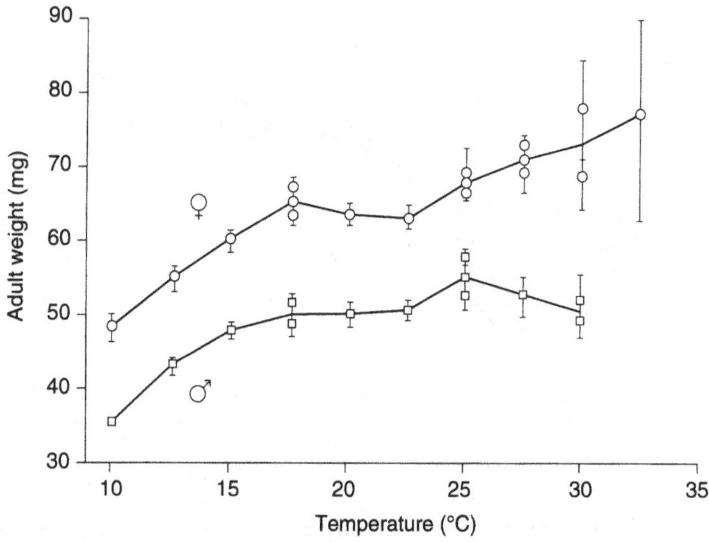

Fig. 3. Average adult weight (+SD) for the beetle *Entomoscelis americana* reared at different temperatures: an exception to the temperature-size rule. (After Lamb & Gerber, 1985.)

in which w is weight, t is time, and a, b, m and n are indices specific to particular combinations of genotype and environment: a, which affects anabolism, is almost constant in response to temperature, but b, which affects catabolism, increases with temperature. The value of m depends at least partly on the way in which the limiting absorptive surface area scales with body mass, whilst n approximates to 1.

Von Bertalanffy (1960) asserted that if catabolism had a higher temperature coefficient (b) than did anabolism then an increase in temperature would increase growth rate and reduce final size.

This simple model potentially applies to a wide range of ectotherms, although simplicity and generality are achieved by summarising many different physiological processes. Alternative interpretations of von Bertalanffy's explanation, however, conclude that a rise in temperature either always (Atkinson, 1994) or at least sometimes (Perrin, 1988) increases catabolism more than anabolism so that the rate of growth will decrease, not increase as he claimed.

In the search for a satisfactory explanation, a range of approaches has been adopted to generate hypotheses. These approaches, which are outlined later, include asking questions about widespread physiological mechanisms and the conditions under which a reduced size at increased developmental temperature would be adaptive. The published literature has then been used to filter out the most promising ideas for future experimental investigation.

Constraints on late growth

While the model of von Bertalanffy (1960) may not always explain the observed reduction in body size at a given developmental stage despite faster average growth, this result may be explained if the physiological limitations at increased temperatures described by von Bertalanffy become evident only late in growth. For instance, Kooijman (1993:48) suggested that the reduction in ultimate size observed at increased temperatures was because the animals fed faster and hence encountered food storage. This argument can be extended to a shortage of other resources and may arise from faster depletion of resources during the experiment or from an increase in the critical concentrations necessary to allow unlimited growth at increased sizes and temperatures. These are discussed next.

Resource depletion

Food depletion seems an improbable (though not impossible) explanation for the same size response observed in many of the studies reviewed by Atkinson (1994) in which food was changed daily, or was described by the authors as being 'in excess', 'fed *ad lib*', or 'more than adequate'. The presence of uneaten food when rations are renewed does not prove that growth is not limited by food concentration as the low rate at which food is encountered may also limit growth, especially for sedentary feeders unable to search for the remaining morsels.

Critical resource levels

At increased temperatures, faster early growth but with subsequent decline at relatively small sizes (Fig. 2) might be explained, even in the absence of food depletion, if food limitation increases as an increasing function of both temperature and size. Growth could thus be initially faster at the higher temperatures until a size is reached at which food starts to become limiting. Although food levels appeared to be maintained at high levels in the studies reviewed by Atkinson (1994), none actually demonstrated that levels were always unlimiting for growth in all treatments. In a study of the effects of food concentration, body size and temperature on the growth rate of two copepod species, Vidal (1980) measured the critical food concentration below which relative growth rate started to decline. This critical concentration increased with both temperature and body weight (Fig. 4). Moreover, a significant reduction in mean size with increased temperature was only observed in the later, larger developmental stages: this is consistent with the hypothesis that reduced size results from increasing limitations on growth by food shortage with increasing size and temperature. Unfortunately, it is not clear from the data presented by Vidal (1980) whether temperature continues to cause a significant reduction in size at food concentrations high enough not to limit growth.

Oxygen may also become limiting with increased size and temperature. One likely contributor to this in aquatic systems is the reduced solubility of oxygen at increased water temperatures (Davis, 1975). Whether the critical oxygen level for growth also increases with temperature and size, in aquatic or terrestrial species, is still not known. In a review of factors affecting fish growth, Cuenco *et al.* (1985) also consider this a serious gap in our knowledge.

Further work is required also to determine the extents to which carbon dioxide, nutrients and water increasingly limit the growth of autotrophs as size and temperatures increase.

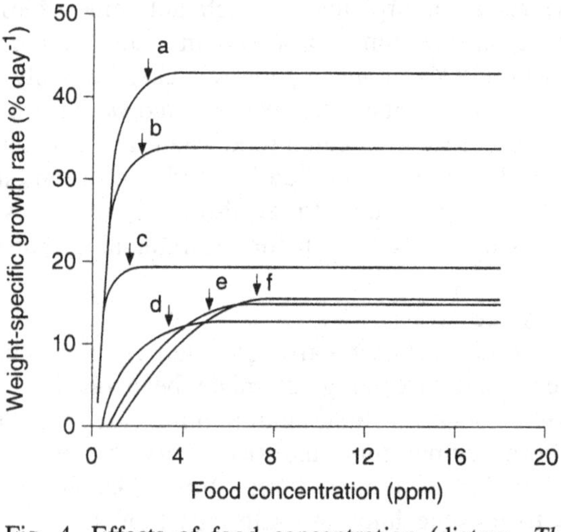

Fig. 4. Effects of food concentration (diatom, *Thalassiosira* sp.) on weight-specific growth rate of the copepod *Calanus pacificus*. Curves a–c are for copepods of 5 μg dry weight; d–f are for those of 125 μg dry weight. Temperatures are 15.5 °C (curves a and f), 12 °C (b and e) and 8 °C (c and d). Arrows indicate critical food concentration for growth: these increase with increased temperature and body size. (Curves are drawn from data of Vidal, 1980.)

Explanations at the cellular level

Another possibility is that the effects of temperature are primarily the result of cellular adaptations or constraints on cell growth and division, and that the effect on the size of the whole organism is merely a correlated response. This might arise if cell size is limited by increased temperature. Atkinson (1994) found no consistent widespread relationship between developmental temperature and cell size, although this does not preclude a dominant role for limitations on cell size in particular species. A second mechanism is that temperature affects the number of stem cells early in development which could influence growth potential and ultimate size, provided that the number of divisions is fixed (see Chapter 7). We do not know whether a reduction in the number of stem cells is a general phenomenon or whether these cells can only divide a set number of times. Further work is clearly needed on the dynamics of stem cell division.

Life-history theory

Another approach is to use life-history theory (for reviews, see Roff, 1992; Stearns, 1992). This comprises optimisation models that predict how different selection pressures will affect life-history traits, including size at maturity, under specified conditions. Of the wide range of functional relationships shown by these models to potentially affect optimal size (Table 2), increased daily mortality resulting from increased temperatures experienced by juveniles is both common among ectotherms and capable of explaining much of the size reduction at increased temperatures (Sibly & Atkinson, 1994). This reasoning is summarised below.

Juvenile mortality rate

Generally, a longer period of juvenile growth will increase body size at the time of first reproduction which in turn will increase fecundity (Caswell, 1989:31). Now consider an organism subject to varying risks of dying because of conditions experienced during the juvenile period. If this organism reduces its juvenile period when it correctly perceives these risks to be high it would more likely survive to be able to reproduce than if it did not accelerate its development. Consequently, high mortality resulting from conditions experienced by juveniles would favour acceleration of development towards adulthood, even at the expense of size at the time of first reproduction. Conversely, when mortality risks were low the costs of spending a longer time growing

Table 2. *Conditions favouring reduced size at a given developmental stage at increased temperatures*

Functional relationship to increased temperature	Applicability[a]	Reference
Increased juvenile mortality	Many cases; less likely when juvenile mortality decreases with parental age and size	Stearns & Koella (1986); Sibly & Atkinson (1994)
Increased population growth rate	Can include isolated populations and those scattered among habitat patches. Not when generation time is fixed	Sibly & Atkinson (1994); Atkinson (1994)
Increased ease of locomotion through water with reduced viscosity, reducing advantages of large size (or at least large locomotory organs or streamlining)	Active (not sedentary) aquatic species	Loosanoff (1959); Atkinson (1994)
Increased risk of sinking which can be reduced by small size	Aquatic plankton	Walsby & Reynolds (1980); Atkinson (1994)
Protection from increased predation by ectotherms favouring large prey	Examples include Cladocera subject to predation from fish but not from *Chaoborus* larvae	Culver (1980); Stibor & Lüning (1994); Atkinson (1995)
Small size may help avoid overheating	Species for which radiation is important for heating and cooling	Stevenson (1985); Atkinson (1994)

[a]A full list of assumptions for each model reviewed can be examined in the original papers. This table indicates major conclusions about the scope of the predictions drawn in the references listed.

would be outweighed by the benefits of larger size. This prediction is robust because it has been made by several models, each with different assumptions about the type of life history or the relationship of the population to the environment to which its members are adapted (e.g. reproductively isolated or scattered among habitat patches) (Sibly & Atkinson, 1994). Only the model of Stearns & Koella (1986) predicted delayed maturation in response to an increased risk of juvenile mortality, but their model included the assumption that juvenile mortality was reduced as parental age and size were increased. Many species do not provide parental care, so this assumption is unlikely to apply widely unless offspring size increases with parental age at maturity (Sibly & Atkinson, 1994).

Mechanisms that have been implicated in causing increased mortality rates at increased temperatures (Atkinson, 1994, 1995, 1996; Sibly & Atkinson, 1994) include:

1. Increased risks of predation (and possibly parasitism) by other ectotherms feeding faster at the higher temperatures;
2. Increased molecular or tissue damage associated with higher metabolic activity;
3. Increased risk of desiccation or drought (terrestrial species);
4. Increased risk of oxygen shortage, or habitat loss by evaporation (aquatic species).

Other functional relationships

Other functional relationships may be worth further investigation, but apply under a more limited range of conditions (Table 2). For example, an increased temperature which increases individual growth rate is also associated with increases in the rate of population growth in some species (for references, see Müller & Geller, 1993; Sibly & Atkinson, 1994). This would favour precocious maturation and reproduction at the expense of body size because in a growing population offspring born early will form a larger fraction of the total than if they were born later, and hence will have a higher Darwinian fitness (Lewontin, 1965). This reasoning does not apply, however, to populations in which generation time is fixed: these lack the demographic pressure to mature especially early at high temperatures as no individual is capable of fitting in additional generations or parts of generations (Meats, 1971).

Other functional relationships listed in Table 2 are associated with particular habitats or predator-size preferences and are discussed by Atkinson (1994).

Further functional relationships that have been predicted to favour a size reduction at a given developmental stage, usually maturity, currently lack a mechanism to link them to increased temperature (Atkinson, 1994; Berrigan & Charnov, 1994; Sibly & Atkinson, 1994) and are not therefore discussed here.

Time constraints

Time constraints on the juvenile period can occur if conditions deteriorate at the end of the season, or if synchronised emergence is required. To complete the life cycle under such time constraints developmental rate would need to be accelerated, leading to earlier maturation at a smaller size at the lower temperature. This can be seen if the period available for the growth of *Chaoborus* larvae (Fig. 2) was limited to about 9 days: on day 9 size is still largest at the highest temperature.

Exceptional responses to temperature

Another method for suggesting casual relationships is to compare niches and conditions for growth of species that are exceptions to the general rule with those which follow the rule, especially species that are otherwise closely related taxonomically and ecologically. Of 109 studies in the review of Atkinson (1994), 61 were of aquatic developmental phases, and of these only six (9.8%) showed an increase in size with increased rearing temperatures. Comparisons between these six exceptions and closely related species that followed the temperature-size rule sought evidence, in both groups, of removal of constraints on growth or factors that would alter the optimal size response to temperature (Atkinson, 1995). Exceptional responses were observed in the pennate diatom *Phaeodactylum tricornutum*, the copepod *Salmincola salmoneus* and four species of mayfly (Table 3).

No satisfactory hypothesis has yet been suggested to explain the exceptional response of *P. tricornutum* (Atkinson, 1995). The copepod *S. salmoneus*, which is parasitic on the gills of Atlantic salmon *Salmo salar*, however, differed from the five other copepods in the review, and indeed from the other 108 cases in Atkinson's (1994) review, in several ways that suggest possible explanations:

1. It was the only gill parasite and the only species for which mechanisms likely to prevent or reduce oxygen shortage at increased temperatures were known, both in the experiment (Johnston & Dykeman, 1987) and in natural populations. Respiratory compensation by the fish at increased temperatures to maintain the amounts of oxygen entering the gills

Table 3. *Aquatic exceptions to the temperature-size rule: a comparison with related species that followed the rule*

Species	Compared with	Differences[a]	Hypotheses[b]
1. *Phaeodactylum tricornutum* (a pennate diatom)	Four other unicellular algae	None found	–
2. *Salmincola salmoneus* (gill parasite of salmon)	Five other copepods	(i) Fish respiratory responses could prevent O_2 limitation at increased temperature (ii) Strongly seasonal life cycle, and temperature effect dependent on photoperiod (iii) No significant risk of ectotherm predation because of large size of host (salmon kelts)	Reduced growth constraints +/or juvenile mortality Time constraints Reduced juvenile mortality
3–5. *Caenis simulans* *Isonychia bicolor*[c] *Tricorythodes atratus*	Five other mayflies	Late summer–autumn emergence (c.f. spring–early summer in other five)	Time constraints
6. *Eurylophella funeralis*	Five other mayflies	Southern species at northern edge of its range	Time constraints

[a]Differences in life history or niche consistent with a reduction in temperature-dependent growth constraints, juvenile mortality; or with increased time constraints are shown.
[b]These hypotheses are discussed in the text.
[c]Study may have included specimens of *Isonychia circe* (B. W. Sweeney, personal communication)
Source: After Atkinson (1995).

would be expected to prevent oxygen shortage for the sedentary copepod, thereby removing or reducing constraints on late growth. Atlantic salmon, which have a high oxygen requirement, also appear in nature to avoid conditions with low levels of dissolved oxygen (Priede *et al.*, 1988). This mechanism would further reduce the chances of growth constraints (or of increased mortality of large individuals due to oxygen shortage) at increased temperature in nature. Increased temperature could therefore be evolutionarily associated with rapid growth without oxygen limitation. Further studies on gill parasites, especially when compared with non-gill parasites on the same fish, would help test whether exceptional responses to developmental temperature are most generally associated with life on gills.

2. The evolution of the developmental response by *S. salmoneus* to increased predation at increased temperature is particularly unlikely, as its host (salmon kelt) is so large that ectotherm predation – which generally increases at increased temperature – is likely to be insignificant. Salmon kelts are typically between 2 and 6 kg in weight and in the experiment were 68–72 cm long, measured to the fork (C.E. Johnston, unpublished results). These were by far the largest animals of the 91 in Atkinson's (1994) review, and therefore one of the least likely to experience increased predation at increased temperatures.
3. The size response of *S. salmoneus* to temperature disappeared when photoperiod was altered. Photoperiod provides an indication of time of year, and as the timing of life history of *S. salmoneus* shows a strong association with the time of year, it might be hypothesised that the increased size results from seasonal time constraints. A reason for doubting this hypothesis, however, is that a lack of size response was associated not with photoperiods corresponding to a period early in the season as would be expected by the 'time constraints' hypothesis, above, but with a photoperiodic regimen which was artificially advanced to reach summer values about 3 months early (Johnston & Dykeman, 1987).

Time constraints may also help explain the exceptional responses observed in the four mayfly species. Adults of three of the four mayfly species that were exceptions to the temperature-size rule emerged from the water during late summer and autumn, whereas five species that followed the rule emerged during late spring and early summer (Sweeney & Vannote, 1978; Vannote & Sweeney, 1980; Atkinson, 1995). This observation is consistent with the hypothesis that time constraints resulting from deteriorating conditions in the autumn would shorten the period available for growth, resulting in smaller size especially at the lower temperatures. The other mayfly species which increased in size with increasing temperature, *Eurylophella funeralis*, was studied at the northern edge of its range where insufficient accumulated temperature may have limited the time available for development. Tests of the hypothesis that perceived time constraints can cause exceptional responses to temperature, alongside tests of alternative explanations proposed by Sweeney & Vannote (1978) and Vannote & Sweeney (1980) have yet to be performed.

Testing the ideas

Tests can be of either the assumptions or the predictions of a hypothesis, and can be correlative or experimental. Aspects of experimental tests are outlined here.

Manipulation of resources

Resource-limited growth can be tested by manipulating amounts of each resource in a factorial experiment, thereby identifying levels which do not limit growth rates, and then recording the effects on size at given developmental stages. Experiments would thus be similar to those of Vidal (1980) but a wider range of resources would be manipulated.

Selection experiments

Selection experiments may be used to test hypotheses predicting that a positive association between developmental temperature and juvenile mortality would select for an adaptive acceleration of development, even at the expense of reduced adult size, at increased rearing temperatures. All experimental treatments would include the same range of varying temperatures but would differ in their correlation between temperature and juvenile mortality. Specific causes of mortality (e.g. ectotherm predation) could be incorporated into the design. Selection experiments could, in principle, also be used to test the idea that a size reduction at reduced temperatures, which was observed in *S. salmoneus* and four species of mayfly, is caused by time constraints on the juvenile period. Again, all experimental treatments would include the same range of varying temperatures but they would differ in the amount of time allowed (by the experimenter) for completion of the juvenile period. In each experiment, after selection over a sufficient number of generations, evolutionary change in reaction norms to temperature should ideally be examined after all populations have been under identical conditions for two generations to remove parental and grandparental effects.

Manipulation of cues

If mortality risks vary within a lifetime, natural selection should favour phenotypically plastic responses to information about these risks. Thus, information indicating seasonal time constraints on the juvenile period (e.g. photoperiod) would be expected to alter the life history, as was found in *S. salmoneus* (Johnston & Dykeman, 1987). The method of manipulating environment cues to test predictions of life-history theory could be applied to mayfly species where they are normally bivoltine or multivoltine and hence likely to be adapted to different periods of adult emergence: experiments would include factorial combinations of several temperatures with photoperiodic regimens corresponding to different times of year.

When using this approach care needs to be taken to separate effects of cues from those of constraints on growth (Atkinson, 1985). For instance, an increased photoperiod may not only provide information about time of year, it may also allow a greater time for feeding each day and thus remove some constraints on growth rate.

Conclusions

Of those effects of rearing temperature that are widespread among ectothermic species, that on body size at a given stage of development is particularly difficult to explain. The present chapter has outlined progress in identifying mechanisms that may explain this enigma, and ways in which these can be tested. Many factors are likely to affect the relationship between temperature and body size, but a combination of theory and analysis of published data identifies three major mechanisms that are particularly widespread and may be capable of explaining much of the observed variation. These are:

1. Constraints on growth resulting from a shortage of resources that is exacerbated by increased size and temperature.
2. A positive association between temperature and juvenile mortality which should select for accelerated juvenile devel-

opment at increased temperatures, even at the expense of size at the time of reproductive maturity.

3. Time constraints that reduce the amount of time for juvenile growth: this will lead to reduced size especially at low temperatures at which growth is slow.

Nonetheless, until these ideas are tested, I concur with Partridge *et al.* (1994: 1275) when they say 'Thermal effects on body size in ectotherms remain both a mystery and a major challenge for future work'.

Acknowledgements

I thank C.E. Johnston and B.W. Sweeney for providing comments and additional information about their studies of *S. salmoneus* and mayflies, respectively. D.L. Trudgill kindly allowed me to see his manuscript prior to publication. Thanks to V. French, R.N. Hughes, I.A. Johnston and L. Partridge for their comments on the manuscript.

References

Atkinson, D. (1985). Information, non-genetic constraints, and the testing of theories of life-history variation. In *Behavioural Ecology*, (ed. R.M. Sibly & R.H. Smith), pp. 99–104. Oxford: Blackwell Scientific Publications.

Atkinson, D. (1994). Temperature and organism size: a biological law for ectotherms. *Advances in Ecological Research* **25**, 1–58.

Atkinson, D. (1995). Effects of temperature on the size of aquatic ectotherms: exceptions to the general rule. *Journal of Thermal Biology* **20**, 61–74.

Atkinson, D. (1996). Matching crops to their environment: Developmental sensitivity versus insensitivity to temperature. In *New Vistas in Plant Genetics* (ed. K.A. Siddiqui). Tando Jam, Pakistan: A.E. Agricultural Research Centre (in press).

Begon, M., Harper, J.L. & Townsend, C.R. (1990). *Ecology: Individuals, Populations and Communities, 2nd edn*, p. 473. Oxford: Blackwell Scientific Publications.

Bělehrádek, J. (1935). *Temperature and Living Matter. Protoplasma Monograph 8*. Berlin: Borntraeger.

Bellinger, R.G. & Pienkowski. R.L. (1987). Developmental polymorphism in the red-legged grasshopper *Melanoplus femurrubrum* (De Geer) (Orthoptera: Acrididae). *Environmental Entomology* **16**, 120–5.

Berrigan, D. & Charnov, E.L. (1994). Reaction norms for age and size in maturity in response to temperature: a puzzle for life historians. *Oikos* **70**, 474–8.

Bertalanffy, L. von. (1960). Principles and theory of growth. In *Fundamental Aspects of Normal and Malignant Growth* (ed. W.N. Nowinski), pp. 137–259. Amsterdam: Elsevier.

Bursell, E. (1974). Environmental aspects in temperature. In *Physiology of the Insecta, 2nd edn* (ed. M. Rockstein), pp. 2–43. New York: Academic Press.

Caswell, H. (1989). *Matrix Population Models*, p. 31. Sunderland, Massachusetts: Sinauer.

Cossins, A.R. & Bowler, K. (1987). *Temperature Biology of Animals*, Chapter 7. London: Chapman and Hall.

Cuenco, M.L., Stickney, R.R. & Grant, W.E. (1985). Fish bioenergetics and growth in aquaculture ponds. II. Effects of interactions among size, temperature, dissolved oxygen, unionized ammonia and food on growth of individual fish. *Ecological Modelling* **27**, 191–216.

Culver, D. (1980). Seasonal variation in sizes at birth and at first reproduction in Cladocera. In *Evolution and Ecology of Zooplankton Communities* (ed. W.C. Kerfoot), pp. 358–66. Hanover: University of New England.

Davis, J.C. (1975). Minimal dissolved oxygen requirements of aquatic life with emphasis on Canadian species: a review. *Journal of the Fisheries Research Board of Canada* **32**, 2295–332.

Hanazato, T. & Yasuno, M. (1989). Effect of temperature in laboratory studies on growth of *Chaoborus flavicans* (Diptera, Chaoboridae). *Archiv fur Hydrobiologie* **114**, 497–504.

Johnston, C.E. & Dykeman, D. (1987). Observations on body proportions and egg production in the female parasitic copepod (*Salmincola salmoneus*) from the gills of Atlantic salmon (*Salmo salar*) kelts exposed to different temperatures and photoperiods. *Canadian Journal of Zoology* **65**, 415–19.

Kooijman, S.A.L.M. (1993). *Dynamic Energy Budgets in Biological Systems*, p. 48. Cambridge: Cambridge University Press.

Lamb, R.J. & Gerber, G.H. (1985). Effects of temperature on the development, growth, and survival of larvae and pupae of a north-temperate chrysomelid beetle. *Oecologia* **67**, 8–18.

Laudien, H. (1973). Changing reaction systems. In *Temperature and Life* (ed. H. Precht, J. Christophersen, J. Hensel & W. Larcher), pp. 355–99. Berlin: Springer-Verlag.

Lewontin, R.C. (1965). Selection for colonizing ability. In *The Genetics of Colonizing Species* (ed. H.G. Baker & G.L. Stebbings), pp. 77–91. New York: Academic Press.

Liu, S.-S., Zhang, G.-M. & Zhu, J. (1995). Influence of temperature variations on rate of development in insects: analysis of case studies from entomological literature. *Annals of the Entomological Society of America* **88**, 107–19.

Loosanoff, V.L. (1959). The size and shape of metamorphosing larvae of *Venus (Mercenaria) mercenaria* grown at different temperatures. *Biological Bulletin* **117**, 308–18.

Meats, A. (1971). The relative importance to population increase of mortality, fecundity and the time variables of the reproductive schedule. *Oecologia* **6**, 223–7.

Müller, H. & Geller, W. (1993). Maximum growth rates of aquatic cilated protozoa: the dependence on body size and temperature reconsidered. *Archiv fur Hydrobiologie* **126**, 315–27.

Partridge, L., Barrie, B., Fowler, K. & French, V. (1994). Evolution and development of body size and cell size in *Drosophila melanogaster* in response to temperature. *Evolution* **48**, 1269–76.

Perrin, N. (1988). Why are offspring born larger when it is colder? Phenotypic plasticity for offspring size in the cladoceran *Simocephalus vetulus* (Müller). *Functional Ecology* **2**, 283–8.

Precht, H., Christophersen, J., Hensel, H. & Larcher, W. (eds) (1973). *Temperature and Life*. Berlin: Springer-Verlag.

Priede, I.G., De L.G. Solbe, J.F., Nott, J.E., O'Grady, K.T. & Cragg-Hine, D. (1988). Behaviour of adult Atlantic salmon, *Salmo salar* L. in the estuary of the River Ribble in relation to variations in dissolved oxygen and tidal flow. *Journal of Fish Biology* **33** (suppl. A), 133–9.

Ratte, H.T. (1985). Temperature and insect development. In *Environmental Physiology and Biochemistry of Insects* (ed. K.H. Hoffman), pp. 34–66. Berlin: Springer-Verlag.

Ray, C. (1960). The application of Bergmann's and Allen's rules to the poikilotherms. *Journal of Morphology* **106**, 85–108.

Roberts, E.H. & Summerfield, R.J. (1987). Measurement and prediction of flowering in annual crops. In *Manipulation of Flowering* (ed. J.G. Atherton), pp. 17–50. London: Butterworths.

Roff, D.A. (1992). *The Evolution of Life Histories: Theory and Analysis*. New York: Chapman and Hall.

Rombough, P.J. (1988). Respiratory gas exchange, aerobic metabolism, and effects of hypoxia during early life. In *Fish Physiology, vol. XI, Part A* (ed. W.S. Hoar & D.J. Randall), pp. 59–161. San Diego: Academic Press.

Shrode, J.B. & Gerking, S.D. (1977). Effects of constant and fluctuating temperatures on reproductive performance of a desert pupfish *Cyprinodon n. nevadensis*. *Physiological Zoology* **50**, 1–10.

Sibly, R.M. & Atkinson, D. (1994). How rearing temperature affects optimal adult size in ectotherms. *Functional Ecology* **8**, 486–93.

Slafer, G.A. & Savin, R. (1991). Developmental base temperature in different phenological phases of wheat (*Triticum aestivum*). *Journal of Experimental Botany* **42**, 1077–82.

Stearns, S.C. (1992). *The Evolution of Life Histories*. Oxford: Oxford University Press.

Stearns, S.C. & Koella, J.C. (1986). The evolution of phenotypic plasticity in life history traits: predictions of reaction norms for age and size at maturity. *Evolution* **40**, 893–913.

Stevenson, R.D. (1985). Body size and limits to the daily range of body temperature in terrestrial ectotherms. *American Naturalist* **125**, 102–17.

Stibor, H. & Lüning, J. (1994). Predator-induced phenotypic variation in the pattern of growth and reproduction in *Daphnia hyalina* (Crustacea: Cladocera). *Functional Ecology* **8**, 97–101.

Summerfield, R.J., Roberts, E.H., Ellis, R.H. & Lawn, R.J. (1991). Towards the reliable prediction of time flowering in six annual crops. I. The development of simple models for fluctuating field environments. *Experimental Agriculture* **27**, 11–31.

Sweeney, B.W. & Vannote, R.L. (1978). Size variation and the distribution of hemimetabolous aquatic insects: two thermal equilibrium hypotheses. *Science* **200**, 444–6.

Sweeney, B.W., Vannote, R.L. & Dodds, P.J. (1986). Effects of temperature and food quality on growth and development of a mayfly, *Leptophlebia intermedia*. *Canadian Journal of Fisheries and Aquatic Science* **43**, 12–18.

Tauber, M.J., Tauber, C.A. & Masaki, S. (1986). *Seasonal Adaptations of Insects*. Oxford: Oxford University Press.

Trudgill, D.L. (1996). A thermal time basis for understanding pest epidemiology and ecology. In *Strategies for Managing Soil-borne Plant Pathogens* (ed. R. Hall). American Phytopathological Society (in press).

Vannote, R.L. & Sweeney, B.W. (1980). Geographic analysis of thermal equilibria: a conceptual model for evaluating the effect of natural and modified thermal regimes on aquatic insect communities. *American Naturalist* **115**, 667–95.

Vidal, J. (1980). Physioecology of zooplankton. I. Effects of phytoplankton concentration, temperature, and body size on the growth rate of *Calanus pacificus* and *Pseudocalanus* spp. *Marine Biology* **56**, 111–34.

Wagner, T.L., Fargo, W.S., Flamm, R.O., Coulson, R.N. & Pulley, P.F. (1987). Development and mortality of *Ips calligraphus* (Coleoptera: Scolytidae) at constant temperatures. *Environmental Entomology* **16**, 484–96.

Walsby, A.E. & Reynolds, C.S. (1980). Sinking and floating. In *The Physiological Ecology of Phytoplankton* (ed. I. Morris), pp. 371–412. Oxford: Blackwell.

Wayne, N.L. & Block, G.D. (1992). Effects of photoperiod and temperature on egg-laying behaviour in a marine mollusk, *Aplysia californica*. *Biological Bulletin* **182**, 8–14.

R.B. HUEY and D. BERRIGAN

Testing evolutionary hypotheses of acclimation

Introduction

Physiologists have long understood that an organism's phenotype is not fixed but rather is partially dependent on the environment experienced during ontogeny. This phenomenon of acclimation, which represents a special type of phenotypic plasticity, offers several research opportunities. Acclimation experiments help elucidate the mechanistic process underlying rapid physiological regulation in response to dynamic environmental fluctuations (Hochachka & Somero, 1984; Prosser, 1986; Cossins & Bowler, 1987). Similarly, acclimation experiments can be used as tools to explore evolutionary adaptation to fluctuating environments (Levins, 1968). Nevertheless, acclimation offers more than a carrot, for it offers a stick as well. The fact of acclimation forces physiologists to control acclimation state as a necessary, if often inconvenient, first step in both mechanistic and evolutionary experiments.

The literature on physiological acclimation is now immense. Inspection, however, suggests that physiologists have been much more rigorous and successful in elucidating the mechanisms underlying acclimation responses than they have in exploring the evolution of these responses. Why have evolutionary explanations in studies of physiological acclimation lacked rigor? (The same question can probably be asked of any field of functional biology (Feder, 1987*a*; Garland & Carter, 1994; Travis, 1994; Bennett, 1996).) In part, this relative lack of rigor may reflect the fact that functional biologists are often more interested in mechanistic issues than in evolutionary ones; but it may also reflect the fact that a need for greater rigor in evolutionary explanations in biology in general has really been evident only in the past 15 years or so (Gould & Lewontin, 1979; Feder, 1987*a*).

The basic evolutionary problem is illustrated with a traditional (if hypothetical) experiment in acclimation. The usual approach is to observe a response, propose a *post hoc* but plausible and adaptive scenario, but not test it. Indeed, a substantial fraction of the literature of acclimation concludes with a variant on the following statement: 'The ability of species X to acclimate to high temperature is adaptive because increased heat tolerance will clearly promote survival in hot environments'. We suspect that rigorous tests of such presumed 'adaptive scenarios' are often deemed unnecessary because of the widespread acceptance of an implicit and appealing assumption: namely, organisms that have the capacity to fine-tune their physiology to the immediate environment should have higher overall performance and thus leave more surviving offspring than would organisms incapable of such fine-tuning. This assumption is the basis for what has recently been called *The Beneficial Acclimation Hypothesis* (Leroi *et al.*, 1994; Zamudio *et al.*, 1995).

Although adaptive scenarios are appealing, their appeal is based only on plausibility, not on rigorous hypothesis testing. Nor surprisingly, the failure to test evolutionary scenarios has been repeatedly and justifiably criticised ever since Gould and Lewontin's provocative attack on the 'adaptationist programme' and the 'Panglossian paradigm'[1] in biology (Gould & Lewontin, 1979). In the last decade several evolutionary physiologists (Feder, 1987*a*; Powers, 1987; Bennett & Huey, 1990; Burggren, 1991; Garland & Adolph, 1991; Garland & Carter, 1994; Bennett, 1996) have constructively interpreted and extended criticism to studies in comparative physiology, and we attempt to do so here with respect to the special case of physiological acclimation.

The general topic of this paper focuses on testing evolutionary hypotheses of acclimation. We begin with a brief discussion of terminology. Next we will review several reasons to doubt the robustness of untested hypotheses of beneficial acclimation. Finally, we will describe several experimental approaches that can be used to test evolutionary hypotheses as well as suggest several new directions for research in this field.

Terminology

Physiologists attempting to study acclimation and related phenomena confront a rather confused terminology (see Rome *et al.*, 1992: 196–7). Moreover, physiologists wishing to interpret their data within the general evolutionary framework of 'phenotypic plasticity' must also confront a suspiciously similar set of terms, which is arguably even more confusing (see Scheiner, 1993). Should one discuss 'acclimation'

[1] Gould & Lewontin (1979) are rightly credited for inspiring many to become more critical of adaptationist explanations; however, J. B. S. Haldane (1958) had earlier noted the problems with such explanations and even referred to naive adaptationism as 'Panglossism' (Haldane, 1958: 21).

and 'acclimatisation', or instead 'phenotypic plasticity', 'reaction norms' and 'polyphenisms'? (This is not the only example of the terminological incompatibility between physiologists and evolutionary biologists – witness how these two groups differentially use the word 'adaptation' (Garland & Carter, 1994: 583).)

Not surprisingly, the independent development of terminology has reinforced the tendency for the literatures of physiology and of evolutionary biology to remain largely non-overlapping. This mutual ignorance is unfortunate, as the conceptual issues underlying acclimation and phenotypic plasticity have much in common. Consequently any attempt to further the dialogue between comparative physiologists and evolutionary biologists must necessarily try to define terms within a common language. Fortunately, Scheiner (1993) has concisely and lucidly simplified a morass of terms used by evolutionary biologists; and here we attempt to incorporate physiological terms into his general framework.

Phenotypic plasticity is a generic term referring to the malleability of an organism's phenotype (physiology, morphology, behaviour) in response to environmental conditions experienced by (or anticipated by) that organism. In other words, phenotypic plasticity describes 'the general effect of the environment on phenotypic expression' (Scheiner, 1993). *Norm of reaction* is a closely related term, which refers to 'the specific form of that effect' (Scheiner, 1993), but is often used interchangeably in the evolutionary literature with phenotypic plasticity.

Several types of phenotypic plasticity can be recognised for the special case of physiological traits.

1. *Acclimation*[2] *responses* are physiological examples of phenotypic plasticity. The time scale of the phenotypic modifications may be minutes to months, and the modifications are more or less reversible within the organism's lifetime. Acclimatory shifts in heat tolerance or metabolic rate in response to photoperiod or temperature are familiar examples. The phenotypic modifications (Levins, 1968) may be either *reactive responses* to experienced environments (for example, to high body temperatures) or *anticipatory responses* to future environments (e.g. to summer conditions, as cued by increasing photoperiod).
2. *Developmental switches* ('polyphenisms') are a special case of plastic responses, in which the phenotype is fixed irrever-

[2] For this discussion we follow Withers (1992) in defining *acclimation* as the process of physiological response to altered environmental conditions in the laboratory and in defining *acclimatisation* as the process of response to alternated environments in nature.

sibly by environmental conditions experienced during a critical phase of development (Levins, 1968; Travis, 1994). Familiar examples include the permanent effects of developmental temperature on sex discrimination in some reptiles (Janzen & Paukstis, 1991), on adult body size in many ectotherms (Atkinson, 1994), bristle number in flies (Gupta & Lewontin, 1982) and on adult colour in many insects (Watt, 1990). Developmental switches may involve either continuous (size) or discontinuous (bristle number) traits (Scheiner, 1993).

3. *Developmental pathologies* are environmentally induced, pathological modifications that occur during development. Familar examples are fluctuating asymmetries (Parsons, 1990; Markow, 1995), supernumerary digits, phenocopies, etc. Such pathologies will not be discussed here.
4. *Labile (transient, acute) effects* are very rapid modifications of phenotypic capacity or performance as a function of an organism's immediate (rather than cumulative) physical or physiological environment (Scheiner, 1993). These effects are often depicted quantitatively by 'performance curves' (e.g. performance or fitness as a function of temperature, Huey & Stevenson, 1979). Performance curves are potentially modifiable by acclimation or by developmental switches. (The evolution of performance curves has been reviewed recently elsewhere (Huey & Kingsolver, 1993; Lynch & Lande, 1993; Gilchrist, 1995) and will not be treated here.)
5. *Cross-generational effects* are maternally or paternally transmitted phenotypic modifications. Cross-generational effects of temperature have recently been described (see below) and are potentially an important, if much understudied, influence on phenotypic plasticity.

Proposing an adaptive hypothesis is not a test of that hypothesis

As mentioned earlier, most of the existing evolutionary interpretations of acclimation patterns are *post hoc* variants of the *Beneficial Acclimation Hypothesis*. The act of proposing a plausible adaptive hypothesis, however, should not be viewed as a test of that hypothesis; rather it can be viewed as merely a test of a physiologist's creativity. Feder (1987*a*) eloquently enumerated many reasons for being critical of *post* *hoc* evolutionary interpretations, and we reiterate a few of his key points. First, even a casual perusal of the literature demonstrates conclusively that at least one adaptive story can readily be concocted for almost any conceivable acclimation response. Second, a given adaptive story, even if plausible, may be wrong. Third, the tendency to suggest only a single evolutionary explanation means that competing hypotheses are ignored. Fourth, even if an adaptive story were in fact true, a full evolutionary exploration requires that any such hypothesis be tested, not just advocated. We will elaborate on the first two points in the remainder of this section and on the second two issues in subsequent sections.

Acclimation of metabolic rate to temperature is a well-known (and appropriately maligned) example of the first problem. Consider an experiment in which an organism has been acclimated to 30 °C and then is shifted acutely to 20 °C (Fig. 1*a*): its metabolic rate will immediately drop (Fig. 1*b*). Continued exposure to 20 °C may lead to one of several different acclimation responses (Precht, 1958; Prosser, 1958; Fig. 1*b*), and each response has been interpreted as adaptive (see Withers, 1992). In complete compensation, the original elevated metabolic rate is restored following continued acclimation to low temperature: complete compensation is thought adaptive because it promotes

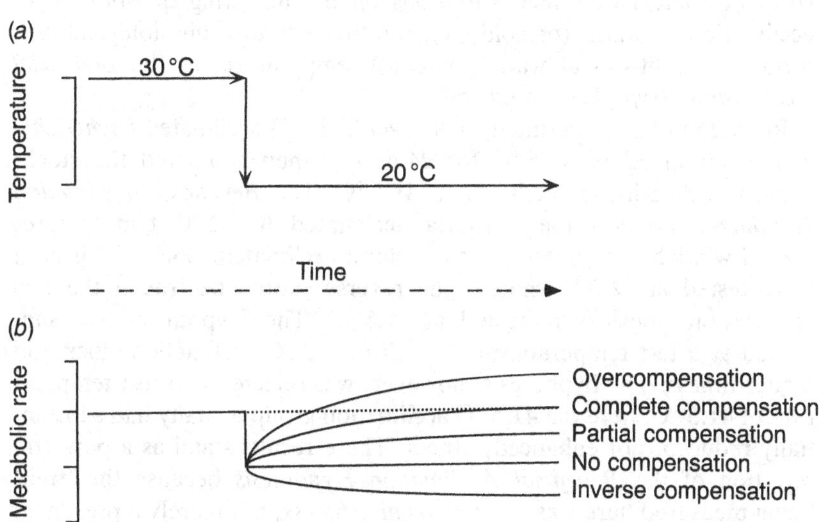

Fig. 1. Possible acclimation responses of metabolic rate (*b*) for an organism that experiences a shift in temperature (*a*). (After Rome *et al.*, 1992 and other sources.)

metabolic homeostasis despite a changing environment. In partial compensation metabolic rate only approaches the original level. Partial compensation may represent a physiological constraint on complete compensation or perhaps an adaptive compromise. In no compensation, no temporal shift in metabolic rate occurs: this lack of response may be adaptive for animals living in constant environments or in a case where the costs outweigh the benefits (Janzen, 1967; Regal, 1977). In inverse compensation, the lowered metabolic rate drops even further following continued acclimation: an inverse response clearly does not promote homeostasis, but it is nevertheless thought adaptive for conserving energy during cold exposure. In over-compensation, the acclimated metabolic rate is elevated above the original level: this might be adaptive in reducing reaction rates at high temperature (Withers, 1992: 179). The observation that any acclimation pattern (Fig. 1*b*) can be 'so explained' should serve as a warning flag that in fact nothing has been explained (Feder, 1987*a*). Adding to that concern is a central lesson emerging from the history of biology: even plausible and reasonable hypotheses are often wrong.

The possibility that plausible adaptive explanations are wrong is not just an academic concern. In fact, three recent studies have tested the *Beneficial Acclimation Hypothesis*, and all reject it. These studies (reviewed later) examined variations on the following question: does acclimation to warm (or cold) temperature enhance physiological performance or fitness at warm (or cold) temperature, as the *Beneficial Acclimation Hypothesis* predicts?

In an important experiment, Leroi *et al.* (1994) acclimated *Escherichia coli* to either 32 or 41.5 °C for 24 h, and then competed the stocks against each other at 32 °C or at 41.5 °C. The *Beneficial Acclimation Hypothesis* predicts that bacteria acclimated to 32 °C temperatures should win when competing with bacteria acclimated to 41.5 °C if both were tested at 32 °C, whereas the reverse should be true if the two acclimation stocks were tested at 41.5 °C. The Hypothesis was supported at a test temperature of 32 °C: the 32 °C acclimation stock had higher fitness. The hypothesis, however, was rejected at a test temperature of 41.5 °C: here the 41.5 °C acclimation group actually had substantially reduced, not enhanced, fitness. These results stand as a powerful rejection of the *Beneficial Acclimation Hypothesis* because the 'trait' being measured here was actually *relative fitness*, not merely a presumed correlate of fitness.

A recent study with *Drosophila melanogaster* yielded similar patterns. Zamudio *et al.* (1995) raised males (egg to adult) at either 18 or 25 °C and then tested territorial success of males in paired contests at 18 or

27 °C. (Note: territorial success is probably a close correlate of a male's fitness, see Zamudio *et al.* (1995).) The *Beneficial Acclimation Hypothesis* predicted that males raised at 18 °C would be dominant in contests at that same temperature, but subordinate in contests at 27 °C. In fact, males raised at 25 °C were dominant at both test temperatures (Fig. 2), again contradicting predictions of the *Beneficial Acclimation Hypothesis*.

Zwaan *et al.* (1992) developed an elegant, full-factorial experiment to test evolutionary hypotheses on aging, but their data are readily re-interpretable in the context of the *Beneficial Acclimation Hypothesis*. These workers raised *D. melanogaster* (egg to adult) at one of three temperatures, and then measured adult longevity (a component of fitness) at the three temperatures in a full 3 × 3 factorial design (Fig. 3). The *Beneficial Acclimation Hypothesis* predicts that flies raised at cool temperatures should live longer at cool temperatures than would flies raised under intermediate or warm temperatures. Similarly, flies raised at intermediate temperatures should live longest at intermediate temperatures, whereas flies raised at warm temperatures should live

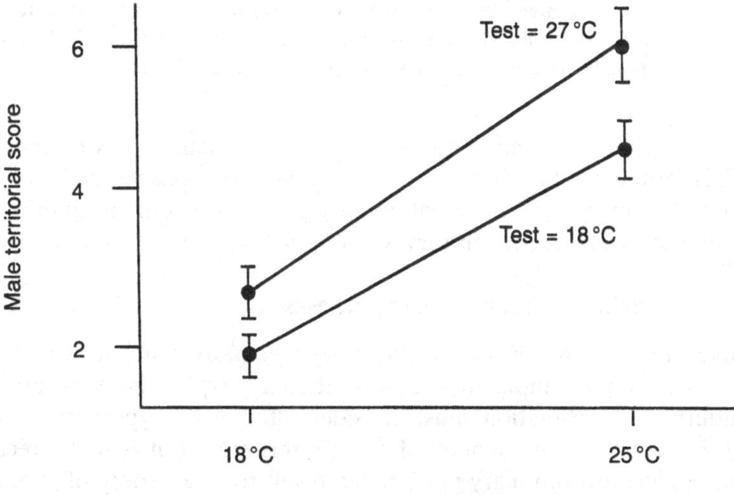

Fig. 2. Territorial scores of male *Drosophila melanogaster* in paired contests. Males raised at 25 °C had higher territorial scores than did males raised at 18 °C, and they did so at test temperatures of both 18 and 27 °C. This result contradicts the *Beneficial Acclimation Hypothesis*, which predicts that males raised at 18 °C should win in contests at 18 °C. (After Zamudio *et al.*, 1995.)

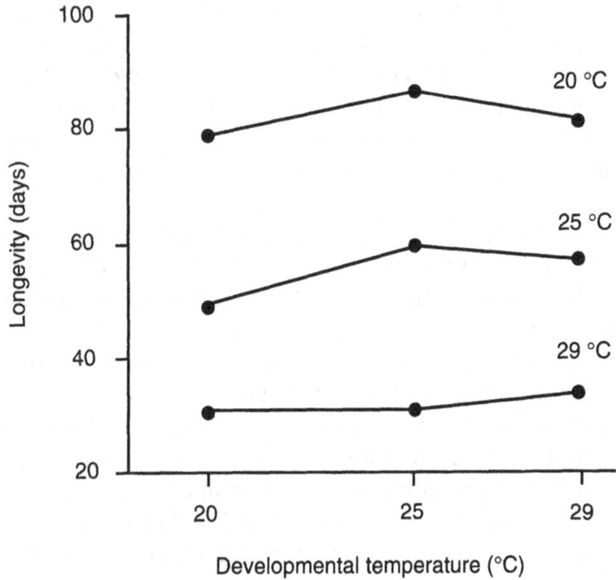

Fig. 3. Longevity as a function of developmental and adult temperature. Patterns contradict predictions of *the Beneficial Acclimation Hypothesis* (see text). (After Zwaan *et al.*, 1992.)

longest at warm temperatures. In fact, our reanalysis of these data (R.B. Huey, G.W. Gilchrist & D. Berrigan, unpublished results) shows adult longevity was maximal following development at intermediate temperatures (Fig. 3), contrary to the *Beneficial Acclimation Hypothesis.*

Seeking alternative hypotheses

Given that the *Beneficial Acclimation Hypothesis* is not invariably supported, then attempts to seek evolutionary explanations of observed patterns of acclimation must consider alternative hypotheses (Feder, 1987*a*; Travis, 1994; Bennett, 1996). In this regard, it is worth recalling that a given evolutionary pattern can result from a variety of evolutionary mechanisms, not just from natural selection. For example, the pattern might reflect random drift (Turelli *et al.*, 1987), adaptation to past rather than contemporary environments (Janzen & Martin, 1982; Dumont & Robertson, 1986; Huey, 1987), or gene swamping from populations adapted to different environments (Stearns & Sage, 1980). Some patterns might not reflect evolution at all and may merely reflect 'developmentally inevitable' (i.e. biophysical) consequences of the effects of environmental factors on physiological systems (Stearns, 1982;

Sultan, 1987). Consequently attempts to explain the evolution of acclimation responses must consider non-adaptive explanations and mechanisms, not just adaptive mechanisms of evolution (Feder, 1987*a*; Garland & Adolph, 1991; Garland & Carter, 1994; Scheiner, 1993; Travis, 1994; Bennett, 1996). (We recognise all too well that doing so is easier said than done.)

Even if a given acclimation pattern does reflect natural selection, the response need not necessarily be adaptive or physiologically 'beneficial'. It could reflect a developmental pathology. Consider again the case of acclimation of metabolic rate to temperature (Fig. 1*b*). A lowering of metabolic rate in response to cold acclimation (inverse compensation, Fig. 1*b*) is, as noted above, typically interpreted as an adaptive mechanism for conserving energy in winter; however, chronic cold acclimation injures and even kills some animals. For example, the lizard *Anolis cristatellus* from Puerto Rico dies within a week of acclimation to 15 °C (Heatwole *et al.*, 1969). Thus a possible reduction in metabolic rate during cold acclimation of such cold-sensitive species is arguably better interpreted as reflecting experimenter-induced injury, rather than adaptive energy conservation. Similarly, partial compensation (Fig. 1*b*) in response to cold acclimation could conceivably reflect increased metabolic costs associated with repair of cold injury, rather than a re-establishment of metabolic homeostasis.

These considerations suggest that a major challenge to evolutionary interpretations of acclimation responses will involve an explicit consideration of non-adaptive factors (see also Feder, 1987*a*; Travis, 1994; Bennett, 1996). This will not always be easy, but it will be rewarding. In a later section, we will return to the issue of multiple hypotheses, when we evaluate strong inference tests of acclimation hypotheses.

Specific ways to test evolutionary interpretations

A variety of approaches have been developed recently to test evolutionary hypotheses. This section highlights a few approaches that seem especially amenable to studies of acclimation and related phenomena. The methods are not equally robust and, of course, no method will be suitable for all organisms or physiological systems of interest. Travis (1994) reviews some other approaches in the context of evolutionary morphology.

Make complex, *a priori* predictions

Given that *post hoc* explanations are of limited interest, a more robust approach would involve making explicit, *a priori* predictions. For example, one could predict both the direction and magnitude of the acclimation response. This is not a great improvement, of course, as one has a 50 : 50 chance of guessing the correct direction of the response (see Garland & Adolph, 1991). Nevertheless, one's confidence can be substantially increased by deriving a *set* of 'complex' predictions. For example, one might predict a particular direction of response for certain species, but a different direction of response for other species. Natural history information (see below, Greene, 1986) may well provide useful information in developing such complex predictions.

Once again patterns of metabolic acclimation to temperature (Fig. 1) serve as a useful example. Feder (1977, 1982) effectively utilised this paradigm of *a priori*, complex predictions in his studies of metabolic acclimation in salamanders and frogs. Feder knew from field data that tropical species generally had less variable body temperatures than did temperate-zone species (Feder & Lynch, 1982). He then proposed *a priori* that the metabolic rate of tropical amphibians should show less pronounced acclimation responses to temperature than would that of temperate-zone species. Extensive comparative studies of acclimation of standard metabolic rates of many tropical versus temperate-zone species strongly supported his prediction (Feder, 1977, 1982), but it remains to be established how observed patterns of acclimation of metabolic rate in these salamanders affect fitness (Feder, 1987*b*).

Tsuji (1988*a, b*) similarly studied the direction and extent of metabolic acclimation in an iguanid lizard (*Sceloporus* spp.). One species (*S. occidentalis*) is distributed over a broad latitudinal gradient in western North America from southern Canada to northern Mexico, and a close relative (*S. variabilis*) occurs only in the tropics. Natural history data revealed that tropical *Sceloporus* are active all year, but that temperate-zone species show latitudinal variation in seasonal activity patterns: lizards in southern California are active during winter, at least on relatively warm days, whereas lizards in Washington to the north hibernate continuously for 5 or 6 months. Drawing on this natural history information, Tsuji (1988*b*) predicted complex patterns of metabolic acclimation in response to a laboratory shift from summer to autumn conditions: (i) the tropical species would show a relatively reduced acclimation response; (ii) the southern California population would show compensatory acclimation, thus promoting activity in midwinter; and (iii) the northern population would show inverse compensation, thus conserving energy in winter. Tsuji (1988*b*) then developed an elegant experimental protocol, using appropriately cycling temperatures, photoperiod and temperature shifts, and a repeated measures design with appropriate controls. The results clearly supported her predictions (Fig. 4) for the summer to autumn period, though patterns

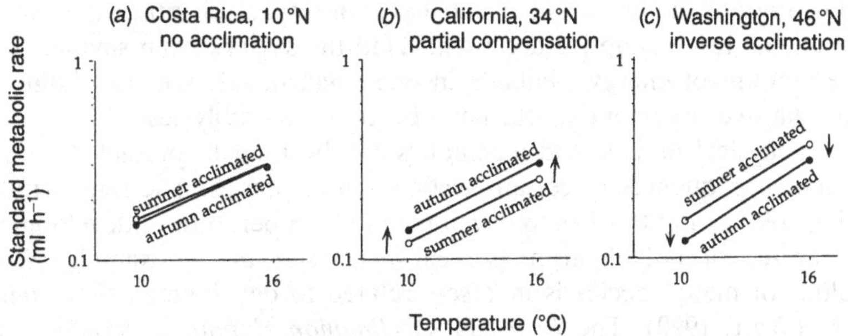

Fig. 4. Geographical variation in pattern of acclimation of metabolic rate as a function of temperature in lizards (*Sceloporus* spp.) from three latitudes. The tropical species shows no acclimation response (i.e. summer-acclimated (circles) versus autumn-acclimated (dots)), the south-temperate population shows partial compensation, and the north-temperate population shows inverse compensation. (After Tsuji, 1988*b*.)

were less clear for acclimation in the spring. Importantly, Tsuji backed up these laboratory experiments with studies of seasonal acclimatisation of the thermal dependence of metabolic rate of *S. occidentalis* taken directly from the field (Tsuji, 1988*a*): these additional studies demonstrate that the laboratory results (Tsuji, 1988*b*) were relevant to patterns actually observed in nature.

Feder's (1977, 1982, 1987*b*) and Tsuji's (1988*a*, *b*) studies serve as exemplars of how comparative and experimental studies, when developed with sensitivity to an organism's natural history, can be developed to test *a priori* predictions. Both Feder and Tsuji chose their species carefully, so that they were able to make a complex set of predictions.

Estimate the 'beneficial' impact of the acclimation response

Another way to evaluate a given acclimation response involves quantifying the presumed benefit of the response, but doing so explicitly in the currency of fitness (or a close correlate). For example, inverse compensation (Fig. 1*b*) will conserve energy during winter, but does it conserve enough energy to benefit the fitness of the animal? Tsuji (1988*a*) asked this question in her analysis of metabolic acclimation of *S. occidentalis* (above). By integrating metabolic and temperature data, she estimated that inverse compensation during winter dormancy would conserve the energy equivalent (in 'lizard currency') of about one egg.

This amount might be enough to have an impact on fitness, as would be crucial to an adaptive argument. Had the savings been smaller (e.g. the amount of energy available in one small meal), the plausibility of an adaptive argument would have been considerably less.

Biophysical models and techniques can be used to predict the functional consequences of certain acclimation responses. The sensitivity of wing colour of butterflies to developmental temperature (a developmental switch or polyphenism, see earlier) serves as an example. Wing colour of many species is inversely related to developmental temperature (Watt, 1990). The *Beneficial Acclimation Hypothesis* would argue that this developmental switch is adaptive in some butterflies (Watt, 1969; Kingsolver & Wiernasz, 1991) because dark wings may provide a thermoregulatory advantage in cool seasons (via increasing heat gain), whereas light wings would be advantageous in warm seasons (reducing heat gain). Other factors, however, (e.g. camouflage, sexual selection and species recognition) influence the evolution of wing colour (Wiernasz, 1989; Wiernasz & Kingsolver, 1991). Consequently, seasonal shifts in colour might potentially reflect seasonal change in selection from predators or for mates, not just in selection from thermoregulatory considerations.

In a series of papers on pierid butterflies, Watt, Kingsolver and colleagues determined the impact of seasonal colour change on thermoregulation (Watt, 1969; Kingsolver & Wiernasz, 1991). First, they used heat transfer models to predict equilibrium body temperatures of butterflies with different wing colours under different environmental scenarios. Second, they then used microclimate models to estimate the impact of colour change on potential activity times of butterflies in different seasons. (Note: Activity time is an important component of fitness in these butterflies (Kingsolver, 1983).) Their analyses supported a '*Beneficial Developmental Switch Hypothesis*'. Dark colours on certain parts of the wing should significantly increase potential activity time in butterflies in spring, whereas light wings should achieve the same in summer (Kingsolver & Wiernasz, 1991).

Watt's and Kingsolver's functional approach is a useful example of a way to determine whether a given plastic response is large enough to have a significant impact, at least potentially, on some relevant aspect of ecological performance. Rather than merely measuring the effect of colour on absorptivity and then speculating about the thermoregulatory significance of colour change, they attempted to estimate the significance of colour change in terms of an important fitness correlate (activity time). Of course such approaches are not direct tests of the fitness consequences. Nevertheless, such quantitative approaches enhance our understanding of complex responses as well as guide future experiments and analyses.

Strong inference in laboratory experiments

A powerful way to explore the evolutionary significance of phenotypic plasticity involves the use of strong inference (Platt, 1964, Feder, 1987*a*) to distinguish among multiple, working hypotheses. Strong inference involves deriving alternative (*a prior*) hypotheses and then designing (and executing) 'crucial' experiments that will exclude one or more of the competing hypotheses. The experiments can be conducted either in the laboratory (this section) or in the field (next section).

To develop the theme of strong inference tests of phenotypic plasticity, we return to the examples of the significance of developmental temperature to adult *D. melanogaster* (Zwaan *et al.*, 1992; Huey *et al.*, 1995; Zamudio *et al.*, 1995). Our goal is to derive a complex experimental design that distinguishes among competing hypotheses. One appropriate design is shown in Fig. 5, in which both developmental and adult temperatures are crossed in a full factorial design.

Figure 6 depicts various scenarios for adult performance (fitness) in three different adult temperature regimens (low, medium, high) as a function of three different developmental temperature regimens (low, medium, high). (Note: we assume that adult performance is curvilinearly related to test temperature, such that performance is maximal at a medium test temperature and thus reduced at high and low test temperature.) These scenarios can be formalised as five hypotheses,

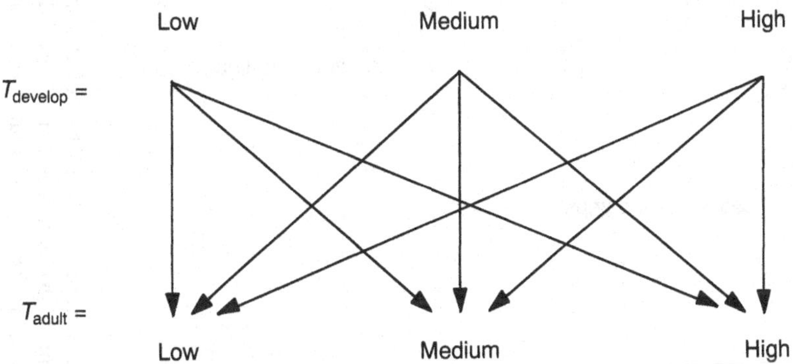

Fig. 5. Factorial experimental design that partitions the effects of developmental and adult temperature. This design enables one to discriminate among competing hypotheses shown in Fig. 6.

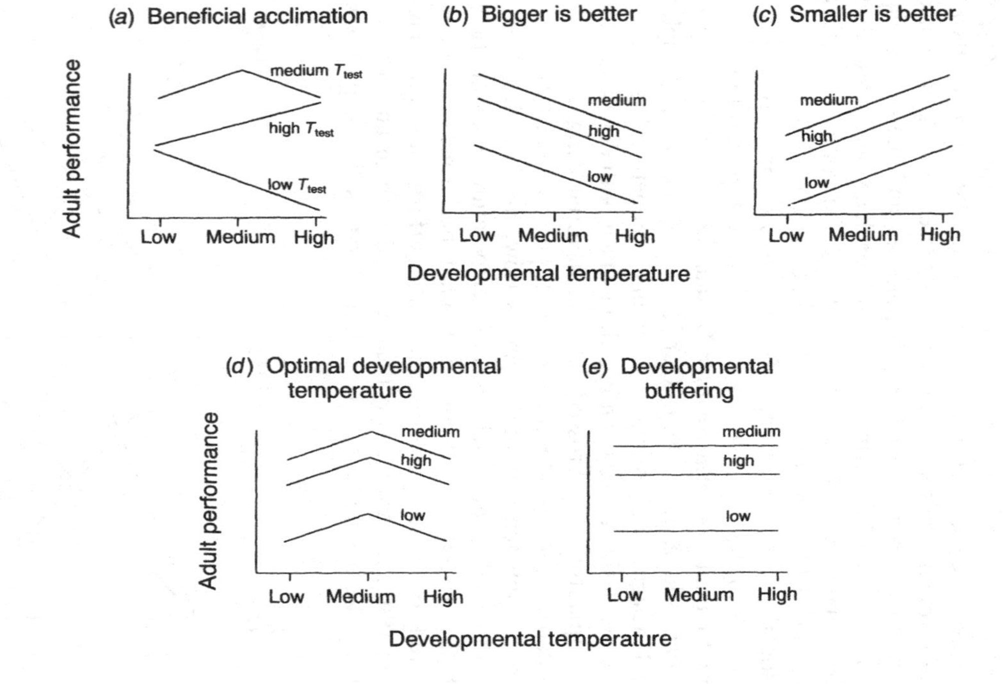

Fig. 6. Various scenarios for adult performance of a hypothetical ectotherm in three different temperature regimens (low, medium, high) as a function of three different developmental temperature regimens (low, medium, high). (Note: Solid lines connect performance of adults tested at a given test temperature (e.g. low). Performance is assumed maximal for adults at medium test temperature, and reduced for adults at high and especially at low test temperature.)

which can be readily distinguished because they make different predictions as to the interactive effects of developmental and adult temperature on adult performance (Fig. 6).

- *Beneficial Acclimation* (Leroi *et al.*, 1994; Zamudio *et al.*, 1995): acclimation is often considered an adaptive physiological mechanism of homeostatic regulation. The *Beneficial Acclimation Hypothesis* predicts that performance is maximal for flies tested as adults in the environment in which they were reared (Fig. 6*a*). Thus adults experiencing high environmental temperatures will have maximal performance if they had developed at high temperatures. (N.B. This particular formulation of the hypothesis assumes that organisms are short lived, such developmental temperatures serve as a direct cue of adult temperatures (see Levins, 1968).)
- *Bigger is Better Hypothesis*: Ectotherms developing at low temperature are phenotypically large (David *et al.*, 1983; Atkinson, 1994; and see pp. 182–204), and large body size often conveys fitness advantages (see pp. 265–92; Partridge *et al.*, 1987; Barbault, 1988; Markow & Ricker, 1992). This hypothesis predicts that organisms raised at low temperature (merely because they are large) will perform better in any adult thermal environment than will organisms raised at higher temperature (Fig. 6*b*).
- *Smaller is Better*: Although large size is usually advantageous, small size appears adaptive in some taxa (McLachlan & Allen, 1987; Aspi & Hoikkala, 1995). Consequently, this hypothesis, which is the counterpoint to *Bigger is Better*, predicts that organisms raised at high temperature, because they are small, will perform well as adults (relative to adults raised at lower temperature) in any thermal environment (Fig. 6*c*).
- *Optimal Developmental Temperature*: Development at intermediate temperature (22–25 °C in *D. melanogaster*) often produces vigorous and healthy individuals (Cohet & David, 1977; Zamudio *et al.*, 1995). Consequently, this hypothesis predicts that organisms developing at intermediate temperature will have the highest performance (relative to those developing at lower or at higher temperature) in any environment (Fig. 6*d*).
- *Developmental Buffering*: Selection may favour the buffering or 'canalisation' of developmental systems against perturbing influences such as temperature (Waddington, 1942; Stearns & Kawecki, 1994). If so, adult performance is independent of developmental temperature (at least for non-extreme temperatures) and thus affected only by adult temperature (Fig. 6*e*).

These five hypotheses fall into two classes. *Bigger is Better* and *Smaller is Better* generate specific predictions based on the general effect of developmental temperature on adult size (see pp. 182–204 and 265–92; Atkinson, 1994; Berrigan & Charnov, 1994) and on whether or not large size promotes adult fitness. In other words, the mechanism underlying these hypotheses is explicitly dominated by body size considerations. In contrast, the remaining hypotheses are non-specific as regards underlying mechanism. For example, *Optimal Developmental Temperature* is based purely on empirical observations (Cohet & David, 1977), not on any established developmental mechanism. Body size, however, could be the mechanism underlying this pattern (Fig. 6*a*) if stabilising selection favours intermediate-sized organisms, as is known in several species (Boake, 1989; Markow & Ricker, 1992).

We can exemplify the power of this approach by reinterpreting data in Zwaan *et al.* (1992), who measured the effects of developmental temperature (18, 25 or 29 °C) on life span of *D. melanogaster* in three adult temperature environments (18, 25 or 29 °C). (Note: Zwaan *et al.* (1992) developed their experiment to test theories of the evolution of aging, not of acclimation, but their data are quite compatible with our approach.) Adult temperature had a major effect on life span, but developmental temperature had minor but significant effects (Fig. 3). Our re-analysis of these data show that longevity is maximal for flies raised at intermediate temperature, thus strongly supporting *Optimal Developmental Temperature* and clearly contradicting *Beneficial Acclimation* (R. Huey, G. Gilchrist & D. Berrigan, unpublished results).

As discussed above, Zamudio *et al.* (1995) used a 2 × 2 factorial design to test competing hypotheses concerning the consequences to male territorial success of development at 18 or 25 °C. Territorial success should be a close correlate of male fitness, because a male that dominates an oviposition site has higher mating success. The 25 °C males, despite being relatively small, were dominant over 18 °C males at both 18 and 25 °C (Fig. 2). These results were consistent with the *Optimal Developmental Temperature Hypothesis* and *Smaller is Better*, but were inconsistent with the three remaining hypotheses.

Zamudio *et al.* (1995) exemplify the need for studying acclimation with at least a 3 × 3 design (Fig. 5): their 2 × 2 design (Fig. 2) could not unambiguously distinguish between the *Optimal Developmental Temperature* and *Smaller is Better Hypotheses*. The difference in size, however, between contestant males had no effect on territorial scores, which is inconsistent with *Smaller is Better* (see Zamudio *et al.*, 1995: 674).

Huey *et al.* (1995) developed a factorial experiment that quantified the effects of different paternal (18 versus 25 °C), maternal (18 versus 25 °C), developmental (18 versus 25 °C), and laying (18 versus 25 °C) temperature on fecundity of *D. melanogaster* early in life. They tested the same set of hypotheses presented above with regard to developmental temperature. Development at 18 versus 25 °C had no significant effect on fecundity early in life, a result that was consistent with the *Developmental Buffering Hypothesis*.

Three studies are instructive with regard to the possible adaptive significance of acclimation. All three studies (plus Leroi *et al.*, 1994) contradict a simple *Beneficial Acclimation Hypothesis*, but support competing hypotheses, thus emphasising the importance of considering competing hypotheses explicitly (Platt, 1964; Feder, 1987*a*).

All these studies measure close correlates of fitness, so the conclusions should be relatively robust with respect to their evolutionary significance. Even so, tests of the adaptive significance of plasticity should ideally rely on direct measures of fitness (Bennett, 1996). In the studies with *E. coli* (see pp. 239–64; Leroi *et al.*, 1994), for example, fitness is directly measured either as the rate of population growth (absolute fitness) or as competitive success (relative fitness). In such experiments, interpretations of patterns shown in Fig. 6 should be relatively straightforward.

Often, however, direct measures of fitness will be impractical or impossible, requiring investigations to use performance measures as a proxy for fitness (e.g. female fecundity, Huey *et al.*, 1995; male territorial success, Zamudio *et al.*, 1995). Even so it is worth remembering that the relationship between performance and fitness may be complex (Huey, 1982; Emerson & Arnold, 1989), especially when temperature or photoperiod is manipulated during development. For example, if developmental conditions predict the adult environment and if the optimal life-history strategy varies among adult environments (Levins, 1968), then the sign of the relationship between a particular performance trait and fitness may change among adult environments (Hoffmann & Parsons, 1991: 182). For example, individuals that develop and live in early summer might achieve highest lifetime fitness by enhancing daily fecundity and sacrificing longevity, whereas the opposite might be true for adults emerging in late fall. In this scenario, the results of a factorial experiment concerning the effects of temperature on fecundity (performance) could superficially support the hypothesis that *Smaller is Better*. In reality, however, these hypothetical organisms are displaying an adaptive response to seasonal environments because fecundity and fitness are positively correlated in early summer but negatively correlated in late fall. Such complications highlight the importance of using natural history information to guide the selection of (and interpretation of) ecologically relevant indices of performance (Huey & Stevenson, 1979; Greene, 1986; Emerson & Arnold, 1989; Bennett & Huey, 1990).

A comment on fixed versus fluctuating temperature regimens

Our discussion has focused on the consequences of development at fixed temperature regimens. However, terrestrial ectotherms developing in nature normally experience temperatures that fluctuate daily as well as seasonally, and fluctuating developmental regimens can have important and complex physiological and life-history effects (Cloudsley-Thompson, 1953; Hagstrum & Milliken, 1991). For example, mosquitoes (*Wyeomyia*) developing at fluctuating temperatures have higher fecundities than do those developing at fixed temperatures (Bradshaw, 1980). Thus, fixed developmental temperatures can be viewed as ecologically 'abnormal' Cloudsley-Thompson, 1953: 187) and potentially even as pathological (Huey, 1982).

Our strong inference approach, which we develop for the simple case of fixed temperature environments, can be extended to include hypotheses involving fluctuating developmental temperatures. For example, Bradshaw's observations (see earlier) suggest the testable hypothesis that development at fluctuating temperatures may enhance adult fitness (in any adult environment) relative to development at constant temperatures.

The choice of specific fluctuating regimens will, however, be difficult: an experimenter will not only have to specify a mean temperature, but also a variance as well as a temporal pattern for the temperature fluctuations. What boundary (upper and lower) temperatures should be selected? Should the fluctuations follow a sine wave? Should photoperiod be manipulated in parallel? Such decisions are probably most appropriately based on observed (see pp. 79–102) or predicted patterns of temperature and photoperiod variation in nature (Bradshaw, 1980; Huey, 1982; Tsuji, 1988*b*).

New evolutionary directions for studies of acclimation

The primary goal of our essay is to encourage more robust studies of the adaptive significance of acclimation responses; however, we would also like to encourage some new directions for future studies. These approaches have been developed primarily by evolutionary biologists studying general aspects of phenotypic plasticity, but the approaches are readily modifiable to study the specific case of physiological acclimation.

Testing the significance of acclimation responses in nature

Controlled laboratory studies are a logical and pragmatic place to begin any analysis of the evolutionary significance of developmental acclimation. Nevertheless, natural environments are inherently more complex than are laboratory ones, and behaviour may enable organisms to circumvent physiological pressures that are unavoidable in simple laboratory environments (see pp. 377–407; Bartholomew, 1964; Chambers *et al.*, 1983; Clarke, 1993; Hoffmann & Parsons, 1991). Consequently, as Laudien (1973: 365) argued over two decades ago, tests of the adaptive significance of physiological responses seen in the laboratory must ultimately be validated in the field. Few physiological studies have implemented Laudien's suggestion, perhaps for two main reasons. First, such field tests usually require techniques of demographic analyses (Bennett & Huey, 1990; Kingsolver, 1995, 1996), which rarely are an integral part of a physiologist's armamentarium. Second, many physiologists may simply feel uncomfortable in giving up their rigid control of the laboratory environments, as such control has traditionally has been viewed a *sine qua non* of good experimental physiology. Moreover, factors (e.g. productivity) other than those considered in the laboratory may dominate seasonal acclimation patterns in nature (Clarke, 1993).

Despite these and other concerns, attempts to test the significance of physiological traits in natural populations (as well as in the laboratory) should become an integral part of evolutionary physiology (Bennett & Huey, 1990). Such field studies will blunt the criticisms of some field biologists, who question the ecological relevance of laboratory studies. Moreover, in some cases, such field trials may even accentuate lab-based patterns. For example, Jaenike *et al.* (1995) recently found that nematode parasites caused greater mortality to field-released *Drosophila* than to laboratory maintained ones, possibly because natural environments are more rigorous than laboratory ones.

With specific respect to acclimation studies, the types of factorial experimental designs (Fig. 5) and associated hypotheses (Fig. 6) described previously for the laboratory can readily be modified for field tests. For example, one could acclimate animals to low versus high temperature, release them into the field in spring and in summer, and then recapture the organisms to determine survival rates in nature as a function of prior acclimation conditions. The *Beneficial Acclimation Hypothesis* would, for example, predict that animals acclimated to low temperature would do relatively well in spring, but that animals acclimated to high temperature would do relatively well in summer. Predictions for the competing hypotheses (Fig. 6) can be similarly developed.

A recent field study by J. Kingsolver (1995) exemplifies this approach. Kingsolver's primary intent was to test whether known seasonal variation in wing colour of a butterfly (*Pontia occidentalis*) was adaptive. As discussed above, thermoregulatory considerations (Kingsolver, 1983; Kingsolver & Wiernasz, 1991) suggested a plausible adaptive hypothesis for the evolution of this 'seasonal polyphenism': generally darker wings (in certain wing regions) in spring might increase rates of heating and hence increase potential activity times, whereas lighter wings in summer might reduce heat gain and hence reduce the risk of overheating in summer.

To test this hypothesis, Kingsolver (1995) raised butterflies in the laboratory under simulated spring versus summer conditions (split-family design), and then released one cohort of butterflies in late spring (1992) and a second cohort in summer (1991, 1992). He then conducted an intensive recapture–release–recapture programme to estimate survival probabilities (as distinct from recapture probabilities) as a function of a butterfly's developmental condition (spring versus summer phenotype). In late spring (1992), developmental condition (i.e. spring versus summer phenotypes) had no significant impact on survival, contrary to the adaptive hypothesis. In the summer, however, butterflies with the summer phenotype often had higher survival. The pattern was more marked in 1992, which was warmer than 1991. Overall, Kingsolver's field test gives only partial support to the adaptive hypothesis at risk.

Kingsolver (1995) is an important study. It develops a rigorous paradigm for testing in nature the adaptive significance, or the lack thereof, of acclimation responses. Furthermore, it raises a caveat. Kingsolver was specially interested in wing coloration, but the rearing conditions (i.e. spring versus summer conditions) he used to manipulate wing colour will undoubtedly affect many aspects of a butterfly's physiology, size, shape, as well as wing colour. Consequently, in the absence of experimental manipulations of wing colour alone, it is risky to attribute the higher survival of the 'summer-phenotype' in summer as a causal function of wing colour alone.

Kingsolver (1996) fully recognised this issue, and so developed an additional set of experiments. He raised butterflies under summer conditions, but used a black (felt) marking pen to darken appropriate parts of the wings of some butterflies, thereby generating butterflies that had the summer phenotype for all traits except wing colour. His release–recapture experiments showed that the blackened male butterflies had markedly reduced survivorship relative to control (normal colour) males. This experiment demonstrates that dark wing colour negatively affects male fitness in summer, probably because dark colour increases the risk of overheating. It does not, however, demonstrate that the higher survival of the 'summer phenotypes' in summer (previous experiments) was attributable only to wing colour: other experimental manipulations or statistical analyses would be necessary to elucidate the possible impact of body size or of summer 'physiology'.

Such field tests of the adaptive significance of phenotypic plasticity can probably be incorporated into many studies of acclimation and other plastic responses (Hoffmann & Turelli, 1985). We are planning to use factorial designs and field release-recapture experiments with *Drosophila* to extend our studies of the adaptive significance of developmental temperature. By repeating these studies in different seasons, we can effectively use our strong-inference approach (Fig. 6) to discriminate against several competing hypotheses concerning developmental effects.

Measuring the costs and trade-offs of acclimation

Robust optimality arguments investigate the costs, not just the benefits, of alternative phenotypes (Alexander, 1982); however, adaptive acclimation hypotheses are typically developed only considering the potential gross benefits of acclimation, not the net benefits. Nevertheless, as Krebs & Loeschcke (1994) and Hoffmann (1995) have recently noted, acclimation responses are likely to involve significant costs, which may well affect the net benefits of alternative acclimation responses. Three kinds of costs are possible. First, maintaining the genetic capacity to acclimate must involve some cost (Dykhuizen & Davies, 1980; Scheiner, 1993), although little is known about it. Second, the transition from one acclimatory state to another must also involve some cost (e.g. enzyme/membrane re-synthesis), and organisms in the transitional states may suffer a transient reduction in performance. Third, acclimation to a new set of conditions may incur trade-offs (e.g. gain in heat tolerance resulting in simultaneous loss in cold tolerance).

Two recent experiments suggest acclimation to high temperature does not invariably enhance fitness and performance at high temperature, contrary to a simple *Beneficial Acclimation Hypothesis*. Leroi *et al.*

(1994) showed that acclimation to high temperature (41.5 °C) enhances the ability of *E. coli* to survive an acute exposure to a higher temperature (50 °C), but this ability comes at the cost of reduced, not enhanced, competitive fitness at 41.5 °C. Similarly, Krebs & Loeschcke (1994) found that a exposure for a few hours to high temperature (36 °C) enhances the ability of *Drosophila melanogaster* to survive a heat shock, but at an apparent cost of reduced fecundity.

In the above two experiments, reduced competitive fitness or reduced fecundity were interpreted as representing a cost of acclimation (Leroi *et al.*, 1994; Krebs & Loeschcke, 1994; Hoffmann, 1995). These reductions might instead reflect a direct and pathological effect of heat exposure on physiology. To distinguish between these alternatives, one needs to develop experimental protocols in which a given acclimation response can be triggered in the absence of a stress (M. E. Feder, personal communication; L. Partridge, personal communication). Using an ingenious approach with *Drosophila*, J. Feder *et al.* (1992) have in fact been able to demonstrate costs associated activation of the heat shock response, and to do so independent of a temperature stress.

Costs and trade-offs are necessarily an integral part of acclimation (Hoffmann, 1995) and other plastic responses (Scheiner, 1993). Consequently, an exploration of the costs and trade-offs involved with acclimation should be a priority of physiological studies. Such costs are directly germane to optimality models, but are also of considerable physiological interest from a mechanistic perspective.

Measuring genetic variation for acclimation responses

Implicit in all arguments concerning the evolution of acclimation responses is the assumption that there is genetic variation (within populations) affecting the direction and magnitude of the acclimation response. Actual studies documenting or suggesting genetic variation with respect to thermal acclimation (plasticity) are rare; and thus studies of this issue would definitely be welcome. Evolutionary geneticists have developed two general ways to assay genetic variation within populations (Falconer, 1989): (i) the demonstration of significant between-family variation in a trait (e.g. in plasticity); and (ii) the successful use of artificial or natural selection to alter the trait.

Several studies have used inter-family variation to infer the existence of genetic variation for phenotypic plasticity in response to temperature. Gupta & Lewontin (1982) showed that the effect of temperature on viability (and several other traits) varied dramatically among genotypes of *Drosophila pseudoobscura* (i.e. different genotypes have ‘crossing reaction norms' with respect to temperature). Sinervo & Adolph (1989) showed that the thermal dependence of growth rates of hatchling lizards (*Sceloporus occidentalis*) differed among families, which suggests that this trait might show heritable variation. On the other hand, Loeschcke *et al.* (1994) found no variation in acclimation to high temperature stress among isofemale lines of *Drosophila buzzatii*. Similarly, Hoffman & Watson (1993) found no interpopulational variation in acclimation responses in *D. melanogaster* or in *D. simulans*.

Selection experiments have also been performed. Scheiner & Lyman (1991) used a family-selection scheme with *D. melanogaster*. They raised some flies from each isofemale line at 19 °C and some at 25 °C. Full-sibs raised at 19 °C were larger than those at 25 °C (control line in Fig. 7), and the difference in mean thorax size (within isofemale lines) measures the degree of plasticity of size. Scheiner & Lyman

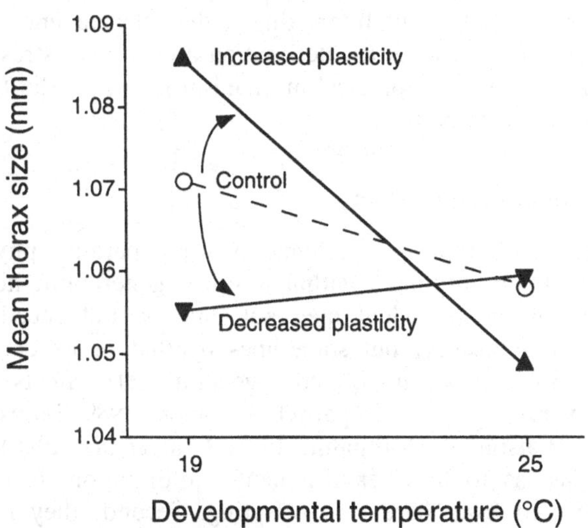

Fig. 7. Twenty generations of artificial selection significantly modify phenotypic plasticity in thorax size as a function of developmental temperature in *Drosophila melanogaster* (Scheiner & Lyman, 1991). Lines were selected for increased plasticity (i.e. increased difference in mean size between full-sibs raised at 19 versus 25 °C) or for decreased plasticity. Relative to the control lines, selection for increased sensitivity of size to temperature showed significantly enhanced plasticity, whereas selection for decreased sensitivity effectively eliminated plasticity. Plotted are the (unweighted) mean thorax sizes for the two replicate lines within each selection regimen (Data courtesy of S. M. Scheiner.)

(1991) selected (replicated) lines for both increased and decreased sensitivity of thorax size to developmental temperature. Both treatments were successful (Fig. 7), demonstrating conclusively that genetic variation exists for the plasticity of this trait with respect to temperature.

McDaniel *et al.* (1995) studied the magnitude of thermal acclimation of thermal preferences of stocks of *Drosophila tripunctata* and *D. robusta* that had been laboratory adapted to a constant temperature for 1, 4, or 7 years. Interestingly, older stocks had smaller acclimation responses; but it is unclear whether this apparent reduction of acclimation response was a result either of selection at constant temperature or of genetic drift. In any case, the response suggests that the magnitude of acclimation shows genetic variation in these species.

Acclimation responses can also evolve as a genetically correlated response to selection. Hoffmann (1990) conducted artificial selection on desiccation resistance in *D. melanogaster*. Resistance increased as a direct response to selection. Interestingly, the selected lines showed reduced acclimation responses (to desiccation and to heat stress), suggesting that a common physiological mechanism may underlie the genetic and acclimatory responses.

Cross-generation effects

The vast majority of studies of acclimation in comparative physiology focus on acclimation responses within a single generation; however, the phenotype of an individual may not only be influenced by its own environmental history, but sometimes by that of its parents or grandparents. Such parental (or cross-generational) effects are of interest for several reasons (Kirkpatrick & Lande, 1989; Mousseau & Dingle, 1991; Watson & Hoffmann, 1995; Crill *et al.*, 1996). First, they raise issues as to how 'environmental information' is mechanistically transferred from parent to offspring. Second, they raise, as was the case for within-generation acclimation, the issue as to how and whether this transfer is adaptive. Third, they can greatly complicate attempts to predict evolutionary responses to environmental change (Falconer, 1989). (For example, A.A. Hoffmann (personal communication) selected for increased cold tolerance in *D. melanogaster*, but found that cold tolerance began to decrease after several generations: this surprising shift proved to be a consequence of a cross-generational effect.) Fourth, they imply that comparative studies may need to raise organisms in a common garden for more than one generation to eliminate environmental effects (Garland & Adolph, 1991).

Maternal effects on life history traits of insects have been intensively studied (Giesel, 1988; Mousseau & Dingle, 1991), but cross-generational effects on physiology have been little studied until recently. Jenkins & Hoffmann (1994) used a two-generation breeding design to discover that the heat tolerance of wild-caught *Drosophila simulans* was strongly influenced by a maternal effect. Watson & Hoffmann (1995) found that either maternal or paternal exposure (*Drosophila* spp.) to cold stress could affect cold resistance of offspring. Huey *et al.* (1995) and Crill *et al.*, (1996) used a factorial design to show significant impact of father's temperature or of mother's temperature on several traits of *D. melanogaster*.

Whether cross-generational effects are adaptive has also been little explored (Kirkpatrick & Lande, 1989; Mousseau & Dingle, 1991). Not surprisingly, adaptive stories can be derived for cross-generational effects of temperature. In principle, cross-generational effects enable an organism to take advantage of environmental information gathered during its parents' lifetimes, not just its own. For example, information that an organism's parents had experienced high temperatures might serve as a cue that an increase in heat tolerance would be an appropriate acclimation response, at least for organisms with short generation times (Levins, 1968). Nevertheless, such speculation should be tested directly (Watson & Hoffmann, 1995; Crill *et al.*, 1996).

Zamudio *et al.* (1995) used male territoriality experiments to test the possible adaptive significance of a cross-generational effect of temperature in *Drosophila*. In these experiments, males were raised only at 25 °C and tested in pairs only at 27 °C. However, one male in a pair had parents raised at 18 °C, whereas the other male in the pair had parents raised at 25 °C. Interestingly, males with 25 °C parents were dominant over males with 18 °C parents. This finding is consistent with cross-generational versions of the *Beneficial Acclimation Hypothesis* or the *Optimal Development Temperature Hypothesis*; but obviously an expanded study (with territoriality scored at additional temperatures, see Fig. 5) is needed to discriminate between these two competing hypotheses. Nevertheless, this study illustrates that the phenomena may be amenable to test.

Concluding remarks

Our primary goal in this paper is to argue that much is to be gained from explicit studies of the evolution of acclimation responses in physiology. Rather than merely invoking the most fashionable adaptive hypothesis, comparative physiologists should endeavour to test that hypothesis, especially in competition with competing explanatory hypotheses. If we have learned anything from the history of biology, it is that nature is inevitably more complex than we could imagine. Consequently, as the Leroi *et al.* (1994) study in particular shows, direct tests of assumptions are likely to enhance (rather than merely complicate) our understanding of the physiology and evolution of the acclimation response.

We do not mean to imply that direct tests of the adaptive significance of acclimation responses should become the primary goal of all those studying acclimation responses. Rather we emphasise that the evolutionary side of acclimation studies has been under-studied in the past and thus offers some exciting research opportunities. Moreover, although we are critical of the practice of asserting an adaptive explanation for some patterns, we recognise that such (untested) suggestions can be valuable, in that they might inspire someone else to develop a test of that hypothesis in a subsequent experiment. Nevertheless, we are critical of those who seemingly assume that the mere ability to concoct an adaptive hypothesis is in itself support for that hypothesis.

Comparative physiologists are in an ideal position to make major contributions to the evolution of acclimation (or phenotypic plasticity). Complete explanations in evolution require an understanding of the mechanism underlying a phenotypic response, not just its phenomenology and potential adaptive significance (Mayr, 1961). As Scheiner (1993: 60) has recently noted, 'A significant advance toward the creation of an evolutionary model (of phenotypic plasticity) with general predictive power would be the ability to predict the shape of reaction norms from physiological first principles'. Studies of acclimation and of related phenomena are well suited subjects for such an advance.

Acknowledgements

We thank I. Johnston and A. Bennett for inviting us to participate in the symposium at St. Andrews. We thank J. Kingsolver, S. Scheiner, J. Tsuji and B. Zwaan for generously giving us access to their data or to unpublished manuscripts. A. Bennett, M. Feder, G. Gilchrist, J. Kingsolver and L. Partridge provided constructive suggestions on drafts of this manuscript. The preparation of this chapter was supported by NSF IBN 9221620 to RBH and NSF Environmental Biology Postdoctoral Fellowship DEB 9393164 to DAB.

References

Alexander, R. McN. (1982). *Optima for Animals.* London: Edward Arnold.

Aspi, J. & Hoikkala, A. (1995). Male mating success and survival in the field with respect to size and courtship song characters in *Drosophila littoralis* and *D. montana* (Diptera: Drosophilidae). *Journal of Insect Behavior* **8**, 67–87.

Atkinson, D. (1994). Temperature and organism size: a biological law for ectotherms. *Advances in Ecological Research* **25**, 1–58.

Barbault, R. (1988). Body size, ecological constraints, and the evolution of life-history strategies. *Evolutionary Biology* **22**, 261–86.

Bartholomew, G.A. (1964). The roles of physiology and behaviour in the maintenance of homeostasis in the desert environment. *Symposium of the Society of Experimental Biology* **18**, 7–29.

Bennett, A.F. (1996). Adaptation and the evolution of physiological characters. In *Handbook of Comparative Physiology* (ed. W.H. Dantzler). Oxford, UK: Oxford University Press (in press).

Bennett, A.F. & Huey, R.B. (1990). Studying the evolution of physiological performance. In *Oxford Surveys in Evolutionary Biology*, vol. 7, (ed. D.J. Futuyma & J. Antonovics), pp. 251–84. Oxford, UK: Oxford University Press.

Berrigan, D. & Charnov, E.L. (1994). Reaction norms for age and size at maturity in response to temperature: a puzzle for life historians. *Oikos* **70**, 474–8.

Boake, C.H.R. (1989). Correlation between courtship success, aggressive success, and body size in a picture-winged fly, *Drosophila sylvestris. Ethology* **80**, 318–29.

Bradshaw, W.E. (1980). Thermoperiodism and the thermal environment of the pitcher-plant mosquito, *Wyeomyia smithii. Oecologia (Berlin)* **46**, 13–17.

Burggren, W.W. (1991). Does comparative respiratory physiology have a role in evolutionary biology (and vice versa)? In *Physiological Strategies for Gas Exchange and Metabolism* (ed. A.J. Woakes, M.K. Grieshaber & C. R. Bridges), pp. 1–13. Cambridge, UK: Cambridge University Press.

Chambers, D.L., Calkins, C.O., Boller, E.F., Ito, Y. & Cunningham, R.T. (1983). Measuring, monitoring and improving the quality of mass reared Mediterranean fruit flies *Ceratitis capitata* Wied. II. Field tests confirming and extending laboratory results. *Zeitschrift für Angewandte Entomologie* **95**, 285–303.

Clarke, A. (1993). Seasonal acclimitization and latitudinal compensation in metabolism: do they exist? *Functional Ecology* **7**, 139–49.

Cloudsley-Thompson, J.L. (1953). The significance of fluctuating temperatures on the physiology and ecology of insects. *Entomologist* **86**, 183–9.

Cohet, Y. & David, J. (1977). Control of adult reproductive potential by preimaginal thermal conditions: A study in *Drosophila melanogaster. Oecologia (Berlin)* **36**, 295–306.

Cossins, A.R. & Bowler, K. (1987). *Temperature Biology of Animals.* New York: Chapman and Hall.

Crill, W.D., Huey, R.B. & Gilchrist, G.W. (1996). Within- and between-generational effects of temperature on the morphology and physiology. *Drosophila melanogaster. Evolution* (in press).

David, J.R., Allemand, R., Herrewege, van & Cohet, Y. (1983). Ecophysiology: abiotic factors. In *The Genetics and Biology of* Drosophila (ed. M. Ashburner, H. L. Carson & J. N. Thompson), pp. 106–69. London: Academic Press.

Dumont, J.P.C. & Robertson, R.M. (1986). Neuronal circuits: an evolutionary perspective. *Science* **223**, 849–53.

Dykhuizen, D. & Davies, M. (1980). An experimental model: bacterial specialists and generalists competing in chemostats. *Ecology* **61**, 1213–27.

Emerson, S.B. & Arnold, S.J. (1989). Intra- and interspecific relationships between morphology, performance, and fitness. In *Complex Organismal Function: Integration and Evolution in Vertebrates* (ed. D.B. Wake & G. Roth), pp. 295–314. New York: Wiley.

Falconer, D.S. (1989). *Introduction to Quantitative Genetics*, 3rd edn. London: Longman.

Feder, J.H., Rossi, J.M., Solomon, J., Solomon, N. & Lindquist, S. (1992). The consequences of expressing hsp70 in *Drosophila* cells at normal temperatures. *Genes and Development* **6**, 1402–13.

Feder, M.E. (1977). Environmental variability and thermal acclimation in neotropical and temperate-zone salamanders. *Physiological Zoology* **51**, 7–16.

Feder, M.E. (1982). Environmental variability and thermal acclimation of metabolism in tropical anurans. *Journal of Thermal Biology* **7**, 23–8.

Feder, M.E. (1987*a*). The analysis of physiological diversity: the prospects for pattern documentation and general questions in ecological physiology. In *New Directions in Ecological Physiology* (ed. M.E. Feder, A. F. Bennett, W.W. Burggren, and R.B. Huey), pp. 38–75. Cambridge, UK: Cambridge University Press.

Feder, M.E. (1987*b*). Effect of thermal acclimation on locomotor energetics and locomotor performance in a tropical salamander, *Bolitoglossa subpalmata. Physiological Zoology* **60**, 18–26.

Feder, M.E. & Lynch, J.F. (1982). Effects of latitude, season, elevation, and microhabitat on field body temperatures of neotropical and temperate zone salamanders. *Ecology* **63**, 1657–4.

Garland, T., Jr. & Adolph, S.T. (1991). Physiological differentiation of vertebrate populations. *Annual Reviews of Ecology and Systematics* **22**, 193–228.

Garland, T., Jr. & Carter, P.A. (1994). Evolutionary physiology. *Annual Reviews of Ecology and Systematics* **56**, 579–621.

Giesel, J.T. 1988. Effects of parental photoperiod on development time and density sensitivity of progeny of *Drosophila melanogaster*. *Evolution* **42**, 1348–50.

Gilchrist, G.W. (1995). Specialists and generalists in changing environments. I. Fitness landscapes of thermal sensitivity. *American Naturalist* **146**, 252–70.

Gould, S.J. & Lewontin, R.C. (1979). The spandrels of San Marco and the Panglossian Paradigm: a critique of the adaptationist programme. *Proceedings of the Royal Society of London* **205**, 581–98.

Greene, H.W. (1986). Natural history and evolutionary biology. In *Predator–Prey Relationships: Perspectives and Approaches from the Study of Lower Vertebrates*, (ed. M.E. Feder & G.V. Lauder) Chicago, IL: University of Chicago Press.

Gupta, A.P. & Lewontin, R.C. (1982). A study of reaction norms in natural populations of *Drosophila pseudoobscura*. *Evolution* **36**, 934–48.

Hagstrum, D.W. & Milliken, G.A. (1991). Modeling differences in insect developmental times between constant and fluctuating temperatures. *Annals of the Entomological Society of America* **84**, 369–79.

Haldane, J.B.S. 1958. The theory of evolution before and after Bateson. *Journal of Genetics* **56**, 11–27.

Heatwole, H., Lin, T.-H., Villalón, E., Muñiz, A. & Matta, A. (1969). Some aspects of the thermal ecology of Puerto Rican anoline lizards. *Journal of Herpetology* **3**, 65–77.

Hochachka, P.W. & Somero, G.N. (1984). *Biochemical Adaptation*. Princeton, NJ: Princeton University Press.

Hoffmann, A.A. (1990). Acclimation for desiccation resistance in *Drosophila melanogaster* and the association between acclimation responses and genetic variation. *Journal of Insect Physiology* **36**, 885–91.

Hoffmann, A.A. (1995). Acclimation: increasing survival at a cost. *Trends in Ecology and Evolution* **10**, 1–2.

Hoffmann, A.A. & Parsons, P.A. (1991). *Evolutionary Genetics and Environmental Stress*. Oxford, UK: Oxford University Press.

Hoffmann, A.A. & Turelli, M. (1985). Distribution of *Drosophila melanogaster* on alternative resources: effects of experience and starvation. *American Naturalist* **126**, 662–79.

Hoffmann, A.A. & Watson, M. (1993). Geographic variation in the acclimation response of *Drosophila* to temperature extremes. *American Naturalist* **142** (suppl.), S93–S113.

Huey, R.B. (1982). Temperature, physiology, and the ecology of reptiles. In *Biology of the Reptilia*, vol. 12 (Physiology C) (ed. C. Gans & F.H. Pough), pp. 25–91. London: Academic Press.

Huey, R.B. (1987). Phylogeny, history, and the comparative method. In *New Directions in Ecological Physiology* (ed. M.E. Feder, A. F. Bennett, W.W. Burggren & R.B. Huey), pp. 76–101. Cambridge, UK: Cambridge University Press.

Huey, R.B. & Kingsolver, J.G. (1993). Evolution of resistance to high temperature in ectotherms. *American Naturalist* **142**, S21–S46.

Huey, R.B. & Stevenson, R.D. (1979). Integrating thermal physiology and ecology of ectotherms: a discussion of approaches. *American Zoologist* **19**, 357–66.

Huey, R.B., Wakefield, T., Crill, W.D. & Gilchrist, G.W. (1995). Effects of parental, developmental, and laying temperature on the early fecundity of *Drosophila melanogaster*. *Heredity* **74**, 216–23.

Jaenike, J., Benway, H. & Stevens, G. (1995). Parasite-induced mortality in myocophagous *Drosophila*. *Ecology* **76**, 383–91.

Janzen, D.H. (1967). Why mountain passes are higher in the tropics. *American Naturalist* **101**, 233–49.

Janzen, D.H. & Martin, P. S. (1982). Neotropical anachronisms: the fruits the gomphotheres ate. *Science* **215**, 19–27.

Janzen, F.J. & Paukstis, G.L. (1991). Environmental sex determination in reptiles: ecology, evolution, and experimental design. *Quarterly Review of Biology* **66**, 149–79.

Jenkins, N.L. & Hoffmann, A.A. (1994). Genetic and maternal variation for heat resistance in *Drosophila* from the field. *Genetics* **137**, 783–9.

Kingsolver, J.G. (1983). Ecological significance of flight activity in *Colias* butterflies: implications for reproductive strategy and population structure. *Ecology* **64**, 546–51.

Kingsolver, J.G. (1995). Fitness consequences of seasonal polyphenism in western white butterflies. *Evolution* **49**, 942–54.

Kingsolver, J.G. (1996). Experimental manipulation of wing pigment pattern and survival in western white butterflies. *American Naturalist 147*, 296–306.

Kingsolver, J.G. & Wiernasz, D.C. (1991). Seasonal polyphenisms in wing-melanin pattern and thermoregulatory adaptation in *Pieris* butterflies. *American Naturalist* **137**, 816–30.

Kirkpatrick, M. & Lande, R. (1989). The evolution of maternal characters. *Evolution* **43**, 485–503.

Krebs, R.A. & Loeschcke, V. (1994). Effects of short-term thermal extremes on fitness components in *Drosophila melanogaster*. *Journal of Evolutionary Biology* **7**, 39–49.

Laudien, H. (1973). Changing reaction systems. In *Temperature and Life* (ed. H. Precht, J. Christophersen, H. Hensel & W. Larcher), pp. 355–99. New York: Springer-Verlag.

Leroi, A.M., Bennett, A.F. & Lenski, R.E. (1994). Temperature acclimation and competitive fitness: an experimental test of the

Beneficial Acclimation Assumption. *Proceedings of the National Academy of Sciences of the USA* **91**, 1917–21.

Levins, R. (1968). *Evolution in Changing Environments.* Princeton, NJ: Princeton University Press.

Loeschcke, V., Krebs, R.A. & Barker, J.S.F. (1994). Genetic variation for resistance and acclimation to high temperature stress in *Drosophila buzzatii. Biological Journal of the Linnean Society* **52**, 83–92.

Lynch, M. & Lande, R. (1993). Evolution and extinction in response to environmental change. In *Biotic Interactions and Global Change*, (ed. P.M. Kareiva, J.G. Kingsolver & R.B. Huey), pp. 234–50. Sunderland, MA: Sinauer.

McDaniel, R., Hostert, E.E. & Seager, R.D. (1995). Acclimation and adaptive behavior of *Drosophila robusta* and *D. tripunctata* adults in response to combined temperature and desiccation stress. *American Midland Naturalist* **133**, 52–59.

McLachlan, A.J. & Allen, D.F. (1987). Male mating success in Diptera: advantages of small size. *Oikos* **48**, 11–14.

Markow, T.A. (1995). Evolutionary ecology and developmental instability. *Annual Review of Entomology* **40**, 105–20.

Markow, T.A. & Ricker, J.P. (1992). Male size, developmental stability, and mating success in natural populations of three *Drosophilia* species. *Heredity* **69**, 122–7.

Mayr, E. (1961). Cause and effect in biology. *Science* **134**, 1501–6.

Mousseau, T.A. & Dingle, H. (1991). Maternal effects in insect life histories. *Annual Reviews of Entomology* **36**, 511–34.

Parsons, P.A. (1990). Fluctuating asymmetry: an epigenetic measure of stress. *Biological Reviews* **63**, 131–45.

Partridge, L., Hoffmann, A. & Jones, S. (1987). Male size and mating success in *Drosophila melanogaster* and *D. pseudoobscura* under field conditions. *Animal Behaviour* **35**, 468–76.

Platt, J.R. (1964). Strong inference. *Science* **146**, 347–53.

Powers, D.A. (1987). A multidisciplinary approach to the study of genetic variation within species. In *New Directions in Ecological Physiology* (ed. M.E. Feder, A.F. Bennett, W.W. Burggren & R.B. Huey), pp. 102–134. Cambridge, UK: Cambridge University Press.

Precht, H. (1958). Theory of temperature adaptation in cold-blooded animals. In *Physiological Adaptation* (ed. C.L. Prosser), pp. 50–78. Washington, DC: American Physiological Society.

Prosser, C.L. (1958). General summary: The nature of physiological adaptation. In *Physiological Adaptation* (ed. C.L. Prosser), pp. 167–180. Washington, DC: American Physiological Society.

Prosser, C.L. (1986). *Adaptational Biology: Molecules to Organisms.* New York: John Wiley.

Regal, P.J. (1977). Evolutionary loss of useless features: is it molecular noise suppression? *American Naturalist* **11**, 123–33.

Rome, L.C., Stevens, E.D. & John-Alder, H.B. (1992). The influence of temperature and thermal acclimation on physiological function. In *Environmental Physiology of Amphibians* (ed. M.E. Feder & W.W. Burggren), pp. 183–205. Chicago, IL: University of Chicago.

Scheiner, S.M. (1993). Genetics and evolution of phenotypic plasticity. *Annual Review of Ecology and Systematics* **24**, 35–68.

Scheiner, S.M. & Lyman, R.F. (1991). The genetics of phenotypic plasticity. II. Response to selection. *Journal of Evolutionary Biology* **4**, 23–50.

Sinervo, B. & Adolph, S.C. (1989). Thermal sensitivity of growth rate in hatchling *Sceloporus* lizards: environmental, behavioral and genetic aspects. *Oecologia (Berlin)* **83**, 228–37.

Stearns, S.C. (1982). The role of development in the evolution of life histories. In *Evolution and Development* (ed. J.T. Bonner), pp. 237–258. Berlin: Springer-Verlag.

Stearns, S.C. & Kawecki, T.J. (1994). Fitness sensitivity and the canalization of life-history traits. *Evolution* **48**, 1438–50.

Stearns, S.C. & Sage, R.D. (1980). Maladaptation in a marginal population of the mosquito fish, *Gambusia affinis*. *Evolution* **34**, 65–74.

Sultan, S.E. (1987). Evolutionary implications of phenotypic plasticity in plants. In *Evolutionary Biology*. (ed. M.K. Hecht, B. Wallace & G.T. Prance), pp. 127–178. New York: Plenum Press.

Travis, J. (1994). Evaluating the adaptive role of morphological plasticity. In *Ecological Morphology: Integrative Organismal in Biology*. (ed. P.C. Wainwright & S.M. Reilly), pp. 99–122. Chicago, IL: University of Chicago Press.

Tsuji, J.S. (1988*a*). Seasonal profiles of standard metabolic rate of lizards (*Sceloporus occidentalis*) in relation to latitude. *Physiological Zoology* **61**, 230–40.

Tsuji, J.S. (1988*b*). Thermal acclimation of metabolism in *Sceloporus* lizards from different latitudes. *Physiological Zoology* **61**, 241–53.

Turelli, M., Gillespie, J.H. & Lande, R. (1987). Rate tests for selection on quantitative characters during macroevolution and microevolution. *Evolution* **42**, 1085–9.

Waddington, C.D. (1942). Canalization of development and the inheritance of acquired characters. *Nature* **150**, 563–5.

Watson, M.J.O. & Hoffmann, A.A. (1995). Cross-generation effects on cold resistance in tropical populations of *Drosophila melanogaster* and *D. simulans*. *Australian Journal of Zoology* **43**, 51–8.

Watt, W.B. (1969). Adaptive significance of pigment polymorphisms in *Colias* butterflies. II. Thermoregulation of photoperiodically controlled melanin variation in *Colias eurytheme*. *Proceedings of the National Academy of Sciences of the USA* **63**, 767–44.

Watt, W.B. (1990). The evolution of animal coloration: adaptive aspects from bioenergetics to demography. In *Adaptive Coloration in Invertebrates* (ed. M. Wicksten), pp. 1–15. College Station, TX: Texas A & M University Press.

Wiernasz, D.C. (1989). Female choice and sexual selection on male wing melanin pattern in *Pieris occidentalis* (Lepidoptera). *Evolution* **43**, 1672–82.

Wiernasz, D.C. & Kingsolver, J.G. (1991). Wing melanin pattern mediates species recognition in *Pieris occidentalis. Animal Behaviour* **43**, 89–94.

Withers, P.C. (1992). *Comparative Animal Physiology*. Fort Worth, TX: Harcourt Brace Jovanovich College Publishers.

Zamudio, K.R., Huey, R.B. & Crill, W.D. (1995). Bigger isn't always better: body size, developmental and parental temperature, and territorial success in *Drosophila melanogaster. Animal Behaviour* **49**, 671–7.

Zwaan, B.J., Bijlsma, R. & Hoekstra, R.F. (1992). On the developmental theory of ageing. II. The effect of developmental temperature on longevity in relation to adult body size in *D. melanogaster. Heredity* **68**, 123–30.

J.A. MONGOLD, A.F. BENNETT and R.E. LENSKI

Experimental investigations of evolutionary adaptation to temperature

Comparative and experimental studies of adaptation

Comparative and experimental analysis

Organismal biologists employ two principal methods in their investigation of the natural world: comparison and experiment. The latter is most familiar in the context of laboratory investigations of functional mechanisms. Experiment is the classic application of the scientific method, including such elements as rigorous and replicated design, controlled manipulation of a single variable of interest and the incorporation of a control group into the study. While experimental science has been crucial to our understanding of how organisms work, to date it has had relatively less application in studying how those organisms came to be the way they are, i.e. in studies of the evolution of organismal characters. In such evolutionary studies, comparative investigations have been by far the dominant methodological tradition.

In the study of evolutionary adaptation to temperature, for example, virtually all our knowledge is derived from comparative studies of different populations, species, or other taxa inhabiting different thermal environments (for reviews, see Precht *et al.*, 1973; Prosser, 1973; Hochachka & Somero, 1984; Cossins & Bowler, 1987). The comparative approach involves the measurement of a character and its correlation with environmental temperature. If the character (e.g. a rate process) is thermally dependent, then it is measured either over a similar range of temperatures or at a single temperature common to the different groups examined. The pattern of character on environmental temperature is then analysed and interpreted, most frequently in an adaptive context (Prosser, 1986; Cossins & Bowler, 1987; Bennett, 1996), and compared with the pattern found for other biological systems inhabiting similar thermal environments. Applications of this approach include, for example, the classical studies of thermal adaptation in different populations of killifish (Powers, 1987), species of barracuda (Graves &

Somero, 1982), and groups of tropical, temperate, and arctic ectotherms (Scholander *et al.*, 1953). These and other such studies have been important in documenting the fit between functional capacity and thermal environment and the associated adaptive shifts in rate processes and thermal niche. Over the past decade, the comparative methodology has incorporated an historical analytical component (Felsenstein, 1985; Brooks & McLennan, 1991; Harvey & Pagel, 1991). Phylogenetically-based comparative studies have the additional advantage of being able to falsify adaptive hypotheses, as well as reconstruct the putative ancestral condition and calculate rates of evolution (Feder, 1987; Huey, 1987; Garland & Adolph, 1994; Bennett, 1996). Examples of such studies examining evolutionary adaptation to temperature include analyses of burst escape speed of lizards (Huey & Bennett, 1987), activity metabolism of anurans (Walton, 1993), and thermoregulation in fish (Block *et al.*, 1993).

Comparative evolutionary analyses, however, have practical limitations and interpretive shortcomings even when they are phylogenetically based (Lauder *et al.*, 1993; LeRoi *et al.*, 1994*b*). For example, they depend crucially on the availability of an independently-derived set of historical relationships and the accessibility of organisms from the groups contained therein. Phylogenies are only tentative hypotheses of relationships, subject to revision (Felsenstein, 1985; Harvey & Pagel, 1991). Conclusions of comparative studies based on phylogenies are consequently also subject to revision, even though the functional observations and measurements of each group remain unaltered (for an example, see the revision of the conclusions of Huey & Bennett, 1987, by Garland *et al.*, 1991, based on newer information concerning branch lengths within the phylogeny). Comparative adaptive studies must, therefore, be tentative, dependent on potential revision of their phylogenetic bases. Furthermore, comparative analytical methodologies all must assume, in one form or another, the principle of parsimony or Occam's razor (Brooks & McLennan, 1991; Harvey & Pagel, 1991; Martins & Garland, 1991): the accepted evolutionary pattern is the one that involves the fewest character transitions. This is a necessary assumption for discrimination among different potential patterns, but it must be recognised that it is only an assumption about how evolution occurs or has occurred. The evolution of any particular character of interest may not have been parsimonious. The intermediate stages of the evolutionary diversification are not observable, so there is no way of knowing for certain whether a parsimony-based analysis is a correct historical description. Comparative evolutionary analyses must therefore remain *ex post facto* interpretations of unobserved events.

Direct experiments studying evolutionary diversification would seem to be a valuable supplement because of the limitations of comparative analyses for studying adaptation. Experimentation could provide exactly those elements lacking in comparative analyses, namely control of the selective environment, direct observation of the ancestral and intermediate (in addition to terminal) stages, and experimental controls and replication. At first consideration, evolutionary experiments might be thought to be impossible, in spite of their desirability. Locating multiple populations, each of large size, and altering the environment of some while leaving others as controls is a daunting experimental challenge. Moreover, the long generation times of most organisms commonly studied by comparative physiologists preclude direct observations of evolutionary change. Some types of organisms, however, are in fact almost ideal for such studies. The genius of comparative physiology has been the recognition and pursuit of the proper type of organism to investigate the question at hand in an expeditious manner (Krogh, 1929; Krebs, 1975). Direct experiments on evolutionary adaptation are in fact quite feasible. It is simply a matter of choosing the best organism for the study.

Experimental evolution

The essence of any experiment is design and control, and experimental studies of evolution are no different in this respect. They require the design and creation of a new series of populations and the subsequent monitoring of genetically-determined phenotypic change in those populations. Experimental evolution involves the imposition of a novel environment on replicated experimental populations, while maintaining control populations for comparison. The control populations permit a statistical evaluation of phenotypic changes attributable to the experimental manipulation, as opposed to random or directional changes in response to factors other than the experimental manipulation. For instance, a character such as body length may be increasing in the experimental populations. Observations on control groups maintained in the original environment permit a determination of whether similar changes are occurring in those populations as well, and may therefore be part of a continuing evolutionary response to some aspect of the environment other than that which was explicitly manipulated. Replication of experimental populations also permits a determination of the directionality of the evolutionary response in contrast to chance divergence (Travisano *et al.*, 1995). Effectively, replication of experimental lines is equivalent to replaying the tape of life, as proposed by Gould

(1989), to evaluate the inevitability of any particular evolutionary change in terms of its direction or underlying mechanism. Finally, in contrast to comparative studies, which are able to observe only evolutionary products, experimental evolutionary studies do not require extrapolation because they can directly measure characters in their initial, intermediate and derived conditions.

Evolution, by definition, involves intergenerational genetic change. Experimental evolutionary studies require the use of organisms with generation times that are relatively short in comparison to the duration of the experiment. Through either selection on variation existing in the original population or selection on new variation arising through recombination or mutation, genetic change is more likely, and is more likely to be detected, with greater numbers of generations. In addition to monitoring the direction of evolutionary change, multi-generational data permit determination of the rate and form of evolutionary change, as well as its trajectory. Experimental evolutionary studies may additionally require that the size of each component population be large, ideally thousands or millions of organisms in each population, to avoid founder effects and genetic drift. Investigators have utilised a variety of laboratory-cultured organisms, including fruit flies, yeast, protists and bacteria (see later) because of these requirements. These organisms can be maintained in discrete populations of very large size in defined and carefully regulated environments. In some of these organisms, measurement of competitive (Darwinian) fitness is also possible, so that the magnitude and rate of adaptation can be determined directly as change in relative fitness.

It is important to recognise that other types of evolutionary studies, while interesting and important, differ fundamentally from experimental evolution. For example, descriptions of the operation of natural selection in the wild (for examples, see Endler, 1986) are not experimental evolution, unless some experimental modification has been introduced, with appropriate replication and controls, into the system examined. Additionally, experimental evolution is not artificial selection. Artificial selection has one or more predefined criteria for the product of the evolutionary process (e.g. a morphological trait). Only organisms that meet the criteria are permitted to remain in the breeding population, and the selective criteria are continually altered to produce organisms with the desired traits. In contrast, experimental evolutionary studies create selective environments and then observe evolutionary change, in whatever form it may take. The approach has consequently been termed 'natural selection in the laboratory' (Rose *et al.*, 1990), although it may also be undertaken in the field [see, for example, the transplant experiments of Reznick and coworkers altering predatory environments of natural populations of guppies (Reznick & Bryga, 1987; Reznick *et al.*, 1990)].

Experimental studies of evolutionary adaptation to temperature

Experimental evolutionary investigations have frequently utilised temperature as the manipulated environmental variable and examined the adaptation of their subject populations to novel thermal regimes. Temperature, and change in temperature, for the ectothermic organisms in these studies is, of course, both environmentally relevant and biologically significant, particularly in view of interest in evolutionary responses to climate change. Even theoretical models of evolution (e.g. Levins, 1968; Lynch & Gabriel, 1987; Pease *et al.*, 1989) frequently use temperature change as an analytical example because of its manifest biological importance and illustrative potential.

The first experimental study of evolutionary adaptation to temperature was undertaken over a century ago by the Rev. W. H. Dallinger (1887). Dallinger, in an attempt to demonstrate that organisms can adapt to environmental change, slowly increased culture temperature for three different species of flagellate protozoans. He was successful in greatly increasing maximal temperatures tolerated in his cultures (from 22 to 70 °C) over a period of several years. He found long periods when no adaptive progress was made, punctuated by times of very rapid improvement in thermal tolerance (perhaps the appearance and selection of favourable mutations), and noted a trade-off in growth at high and low temperatures of the adapted lines. Although his experiments lacked certain features, such as thermal and contamination controls, they were extraordinary for their time. In a letter to Dallinger, Darwin declared that he found them to be 'extremely curious and valuable' and the adaptive responses of the protists to be 'very remarkable' (Dallinger, 1887).

More recent investigations of experimental evolution in different thermal environments have utilised a variety of organisms. Fruit flies (*Drosophila* spp.) have been particularly popular as experimental subjects because of the ease of their long-term laboratory culture, the obvious impact of temperature on their functional capacities, and the wealth of information about their genetics and other aspects of their biology. Several laboratories have now maintained replicated populations of flies at different temperatures for over a hundred generations and have examined the evolution of such characters as heat tolerance, body size, and shift in thermal niche (for examples, see Cavicchi *et al.*, 1989; Huey *et al.*, 1991; Partridge *et al.*, 1995). In addition, similar experiments are feasible using many other types of eukaryotic organisms, including nematodes (Grewal *et al.*, 1994), protists (Walton *et al.*, 1995), and fungi (Jinks & Connolly, 1973). The use and further development of these and other model systems for experimental evolutionary studies is certain to occur in the near future. In our own studies of evolutionary thermal adaptation, we have chosen to use bacteria as experimental subjects. We review below the utility of bacteria for these types of studies, the structure of our experiments and conclusions, and their implications for evolutionary adaptation to temperature.

Experimental studies of temperature adaptation in bacteria

Bacteria as experimental organisms

Bacteria possess all of the characteristics mentioned in the foregoing discussion of desirable attributes for subjects of experimental evolutionary studies. In addition, some bacteriological techniques are particularly useful for such studies. For example, bacterial populations can be preserved in a frozen state and resuscitated later for analysis. This means that direct comparisons can be made between the ancestral and derived genotypes under identical environmental conditions. Bacteria are also easily cloned, allowing the experimenter to found replicate populations with initially identical genetic composition. Finally, the wealth of information on the biochemistry, molecular biology and genetics of certain bacterial species, most notably *Escherichia coli*, may allow one to dissect measurable fitness changes into their underlying physiological adaptations and elucidate the genetic mechanisms that operate under a natural selection regime.

Given the practical advantages that these organisms offer, we chose to use bacterial populations to study thermal adaptation and to address some general questions regarding the process of evolutionary adaptation of organisms to the environment. How rapid is the evolutionary response to a change in temperature? How specific is the adaptive response to a particular environmental change? Does the direction or magnitude of change in temperature influence the dynamics of the evolutionary response? Does adaptation to a novel thermal environment imply trade-offs in performance in the ancestral environment or at other temperatures? Will replicate populations diverge as a result of finding unique solutions to identical environmental conditions, or will they repeatedly arrive at the same physiological adaptations to a common environment?

In the following sections, we review the structure of our experiments, their outcomes, and their implications for adaptive evolution. Detailed methods and analyses can be found in the original papers (Bennett *et al.*, 1990, 1992; Lenski & Bennett, 1993; LeRoi *et al.*, 1994*a*; Travisano *et al.*, 1995; Bennett & Lenski, 1996; Mongold *et al.*, 1996).

Overview of the experimental system

In our studies of evolutionary adaptation to temperature, we utilised a lineage of *Escherichia coli* B that lacks both plasmids and bacteriophage and is therefore strictly asexual. Thus we were able to establish a series of genetically identical, asexual populations all derived from a common ancestor, place them in different thermal environments and observe the adaptive responses and consequences which arose through de novo mutation and natural selection. The large population sizes and replication of populations maintained in each experimental environment make genetic drift an unlikely explanation for any group effects which may be observed. Immigration of naturally-occurring *E. coli* into our populations was ruled out by monitoring a series of genetic markers which are not typically found in wild strains. Migration between experimental populations was also monitored and ruled out by introducing a neutral genetic marker into the ancestral genotype and interspersing replicate populations with alternate states of this marker.

At the beginning of the experiment, and at subsequent intervals, samples from the evolving populations were removed and stored at −80 °C. These samples can be stored indefinitely and subsamples can then be thawed and used in future analyses. This feature of bacterial cells means that direct comparisons of the ancestral and derived genotypes may be conducted under identical environmental conditions. This is an extremely important factor in the experimental design because it allows heritable, genetic changes to be distinguished from physiological responses as a result of acclimation to different environments.

Our experimental populations were maintained and evolved in serial transfer culture. Propagation involved transferring a fraction of each population to fresh media on a daily basis. After an initial lag phase, the populations grow exponentially until the limiting nutrient (glucose in our experiments) is depleted and growth ceases. The bacteria then enter a quiescent stationary phase, which is maintained until new nutrients become available. This 'feast or famine' availability of resources and consequent cycling of growth phase may more closely approximate the condition of bacteria in nature than the continuous growth and stable (but low) resource levels provided in chemostats. Additionally, the multiple population growth phases afforded by serial transfer culture may allow a greater number of potential adaptive mechanisms in novel thermal environments. The length of the cycle period and size of the fraction transferred simply represent a constant level of random mortality in the environment. Although the growth rate differs with temperature, the experimental populations are limited by resource availability and, therefore, undergo the same number of generations per cycle in all of the thermal environments.

We used two experimental measurements of fitness to assess the pattern of adaptation in our experimental populations. The first is the Malthusian parameter, or absolute fitness (Bennett & Lenski, 1993), of a particular genotype in a particular environment. The absolute fitness is defined as the slope of the natural logarithm of population density regressed over time. This may be estimated for each genotype in pure culture under identical environmental conditions at varying temperatures. This provides an operational definition of the bacterium's thermal niche. The thermal niche encompasses the range of temperatures over which a genotype can grow fast enough to maintain a constant population size. To persist in our serial transfer environment, a population must be capable of increasing its population size 100-fold within a 24-h period. Outside of its thermal niche, a genotype would be incapable of maintaining a constant population size in the face of daily serial dilution, and its density would decline over time leading to eventual extinction. A negative Malthusian parameter therefore indicates that the population is growing more slowly than the dilution rate imposed by the environment.

The second measurement is the Darwinian fitness of a derived genotype relative to its ancestor. This quantity provides a measurement of a particular genotype's competitiveness in the environment of interest and, from an evolutionary point of view, is the most important property of an organism. Relative fitness is defined as the ratio of the Malthusian parameters of an evolved line and its ancestor under conditions of direct competition (Lenski *et al.*, 1991). Populations of the lines that are to compete are first separately acclimated to the environment of interest. After this preconditioning, the populations are mixed in fresh media and their relative abundances determined at the beginning and end of a specified time interval. An easily scorable genetic marker that is neutral in the assay environment is used to differentiate the two competitors in mixed culture. The rate of increase, or Malthusian parameters, for each competitor can then be calculated and relative fitness obtained from their ratio. By making replicated measures of relative fitness in a controlled environment, one can calculate the measurement error inherent in the experimental procedure and determine the significance of even small differences in the mean fitness of a population as it changes through time. Similarly, the ability to found replicate populations with genetically identical individuals means that the genetic variation measured among populations can be partitioned into variation as a result of: (i) chance differences in the beneficial mutations that have arisen in populations maintained under identical conditions; (ii) differences among treatment groups experiencing different environments; and (iii) genotype by environment interactions. This type of analysis is the same as the approach taken in comparative studies of adaptations among phylogenetically related taxa inhabiting similar and dissimilar habitats. The difference is that in laboratory evolution experiments, the phylogenetic relationships of the populations and their environmental histories are completely known and under the control of the experimenter. The phylogeny and thermal history of the experimental lineages in our study are illustrated in Fig. 1.

Evolutionary responses to novel thermal environments

In the experimental study outlined in Fig. 1, 30 populations were founded as clones from a single ancestral genotype. This ancestor was the product of a lineage of *E. coli* B that had been maintained in the laboratory under constant environmental conditions (serial transfer culture in minimal glucose medium at 37 °C) for 2000 generations (Lenski *et al.*, 1991). During this period, the lineage underwent extensive genetic adaptation to the laboratory culture conditions, but experienced relatively little further adaptive change during subsequent culture (Lenski & Travisano, 1994).

The 30 new experimental populations were divided into five treatment groups with six replicate lines each. These five groups were propagated for an additional 2000 generations under similar culture conditions, the only difference being the temperature at which the populations were incubated (20, 32, 37 and 42 °C, and daily alternation between 32 and 42 °C). The groups are named according to the temperature at which they were propagated, as indicated in Fig. 1. Using an ancestor from a lineage already well adapted to the culture environment was presumed to increase the likelihood that any further adaptation would be a specific response to a novel thermal environment rather than further adaptation to the general culture conditions. The 37 °C group served as a control to estimate the extent of further adaptation to general

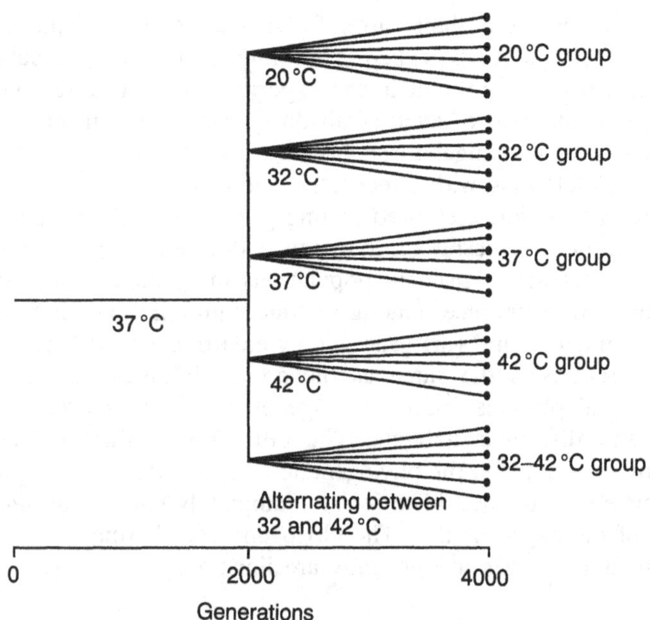

Fig. 1 Phylogeny and thermal history of the experimental lines and groups being described (derived from Mongold *et al.*, 1996.) Culture temperatures are shown below the designated lines, and group names are given at the right.

culture conditions because 37 °C was the ancestral temperature. An equal increase and decrease of 5 °C from the ancestral temperature were experienced by the 42 and 32 °C groups, respectively; however, 42 °C is also within 1 °C of the upper limit of the ancestral thermal niche, above which the ancestor cannot maintain itself in serial dilution culture. The corresponding lower thermal niche boundary of the ancestor is just below 20 °C, very close to that experienced by the 20 °C group. These four experimental groups were exposed to constant thermal environments, while the 32–42 °C group experienced and evolved in a variable thermal environment.

The relative fitness of clones isolated from each experimental line was measured relative to the common ancestor of all thirty lines (Fig. 2). The measurements of relative fitness were conducted at the temperature at which each group had evolved. For the 32–42 °C group, relative fitness was separately measured at 32 and 42 °C to assess the extent of adaptation to each component of the variable environment. By definition, the relative fitness of all lines was 1.0 at the beginning

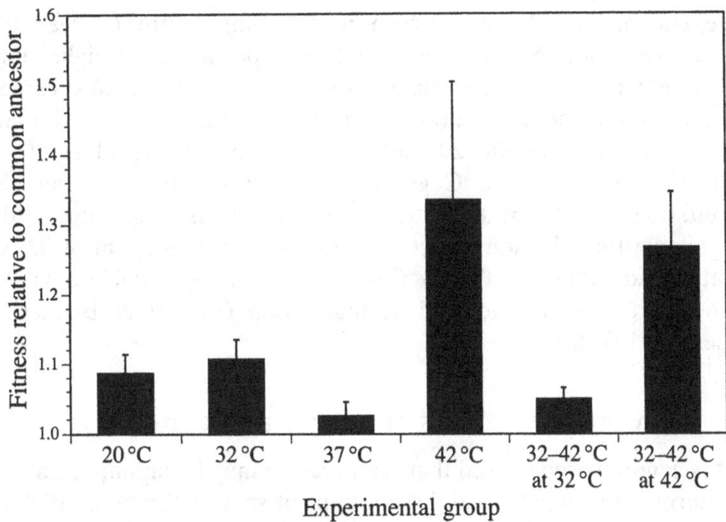

Fig. 2. Mean relative to their common ancestor for groups of *E. coli* populations that were propagated for 2000 generations at constant 20, 32, 37, or 42 °C or with daily alternation between 32 and 42 °C. Fitnesses were assayed at each group's selective temperature, or as indicated. The error bars represent 95% confidence intervals based on the six replicate lines in each group using a two-tailed t-distribution with $n-1 = 5$ degrees of freedom. (After Bennett & Lenski, 1996.)

of the experiment. The rate of adaptation was estimated by regressing relative fitness measured for each line over time (in generations). This rate of adaptation was significantly higher for all four of the groups experiencing novel thermal environments than for the control (37 °C) group, indicating that most of the adaptation that occurred was in response to the new thermal environments. The magnitude of the adaptive response was also highly variable between groups (Fig. 2). While there was no significant difference between the rates of adaptation to the environments below the ancestral temperature (i.e. 20 and 32 °C), the rate of adaptation to the high temperature environment (42 °C) was markedly greater. Similarly, in the 32–42 °C group, the rate of adaptation to the 42 °C component of the variable environment was substantially greater than that to the 32 °C component and was not significantly different from that of the 42 °C group to 42 °C. Therefore, the heterogeneity of the adaptive response to the different thermal regimens is not simply a function of the magnitude of the change from the ancestral temperature. An alternative explanation is that the dependence of the mutation rate on temperature (Savva, 1982) is responsible for the increased adaptive response to the higher temperature environments. The data, however, do not fit this explanation either. First, there was no difference in the magnitude of the adaptive response between the 20 and 32 °C groups (Mongold *et al.*, 1996). Second, in the 32–42 °C group, variation generated by an elevated mutation rate during the days spent at the high temperature would be available for selection to act on it during the days spent at 32 °C. The rate of adaptation to the 32 °C component of the variable environment, however, was not elevated in that group (Lenski & Bennett, 1993; Lenski, 1995).

Correlated effects at other temperatures

It is generally recognised that temperature may be an important variable determining geographical distributions of species (Somero, 1995). What is less clear are the factors constraining species from evolutionary expansion of their ranges, as most natural populations have abundant genetic variation and hence apparent potential for adaptation to novel environments. One common view is that adaptation to novel environments is constrained by trade-offs in performance (Futuyma & Moreno, 1988). That is, genetic correlations among performance traits are assumed to be important across an environmental gradient, such that specialisation for performance in a novel environment may be associated with a decrement in performance in the ancestral environment or other environments. On the other hand, depending on the nature of the correlations, improvement in performance in a marginal environment might carry over to environments beyond the range of experience and therefore actually extend an organism's potential thermal niche. Although thermal trade-offs and other correlations between environments are widely assumed (Levins, 1968; Lynch & Gabriel, 1987; Pease *et al.*, 1989), there is little empirical support for their existence (for example, Huey & Hertz, 1984; but see Gilchrist, 1996). In our bacterial experiments, however, we were able to examine directly whether evolutionary (genetic) trade-offs occurred within the original ancestral thermal niche during adaptation. Furthermore, we could also examine whether a shift in thermal niche limits occurred during this thermal adaptation.

Improvement in relative fitness within the original thermal limits of the bacterium was very specific to the temperature in which the particular lines evolved. Among the groups maintained at a constant temperature (i.e. the 20, 32, 37 and 42 °C groups), the range of temperatures over which each group improved relative to the ancestor usually extends only a few degrees in either direction from the temperature at which that group had evolved (Fig. 3). These groups may therefore be considered thermal specialists. The group maintained in a variable environment (32–42 °C group), on the other hand, showed significant improvement across a broad range of temperatures between its maximum and minimum daily temperatures, even extending to temperatures to which the lines had never been exposed (Fig. 3) (Bennett *et al.*, 1992). This group therefore comprises thermal generalists.

Although all of the specialist groups adapted to the greatest degree in their own thermal environment, the association of trade-offs in performance at other temperatures was highly asymmetrical with respect to the temperature of adaptation. In spite of the extensive adaptation of the groups which evolved at the higher temperature (42 °C), they suffered no significant loss in fitness at lower temperatures, even those below the ancestral temperature (Fig. 4). In contrast, adaptation to a much lower temperature (20 °C) was significantly correlated with a loss in relative fitness at higher temperatures. Similarly, the limits of the thermal niche, as measured by the temperature at which absolute fitness

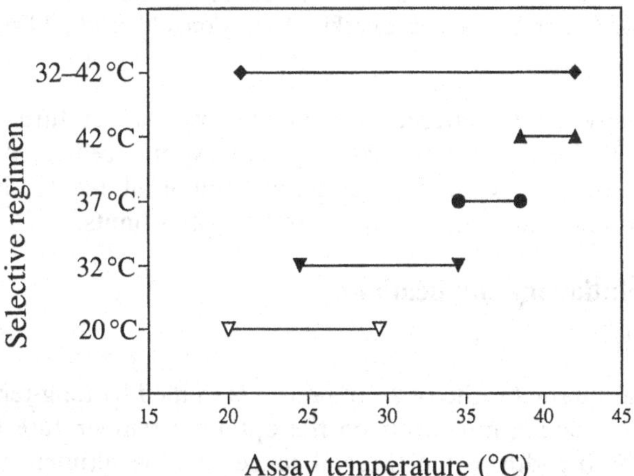

Fig. 3 Specificity of genetic adaptation with respect to environmental temperature in groups of *E. coli* populations that were propagated for 2000 generations at constant 20, 32, 37, or 42 °C or with daily alternation between 32 and 42 °C. Each solid line indicates the approximate range of temperature over which the mean fitness of a group was improved significantly relative to the common ancestor ($p < 0.05$). (After Bennett & Lenski, 1993; Mongold *et al.*, 1996)

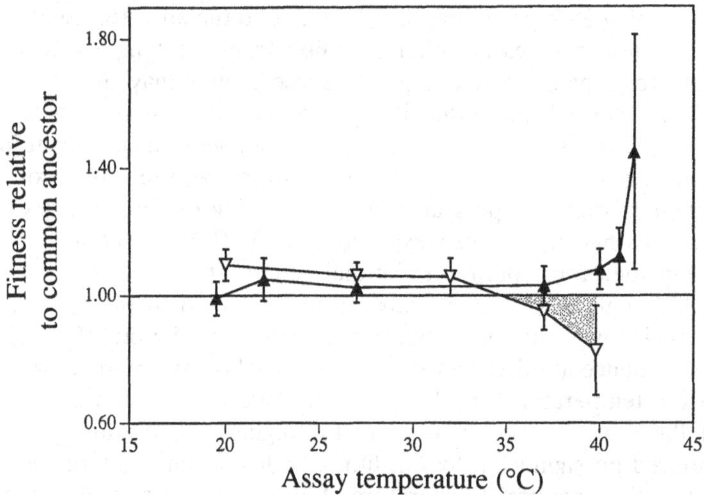

Fig. 4. Mean fitness of the 20 (▽) and 42 °C (▲) groups, relative to their common ancestor, measured across their thermal niche. The error bars represent 95% confidence intervals based on the six replicate lines in each group ($n-1 = 5$ d.f.). Note the trade-off in fitness of the 20 °C group when assayed at the higher temperatures (shaded area). (After Bennett & Lenski, 1993; Mongold *et al.*, 1996.)

became negative, were altered only in the low temperature adapted lines (Fig. 5): the lines of the 20 °C group experienced a downward shift in both their upper and lower thermal limits of 1–2 °C. None of the other groups had significantly altered thermal limits.

Evolutionary implications

Niche evolution

The potential thermal niche of an organism is shaped by long-term evolutionary forces acting in concert on the optimum temperature for performance, the breadth of the thermal range and the ultimate limits of thermal tolerance. An understanding of the correlations between these traits is fundamental to understanding the diversity of ecological types found in nature. For example, does evolutionary adaptation to a higher temperature result in a correlated shift in the entire range of thermal tolerance, a shift only in the optimum temperature for performance, or a broader niche with performance within the original thermal range being unaffected? Our experiments enable us to directly examine the effects of adaptation to altered temperatures on niche evolution (Fig. 5). Adap-

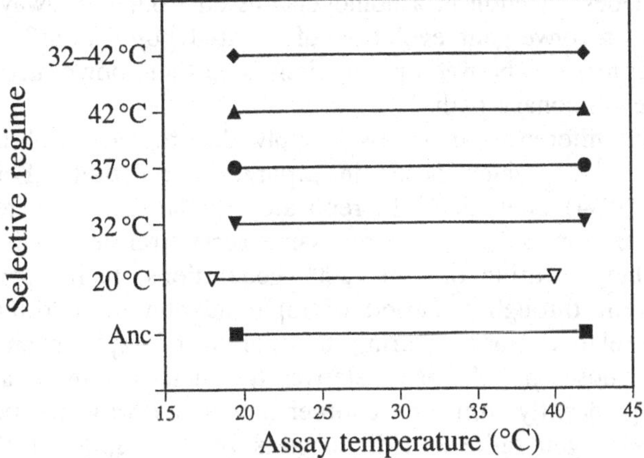

Fig. 5. Thermal niche of the ancestor and five experimental groups. Each solid line indicates the approximate range of temperatures over which the mean Malthusian parameter for a group was not negative, i.e. those temperatures where the lines were able to maintain a constant population density by replicating sufficiently to offset the daily serial dilution. The ancestral niche has been modified only in the 20 °C group. (After Bennett & Lenski, 1993; Mongold *et al.*, 1996.)

tation to 20 °C, very near the lower thermal niche boundary, resulted in both the lower and upper limits of thermal tolerance being shifted 1–2 °C lower. This group not only adapted to the selective temperature, but also became preadapted to even lower temperatures.

Performance at high temperatures was reduced in the process. In contrast, adaptation to a temperature very near the upper thermal limit, 42 °C, while greatly increasing fitness at that temperature (Fig. 2), did not appreciably shift either the upper or lower thermal limits from their ancestral condition. These results demonstrate the complexity of evolutionary changes in the niche. Starting with a single genotype of this single species, selection at or near the upper and lower niche limits produced completely different patterns of niche evolution. It is highly unlikely that any universal or predictable pattern of niche evolution will emerge for organisms in general if such contrasting effects can be observed within a single experimental system.

Origin and maintenance of diversity

The foregoing discussion examined how an organism's ecological potential might be affected by adaptation to a specific environment. Also of interest is the effect of adaptation on an organism's evolutionary potential. Does selection in a homogeneous environment always result in parallel or convergent evolution of isolated populations? Or, will historical differences between populations send them down increasingly divergent evolutionary paths?

Historical differences may arise simply due to chance differences in the mutations which occur in separate populations. Lenski & Travisano (1994) maintained 12 replicate populations of bacteria for 10 000 generations at 37 °C in the same serial transfer environment described here. During the first 2000 generations, their populations similarly went through a period of rapid adaptation followed by a period of relative stasis. During the period of rapid change, the fitness of those populations relative to their common ancestor diverged significantly from one another and that divergence persisted for over 8000 generations. This suggests that, in spite of the fact that they experienced identical environmental conditions, the replicate populations acquired distinct adaptations with unequal effects on fitness.

Historical differences can also arise between populations which have adapted to different environmental conditions. In our study, we had groups of populations with 2000 generations of evolutionary history in different thermal environments. We could then ask whether historical contingencies had arisen which would constrain their future adaptive potential and promote continued divergence even if the populations subsequently experienced identical environments. To test this, the experimental lines previously adapted to 32, 37, 42 and 32–42 °C were moved to 20 °C and propagated in the same manner for an additional 2000 generations. After that period of time, all groups significantly increased their relative fitnesses at 20 °C, but there were no significant differences in the magnitude of this increase among the four groups with different thermal histories (Travisano *et al.*, 1995; Mongold *et al.*, 1996). Thus, in this case, specialisation for growth in different thermal environments had no directional impact on future potential for adaptation to this novel thermal environment. This conclusion must be qualified, however, as we do not know whether the outcome might have been different if the populations had been allowed to diverge for longer in their original selective environments.

Evolutionary response to stress

It has been proposed that physiological stresses, such as those caused by changes in climate, may accelerate the rate of adaptive evolution

(Parsons, 1987; Hoffman & Parsons, 1991). This proposition is based on observations of rapid divergence of local populations inhabiting marginal habitats on the edge of a species' geographical range or in semi-toxic environments (e.g. mine tailings with high concentrations of heavy metals) (Hoffmann & Parsons, 1991; Howarth, 1993). The proximate cause hypothesised for this rapid evolution is that physiological stress exposes a greater fraction of existing genetic variation to the action of selection.

A problem in testing this proposal is the difficulty in operationally defining and quantifying 'stress' in most biological systems (see Hoffmann & Parsons, 1991; Lenski & Bennett, 1993). Many of the criteria proposed for stress, both biochemical (e.g. stress protein formation) and ecological (e.g. growth rate and yield depression), can, however, easily be measured in bacteria. Our experimental system can therefore be used to illuminate the correlations between stress and evolutionary adaptation. Both 20 and 42 °C are very near the lower and upper thermal limits of population persistence of our ancestral bacterium. Environments near a niche edge are therefore expected to be stressful, and hence considered to be marginal. The intensity of selection might be expected to be equally intense for beneficial mutations occurring in both 'edge' environments. On the other hand, the biology of the bacteria at the two thermal boundaries is very different. The upper niche boundary is extremely sharp and characterised by a sudden shift from rapid growth to marked death between 42 and 44 °C. By contrast, the lower niche boundary is characterised by a gradual reduction in growth rate, which at about 19 °C becomes insufficient to offset the losses due to serial dilution. Lower temperatures are not lethal, but sufficiently inhibiting to slow growth rate. Growth rate is depressed at both 20 and 42 °C in comparison to 37 °C; in contrast, growth yield, biomass formed from available nutrients, is depressed only at 42 °C, not at 20 °C. Both temperatures induce a suite of 'stress response' genes in *E. coli* (Jones *et al.*, 1992; Craig *et al.*, 1993). At least 20 proteins are preferentially induced by shifts up in temperature (Neidhardt *et al.*, 1984; Delaney *et al.*, 1992). The functions of all of these proteins are not yet known, but they are believed to act, among other things, as molecular chaperones (Ellis & van der Vies, 1991; Martin *et al.*, 1991; Craig et al., 1993), aiding in the correct folding and oligomerisation of proteins. Following a downshift of 13 °C or more, the major cold shock protein, F10.6, is induced (Jones *et al.*, 1992). Induction of the cold shock response is believed to be negatively regulated by the level of (p)ppGpp and perhaps has some connections with the stringent response network (Jones *et al.*, 1992).

Therefore, by many criteria (e.g. niche edge, stress protein formation, growth rate depression), both 20 and 42 °C are stressful environments, although they differ with respect to induced mortality. The evolutionary response to those environments, however, was quite different. Adaptation was far more rapid and extensive in the high temperature environment, both in the 42 °C group and in the 32–42 °C group in the 42 °C environment (Fig. 2). In this environment, the expected match between stress and rapid evolution was indeed observed. The rate of adaptation to 20 °C, however, was considerably slower and no different from that at 32 °C, a very benign thermal environment that meets almost none of the criteria for stress. Stressful environments, therefore, do not invariably produce rapid rates of evolutionary adaptation.

Interestingly, adaptation to the lower temperature resulted in more far-reaching effects with regard to performance in other thermal environments. It entailed extensive trade-offs within the ancestral thermal niche and a downward shift of both upper and lower limits of the thermal niche, whereas adaptation to the high temperature was very temperature specific but produced neither trade-offs nor a shift in which limits (Figs. 4 and 5). Thus, rapid evolution associated with stress does not necessarily entail extensive changes in niche structure, but may be highly specific to a single environment.

The observations of an elevated rate of adaptation to high temperature, and the asymmetrical nature of the correlated responses associated with adaptation to high versus low temperature, may provide some insight into the targets of selection in these thermal environments. It appears that many of the traits that are important for performance at lower temperatures are functional across the thermal spectrum and are very temperature sensitive in terms of optimal performance. At high temperature, however, other functions may come into play that are not even expressed at lower temperatures. If these are the targets of selection, then their alteration may have relatively little impact on performance in other thermal environments.

Preadaptation

Adaptation entails the evolution of specific mechanisms that improve performance in a particular selective environment. These mechanisms may impact not only the functions selected but a host of correlated responses. These correlated responses may have only minor, or even no, functional consequences in the selective environment. If the population subsequently colonises new environments or the original environment changes, however, then these formerly unimportant traits may differentially enable or disable the population in the new environment. Specifically, they may preadapt the population to the new circumstances, increasing fitness or even permitting persistence when it would otherwise be impossible. Alternatively, they may hinder performance in the new environment perhaps even dooming the population to eventual extinction if it cannot cope with the environmental change.

Although the concept of preadaptation is intuitively appealing, demonstrating its occurrence and generality, and investigating its properties can be problematic or impossible. Our bacterial system, however, demonstrates unequivocally that adaptation to one well-defined and characterised environmental factor, temperature, can entail widespread, genetically-determined divergence in functional capacities of other traits of no current selective value. Specifically, capacities to utilise the nutrient maltose, which was not present in the selective environment for at least 4000 generations and probably considerably longer, were greatly and differentially altered during temperature adaptation. Glucose was the only nutrient supplied during this evolutionary experiment, and glucose and maltose are transported across the outer and inner membranes by completely different pathways (Nikaido & Saier, 1992). Many of the adaptive mechanisms in our diverse thermal environments apparently involved changes in glucose transport, and these had a variety of correlated effects on the ability of the bacteria to utilise maltose as a nutrient. All possible responses were seen among the experimental lines (Fig. 6). In some, fitness in maltose was unaltered from the ancestral condition; in others, fitness declined; in still others, gains in fitness in maltose actually equalled or even exceeded those in glucose. In the 42 °C group, for instance, all individual lines had improved fitness in maltose, increasing the average fitness of the group in maltose by 55%. Some lines had as much as doubled fitness in maltose. Similar improvements in performance in maltose occurred in the 32–42 °C group at 42 °C.

For whatever mechanistic reasons, adaptation to a high temperature–glucose environment preadapted these experimental lineages to environments with maltose as the nutrient. If these lineages should now secondarily encounter maltose-containing environments, they will be more fit in these new environments than their ancestor as a result of temperature adaptation. Such preadaptation was not, however, an inevitable outcome of adaptation to novel temperatures: some individual lines of the 20 and 32 °C groups declined in fitness in maltose and were less able than their common ancestor to prosper competitively in

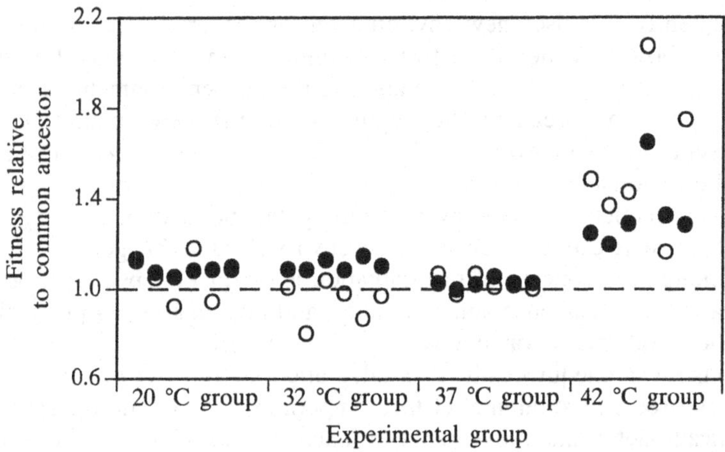

Fig. 6. Fitnesses of the individual lines in each of the four experimental groups selected at constant temperature (20, 32, 37 and 42 °C) relative to their common ancestor when grown in glucose (●) and maltose (○). (After Bennett & Lenski, 1993.)

those environments (Fig. 6). The differences in these correlated characters must depend on the exact mechanistic nature of the thermal adaptation in each lineage. They remain covert and unimportant, until the organisms find themselves tested in new environments.

Conclusion and perspectives

The study of evolutionary adaptation, as both process and product, is proving to be a fruitful field for interaction between organismal and evolutionary biology. The former provides a detailed understanding of the functional basis of adaptation, an appreciation for the integration of adaptive traits into all aspects of organismal function and a rich tradition of study of organism–environment interaction. Evolutionary biologists contribute a firm foundation in genetics and population biology and a rigorous theoretical and analytical approach to analyses of adaptation. Together, these formerly distinct fields can build a much richer and more comprehensive understanding of evolutionary adaptation than either could possibly accomplish separately. We have attempted to combine approaches and viewpoints from both traditions in our studies.

What specifically can be learned from our studies on evolutionary adaptation to temperature in bacteria? Our study system has strict limitations: it was founded from a single strain of bacteria, indeed by a single genotype of that strain. Its populations were originally genetically homogeneous. Reproduction was strictly asexual, without the possibility of sexual recombination, so that all novelty had to arise *de novo* by mutation. This system cannot, therefore, serve as a model for the initial stages of adaptive evolution in genetically heterogeneous, sexually reproducing populations. In these, exposure to a new environment might favour a portion of the existing variability within the population, and adaptive novelty might arise through genetic recombination and be selected subsequently. Ultimately, however, the generation of novelty depends on mutation. Comparative studies on natural populations and species demonstrate that evolutionary adaptation to different thermal regimens depends on the origin of different alleles (examples reviewed by Hochachka & Somero, 1984; Powers, 1987; Somero, 1995). Our system permits the explicit study of the generation and consequences of such mutational novelty. This is not a general model for thermal adaptation in all biological systems: no single organism ever could be such, be it bacterium, plant, fly or mouse. The generality of the specific patterns obtained in our studies can be determined only by similar analyses on other systems.

The outstanding utility of this study system is in its ability to address and test a host of different general assertions about biological adaptation. *E. coli* is a typical mesophilic bacterium with a broad thermal niche and is not specialised for function in extreme thermal environments. It is therefore an excellent organism in which to examine the evolutionary responses to both moderate and stressful environmental change. Using this system, we have been able to conclude the following properties of evolutionary temperature adaptation:

1. Rates of adaptation (fitness improvement) can vary significantly among different novel thermal environments.
2. Adaptation can be highly temperature specific, often to a range of only a few degrees.
3. Adaptation and specialisation do not necessarily involve trade-offs in other environments.
4. Adaptation, even to niche extremes, does not necessarily involve a change in thermal niche.
5. Adaptation to stressful environments is not necessarily more rapid or extensive than adaptation to non-stressful environments.
6. Adaptation to an historical environment does not necessarily impede the rate of adaptation to a novel environment.

7. Correlated consequences of thermal adaptation may result in extensive preadaptation to other novel environments.

Several of these conclusions contradict widely held assertions (e.g. Levins, 1968; Hoffmann & Parsons, 1991) about patterns of evolutionary adaptation. If such evolutionary patterns are not found in this first detailed experimental test of their assertions, they are certainly not universal and are unlikely to be general. This is the particular utility of experimental evolution: it permits us to test propositions that we could formerly only assert.

Acknowledgements

We thank the Society for Experimental Biology and NSF Grant IBN-9400659 for support to attend this symposium. Our original research was supported by an Irvine Faculty Research Fellowship and NSF Grants (DEB-9208662 and IBN-9507416) to A.F. Bennett and R.E. Lenski. Further support was provided by the NSF Center for Microbial Ecology (BIR-910006) at Michigan State University.

References

Bennett, A.F. (1996). Adaptation and the evolution of physiological characters. In *Handbook of Comparative Physiology*, ed. W.H. Dantzler. New York: Oxford University Press (in press).

Bennett, A.F., Dao, K.M. & Lenski, R.E. (1990). Rapid evolution in response to high temperature selection. *Nature* **346**, 79–81.

Bennett, A.F. & Lenski, R.E. (1993). Evolutionary adaptation to temperature. II. Thermal niches of experimental lines of *Escherichia coli. Evolution* **47**, 1–12.

Bennett, A.F. & Lenski, R.E. (1996). Evolutionary adaptation to temperature. V. Adaptive mechanisms and correlated responses in experimental lines of *Escherichia coli. Evolution* **50**, 493–503.

Bennett, A.F., Lenski, R.E. & Mittler, J.E. (1992). Evolutionary adaptation to temperature. I. Fitness responses of *Escherichia coli* to changes in its thermal environment. *Evolution* **46**, 16–30.

Block, B.A., Finnerty, J.R., Stewart, A.F.R. & Kidd, J. (1993). Evolution of endothermy in fish: Mapping physiological traits on a molecular phylogeny. *Science* **260**, 210–14.

Brooks, D.R. & McLennan, D.A, (1991). *Phylogeny, Ecology, and Behavior*. Chicago: University of Chicago Press.

Cavicchi, S., Guerra, V., Natali, V., Pezzoli, C. & Giorgi, G. (1989). Temperature-related divergence in experimental populations of *Drosophila melanogaster*. II. Correlation between fitness and body dimensions. *Journal of Evolutionary Biology* **2**, 235–51.

Cossins, A.R. & Bowler, K. (1987). *Temperature Biology of Animals*. New York: Chapman and Hall.

Craig, E.A., Gambill, B.D. & Nelson, R.J. (1993). Heat shock proteins: molecular chaperones of protein biogenesis. *Microbiological Review* **57**, 402–14.

Dallinger, W.H. (1887). The president's address. *Journal of the Royal Microscopical Society*. April, 185–99.

Delaney, J.M., Ang, D. & Georgopoulos, C. (1992). Isolation and characterization of the *Escherichia coli htrD* gene whose product is required for growth at high temperature. *Journal of Bacteriology* **174**, 1240–7.

Ellis, R.J. & van der Vies, S.M. (1991). Molecular chaperones. *Annual Review of Biochemistry* **60**, 321–47.

Endler, J.A. (1986). *Natural Selection in the Wild*. Princeton: Princeton University Press.

Feder, M.E. (1987). The analysis of physiological diversity: the prospects for pattern documentation and general questions in ecological physiology. In *New Directions in Ecological Physiology* (ed. M.E. Feder, A.F. Bennett, W.W. Burggren & R.B. Huey, pp. 38–75. Cambridge: Cambridge University Press.

Felsenstein, J. (1985). Phylogenies and the comparative method. *American Naturalist* **125**, 1–15.

Futuyma, D.J. & Moreno, G. (1988). The evolution of ecological specialization. *Annual Review of Ecological Systematics* **19**, 207–33.

Garland, T., Jr. & Adolph, S.C. (1994). Why not to do two-species comparative studies: limitations on inferring adaptation. *Physiological Zoology* **67**, 797–828.

Garland, T., Jr., Huey, R.B. & Bennett, A.F. (1991). Phylogeny and thermal physiology in lizards: a reanalysis. *Evolution* **45**, 1969–75.

Gilchrist, G.W. (1996). A quantitative genetic analysis of thermal sensitivity in the locomotor performance curve of *Aphidius ervi*. *Evolution* (in press).

Gould, S.J. (1989). *Wonderful Life: The Burgess Shale and the Nature of History*. New York: Norton.

Graves, J.E. & Somero, G.N. (1982). Electrophoretic and functional enzymatic evolution in four species of eastern Pacific barracudas from different thermal environments. *Evolution* **36**, 97–106.

Grewal, P.S., Selvan, S. & Gaugler, R. (1994). Thermal adaptation of entomathogenic nematodes: niche breadth for infection, establishment and reproduction. *Journal of Thermal Biology* **19**, 245–53.

Harvey, P.H. & Pagel, M.D. (1991). *The Comparative Method in Evolutionary Biology*. Oxford: Oxford University Press.

Hochachka, P.W. & Somero, G.N. (1984). *Biochemical Adaptation*. Princeton: Princeton University Press.

Hoffmann, A.A. & Parsons, P.A. (1991). *Evolutionary Genetics and Environmental Stress*. Oxford: Oxford University Press.

Howarth, F.G. (1993). High-stress subterranean habitats and evolutionary change in cave-inhabiting arthropods. *American Naturalist* **142**, S65–S77.

Huey, R.B. (1987). Phylogeny (history, and the comparative method. In *New Directions in Ecological Physiology* (ed. M.E. Feder, A.F. Bennett, W.W. Burggren & R.B. Huey), pp. 76–98. Cambridge: Cambridge University Press.

Huey, R.B. & Bennett, A. F. (1987). Phylogenetic studies of coadaptation: preferred temperatures versus optimal performance temperatures of lizards. *Evolution* **41**, 1098–115.

Huey, R.B. & Hertz, P.E. (1984). Is a jack-of-all-temperatures a master of none? *Evolution* **38**, 441–4.

Huey, R.B., Partridge, L. & Fowler, K. (1991). Thermal sensitivity of *Drosophila melanogaster* responds rapidly to laboratory natural selection. *Evolution* **45**, 751–6.

Jinks, J.L. & Connolly, V. (1973). Selection for specific and general response to environmental differences. *Heredity* **30**, 33–40.

Jones, P.G., Cashel, M., Glaser, G. & Neidhardt, F.C. (1992). Function of a relaxed-like state following temperature downshifts in *Escherichia coli. Journal of Bacteriology* **174**, 3903–14.

Krebs, H.A. (1975). The August Krogh principle: 'For many problems, there is an animal on which it can be most conveniently studied'. *Journal of Experimental Zoology* **194**, 221–6.

Krogh, A. (1929). Progress of physiology. *American Journal of Physiology* **90**, 243–51.

Lauder, G.V., LeRoi, A.M. & Rose, M.R. (1993). Adaptations and history. *Trends in Ecology and Evolution* **8**, 294–7.

Lenski, R.E. (1995). Evolution in experimental populations of bacteria. In *Population Genetics of Bacteria* (ed. S. Bamberg, J.P.W. Young, S.R. Saunders & E.M.H. Wellington). Cambridge: Cambridge University Press.

Lenski, R.E. & Bennett, A.F. (1993). Evolutionary response of *Escherichia coli* to thermal stress. *American Naturalist* **142**, S47–S64.

Lenski, R.E., Rose, M.R., Simpson, S.C. & Tadler, S.C. (1991). Long-term experimental evolution in *Escherichia coli.* I. Adaptation and divergence during 2000 generations. *American Naturalist* **138**, 1315–41.

Lenski, R.E. & Travisano, M. (1994). Dynamics of adaptation and diversification: A 10 000-generation experiment with bacterial populations. *Proceedings of the Naturalist Academy of Sciences* **91**, 6808–14.

LeRoi, A.M., Lenski, R.E. & Bennett, A.F. (1994*a*). Evolutionary adaptation to temperature. III. Adaptation of *Escherichia coli* to temporally varying environment. *Evolution* **48**, 1222–9.

LeRoi, A.M., Rose, M.R. & Lauder, G.V. (1994*b*). What does the comparative method reveal about adaptation? *American Naturalist* **143**, 381–402.

Levins, R. (1968). *Evolution in Changing Environments*. Princeton: Princeton University Press.

Lynch, M. & Gabriel, W. (1987). Environmental tolerance. *American Naturalist* **129**, 283–303.

Martin, J., Horwich, A.L. & Hartl, F.-U. (1991). Prevention of protein denaturation under heat stress by the chaperonin Hsp60. *Science* **258**, 995–8.

Martins, E.P. & Garland, T., Jr. (1991). Phylogenetic analyses of the correlated evolution of continuous characters: a simulation study. *Evolution* **45**, 534–57.

Mongold, J.A., Bennett, A.F. & Lenski, R.E. (1996). Evolutionary adaptation to temperature. IV. Selection at a niche boundary. *Evolution* **50**, 35–43.

Neidhardt, F.C., Vanbogelen, R.A. & Vaughn, V. (1984). The genetics and regulation of heat-shock proteins. *Annual Review of Genetics* **18**, 295–329.

Nikaido, H. & Saier, M.H., Jr. (1992). Transport proteins in bacteria: common themes in their design. *Science* **258**, 936–42.

Parsons, P.A. (1987). Evolutionary rates under environmental stress. In *Evolutionary Biology*, vol. 21 (ed. M.K. Hecht, B. Wallace & G.T. Prance), pp. 311–47. New York: Plenum Press.

Partridge, L., Barrie, B., Fowler, K. & French, V. (1995). Evolution and development of body size and cell size in *Drosophila malanogaster* in response to temperature. *Evolution* **48**, 1269–76.

Pease, C.M., Lande, R. & Bull, J.J. (1989). A model of population growth, dispersal, and evolution in a changing environment. *Ecology* **70**, 1657–64.

Powers, D.A. (1987). A multidisciplinary approach to the study of genetic variation within species. In *New Directions in Ecological Physiology* (ed. M.E. Feder, A.F. Bennett, W.W. Burggren & R.B. Huey), pp. 102–30. Cambridge: Cambridge University Press.

Precht, H., Christofersen, J., Hensel, H. & Larcher, W. (1973). *Temperature and Life*. Berlin: Springer-Verlag.

Prosser, C.L. (1973). *Comparative Animal Physiology*, 3rd edn. Philadelphia: Saunders.

Prosser, C.L. (1986). *Adaptational Biology: Molecules to Organisms*. New York: John Wiley.

Reznick, D.N. & Bryga, H. (1987). Life-history evolution in guppies. 1. Phenotypic and genotypic changes in an introduction experiment. *Evolution* **41**, 1370–85.

Reznick, D.N., Bryga, H. & Endler, J.A. (1990). Experimentally-induced life-history evolution in a natural population. *Nature* **346**, 357–9.

Rose, M.R., Graves, J.L. & Hutchison, E.W. (1990). The use of selection to probe patterns of pleiotropy in fitness characters. In *Insect Life Cycles: Genetics, Evolution and Co-ordination* (ed. F. Gilbert), pp. 29–42. New York: Springer-Verlag.

Savva, D. (1982). Spontaneous mutation rates in continuous cutlures: the effect of some environmental factors. *Microbios* **33**, 81–92.

Scholander, P.F., Flagg, W., Walters, V. & Irving, L. (1953). Climatic adaptation in arctic and tropical poikilotherms. *Physiological Zoology* **26**, 67–92.

Somero, G.N. (1995). Proteins and temperature. *Annual Review of Physiology* **57**, 43–68.

Travisano, M., Mongold, J.A., Bennett, A.F. & Lenski, R.E. (1995). Experimental tests of the roles of adaptation, chance, and history in evolution. *Science* **267**, 87–90.

Walton, B.M. (1993). Physiology and phylogeny: the evolution of locomotor energetics in hylid frogs. *American Naturalist* **141**, 26–50.

Walton, B.M., Gates, M.A., Kloos, A. & Fisher, J. (1995). Interspecific variability in the thermal dependence of locomotion, population growth, and mating in the ciliated protist *Euplotes vannus*. *Physiological Zoology* **68**, 98–113.

L. PARTRIDGE and V. FRENCH

Thermal evolution of ectotherm body size: why get big in the cold?

Introduction

Body size has profound consequences for animal life history and ecology (Bonner, 1965; Peters, 1983; Calder, 1984; Schmidt-Neilsen, 1984; Damuth, 1987), so it is very important to understand exactly how natural selection acts in the evolution of this character. Temperature is a crucial aspect of the environment that appears to influence body size in two ways. First, temperature may be an agent of natural selection in producing evolutionary (genetic) changes in the developmental mechanisms that control growth rate and adult size and, second, the thermal conditions during an individual's development may affect its final adult size. Thermal evolution of body size has been discussed mainly in the context of endotherms (principally mammals), where the relationships between the surface area and volume of the body are important in determining rates of heat production and dissipation, and hence in maintaining the standard body temperature. It has long been maintained that endotherms tend to evolve larger body size in colder conditions (Mayr, 1963): an idea traditionally termed 'Bergmann's rule' (Bergmann, 1847). This generalisation, however, has since been questioned (Ralls & Harvey, 1985), and the relationships between body size and thermal ecology within and between endothermic species await rigorous comparative analysis.

Thermal evolution of body size is not restricted to endotherms, however, as several species of ectotherm show clear geographical clines in body size, with the larger individuals found in populations derived from higher latitudes, even when all are reared in standard conditions. Such clines have been demonstrated in the copepod crustacean *Scottolana canadensis* (Lonsdale & Levinton, 1985) and, particularly, in insects such as the honey bee *Apis mellifera* (Alpatov, 1929), the house fly *Musca domestica* (Bryant, 1977) and several species of the fruit fly *Drosophila* (Stalker & Carson, 1947; Prevosti, 1955; David & Bocquet,

1975; Lemeunier *et al.*, 1986; Coyne & Beecham, 1987; Imasheva *et al.*, 1994; James *et al.*, 1995, but see Long & Singh, 1995). The repeatability of these size clines, in different species and in the same species across different continents, suggests that the genetic differences underlying them are caused primarily by natural selection rather than by drift and dispersal (Endler, 1977).

Many ecological variables change with latitude (e.g. levels of competition and parasitism, food supply, day length, rainfall and temperature), but two lines of evidence suggest that temperature may be particularly important with respect to the clines in body size. First, genetic changes conferring increased body size have also been reported with increasing altitude in *Drosophila* (Stalker & Carson, 1948) and in an anuran *Rana sylvatica* (Berven, 1982), and in the cooler periods of the breeding season in *Drosophila* (Stalker & Carson, 1949; Tantawy, 1964). Stronger evidence comes from results of laboratory studies of thermal evolution in *Drosophila*, where replicated populations have been kept in long-term culture at different temperatures but in otherwise similar conditions. As with the field clines, replication is important to rule out genetic drift as a cause of genetic divergence between populations. In this way, both *D. pseudoobscura* (Anderson, 1966, 1973) and *D. melanogaster* (Partridge *et al.*, 1994*a*) have been shown to increase significantly in body size in response to several years' thermal selection at lower temperature.

Temperature is therefore strongly implicated as a selective agent influencing *Drosophila* body size, although its mechanisms of action are far from clear and may involve other, intervening variables. Resistance to desiccation is unlikely to be an issue as the direction of change in body size is opposite to that predicted. Rates of heat exchange and equilibrium body temperatures are also unlikely to be important because several of the organisms that show thermal evolution of body size, including *Drosophila*, are in a size range where they very rapidly adopt the temperature of their surroundings (Stevenson, 1985).

It has long been known that rearing temperature can influence not only growth rate but also the final adult size of ectotherms (Ray, 1960; von Bertalanffy, 1960). In a careful and extensive review of the literature, Atkinson (1994) showed that, in over 80% of ectotherm species studied, decreased rearing temperature resulted in an increase in size. This was a consistent pattern across protists, plants and animals, across animal phyla and across the many studies of insects. Recently there has been considerable theoretical interest in the evolution of this pattern of response (or reaction norm) to rearing temperature (see pp. 183–204; Atkinson, 1994; Berrigan & Charnov, 1994; Sibly & Atkinson,

1994). The similarity of the evolutionary and developmental responses of body size to temperature suggests that they may have similar underlying explanations.

Here we discuss the thermal evolution of body size in ectotherms, concentrating on insects, and particularly on *Drosophila*, because the phenomenon is much better documented there. Also, we have much more experimental evidence relating to the different selection pressures that may be involved, in the larval and in the adult parts of the life history. Body size may evolve in response to temperature, both as a result of physiological effects within individual organisms and because of ecological effects on the dynamics of the populations of which the individuals are members. We discuss both of these. *Drosophila* has proved very useful for tackling the problem, because studies of ecology, genetics and physiology can be combined. In particular, evolutionary experiments in the laboratory are feasible on a reasonable time scale. An evolutionary change in body size in response to temperature must occur by modification of the development mechanisms that control growth and its cessation. We start by discussing these mechanisms, which are better understood in insects than in other animals.

Developmental control of insect growth and size

The insect body is contained within the cuticular exoskeleton which is secreted by the surface epidermis and, in general, growth is possible only during replacement of the cuticle in the periodic moult cycles of pre-adult development. The moult cycle is characterised by sequential changes in the epidermal cells which enlarge, separate from the old cuticle, divide to form a larger corrugated cell sheet, and then secrete a new cuticle on their outer surface. At moulting, the old cuticle is split and shed, and then the soft new cuticle flattens out and hardens. In this way, size increases by characteristic increments between the stages (instars) until the adult appears.

The progress of the moult cycle is controlled by hormones (for reviews, see Sehnal, 1985; Nijhout, 1994) and, in particular, by the changing concentration of ecdysone which is secreted from the prothoracic gland. Ecdysone secretion is triggered by the release of prothoracicotropic hormone (PTTH) from the brain. In both hemimetabolous (e.g. bugs) and holometabolous (e.g. moths and flies) insects, the release of PTTH is associated with a period of feeding and, particularly, with growth to a 'critical weight' (Nijhout, 1994). If starved before reaching the critical weight for that instar, the larva will fail to initiate the moult cycle, whereas starvation just after this point permits moulting. As size depends on the feeding (and resulting growth in the epidermis) that occurs after the critical weight, the starved insect will be very small in the next instar. In several insect species, critical weight has been shown to depend on the size at the beginning of that instar, suggesting that distension of body may directly stimulate PTTH release and, in two species of bug, the abdominal pressure receptors have now been identified (Nijhout, 1994). This may not be a general mechanism, however: in other insects, feeding before the critical point may act through raising nutrient levels in the blood (Sehnal, 1985). PTTH release and progress through the moult cycle frequently also depend on environmental stimuli, such as photoperiodic phase, and can be delayed by regeneration following injury or limb amputation (Bulliere & Bulliere, 1985; Sehnal & Bryant, 1993). Completion of the last larval instar determines the final (adult) size of the insect, and this instar is characterised by an early fall in concentration of juvenile hormone (JH). The elimination of JH also seems to be associated with attaining a specific larval size (weight) at the start of the instar but the mechanism is not known. In many insects, this specific size (and hence the final adult size) can be influenced by the quality of larval nutrition and by other environmental stimuli (Sehnal, 1985; Nijhout, 1994).

Flies, such as *Drosophila*, have a dramatic metamorphosis in which the last (third) larval instar moults to the pupal and then adult stages. The hormonal changes leading to metamorphosis are initiated early in the third larval instar: as in other insects, starvation before this critical point blocks further development (Bakker, 1959; Robertson, 1963; Partridge *et al.*, 1994*b*). After the critical point, a period of feeding normally results in continued growth of the larva which then wanders from the food, empties its gut, contracts and hardens its persisting larval cuticle (pupariation), inside which it moults into the pupa. During metamorphosis one cell population is largely replaced by another. The larval cells do not divide postembryonically, but most of them become very large and polyploid and then die at metamorphosis. The epidermis (and most internal tissues) of the adult derive from groups of 'imaginal' cells which grow and divide within the larva (or, in some cases, the pupa). The adult wing and surrounding thorax, for example, come from about 20 cells lying beneath the thoracic epidermis of the hatching larva. These cells divide through larval life to form the mature imaginal wing disc (an invaginated pouch of around 50 000 cells) which then evaginates during metamorphosis to form the dorsal mesothorax (Fig. 1).

Cell division in the imaginal disc normally slows and ceases around the time of pupariation, but this appears not to be controlled only by

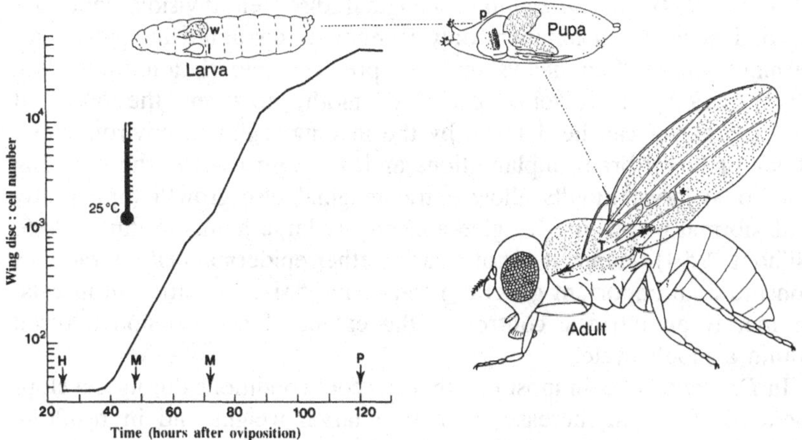

Fig. 1. Development of the adult in *Drosophila melanogaster*. Graph shows the approximate cell number in the wing imaginal disc between hatching (H) and pupariation (P). (After Bryant & Levinson 1985; Bryant, 1987). There are three larval instars, separated by moults (M), and the mature larva develops inside the puparium (p) into the pupa and then the adult. The wing (w) and mesothoracic leg (l) imaginal discs are shown at larval stage, and the epidermal structures derived from the wing disc are shown by stippling on the pupa and adult. Measurements of thorax length (T) and wing area (distal to the small arrows) are indicated on the adult, and the asterisk marks the location of cell density counts on the wing.

the hormonal changes. If pupariation is delayed by early damage to some of the discs, no part of the resulting fly becomes abnormally large, indicating that undamaged discs do not grow beyond their normal size, despite the extension of larval life (Simpson *et al.*, 1980). Similarly, immature imaginal discs removed and cultured *in vivo* in an adult abdomen will continue to grow, but will not exceed approximately the normal disc cell number (Bryant & Levinson, 1985). Hence growth termination, which determines the size of the fly, may be largely intrinsic to the imaginal discs, given adequate larval nutrition (Bryant & Simpson, 1984). Furthermore, studies of regeneration in imaginal discs (and in larval epidermis of other types of insect) show that cell division is linked to the specification of spatial patterns of cell fate (e.g. the location of bristles, veins, etc.) and both are controlled by short-range cell interactions. Hence the disc-intrinsic termination of growth is likely to operate via local interactions, rather than by assessment of a 'target' total cell number for the disc (Bryant & Simpson, 1984; Bryant, 1987;

French, 1989). In *Drosophila*, imaginal disc cell division continues beyond normal cell numbers only in mutants disrupted in spatial patterning, and it then delays or even prevents pupariation (Sehnal & Bryant, 1993). In several species of moth, however, the onset of metamorphosis can be delayed by the feeding regimen, environmental shock or larval brain implantations and, in these insects, the resulting additional larval moults allow extra imaginal disc growth (apparently with normal patterning) to give abnormally large adults (Sehnal, 1985; Nijhout, 1994). Hence it is not clear whether epidermal cell interactions constitute an important general mechanism of size limitation in insects, or merely an intrinsic control of the extent of cell division allowed within a moult cycle.

In *Drosophila*, as in most ectotherms, cool conditions during development result in an increase in mature larval weight and in resulting adult size (e.g. Alpatov, 1930; see later). Adult size may also be altered genetically by thermal selection or by direct selection for larger or smaller adults. Unfortunately, in insects (and indeed in other ectotherms) there is no general understanding of the way in which size may change through influences on the level of nutrition, on the allocation of nutrients or on the hormonal and tissue-intrinsic controls of cell division (see later).

Selection on *Drosophila* body size

Large body size appears to be an unconditional advantage for adult *Drosophila*. The trait shows a narrow sense heritability of between 0.2 and 0.6 in the laboratory and of 0.2–0.3 in nature, with different measures of body size such as wing and thorax length showing strong genetic correlations (e.g. Robertson & Reeve, 1952; Coyne & Beecham, 1987; Robertson, 1987; Prout & Barker, 1989; Cowley & Atchley, 1990; Wilkinson *et al.*, 1990; Partridge & Fowler, 1993). Several laboratory and field studies have demonstrated a correlation between large adult size and fitness components such as fecundity and survival in both females and males (for references, see Partridge & Fowler, 1993). These phenotypic correlations are not a strict demonstration of natural selection on the genetic component of variance for size, however, as large adult size and increased adult fitness could be independent consequences of favourable larval growth conditions. Longevity, male mating success, male fertility and female fecundity have all been shown to be correlated genetically with large body size (for references, see Partridge & Fowler, 1993; James *et al.*, 1995). As genetically-based large adult size is beneficial, but the flies do not appear to be getting larger over time, there must be counterbalancing selection at some other stage in the life history.

When replicated lines of *D. melanogaster* were artificially selected over many generations for large, control and small thorax size, large adult size was associated consistently with an extended period of larval development (Partridge & Fowler, 1993; Santos *et al.*, 1992, 1994). Direct examination of larval growth (Fig. 2) showed no significant increase in growth rate: larvae from the 'large' size lines grew at the control rate, but for longer, to attain a higher larval weight before pupariation (L. Partridge, R. Langelan, K. Fowler & V. French, unpublished results). It appears that the alternative strategy for achieving large size, of growing faster, is not adopted, possibly because larval

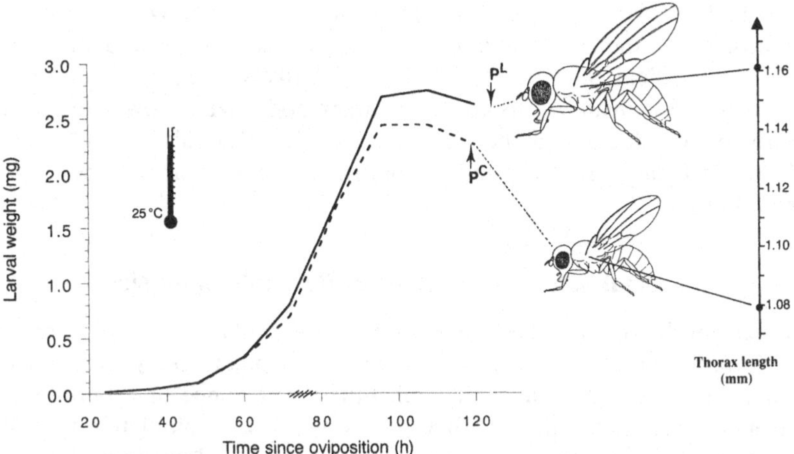

Fig. 2. Response of *Drosophila* to selection for thorax length. The increase in larval weight of 'Large' (—) and 'Control' (- - -) size-selected lines of *D. melanogaster* reared at 25 °C, obtained by weighing individual female larvae. Shading on the time axis indicates the approximate time that larvae achieve their 'critical weight' for pupariation, as determined by starving timed and weighed larvae. P^L and P^C are mean pupariation times for the lines, and the scale gives the mean thorax lengths for the resulting female flies (both of these measurements come from low density 25 °C cultures reared in conditions slightly different to those used to monitor larval growth). The data (After Partridge *et al.*, unpublished results) were obtained from crosses between two replicate lines, to remove any effects of inbreeding depression. Only females are illustrated; males behaved similarly but were smaller, as larvae and as adults.

growth rate has been subject to strong directional selection. Hence the response to selection for rapid larval development is correlated with a decrease in adult body size (Zwaan, 1993), and the response to selection for increased larval feeding rate is not accompanied by a correlated increase in larval growth rate (Burnet *et al.*, 1977). The extended larval development in the 'large' size-selected lines was associated with increased pre-adult mortality, particularly in high-density larval cultures (Partridge & Fowler, 1993; Santos *et al.*, 1992, 1994).

The picture, therefore, is of antagonistic selection pressures on *Drosophila* body size: genetic variants that increase adult size have negative effects on pre-adult survival and age at first breeding, but positive effects on the lifespan and reproductive success of the adults. These conflicting selection pressures on different life-history stages may be taxonically widespread determinants of adult size (Stearns, 1992), so the findings for *Drosophila* are likely to be general. *Drosophila* size evolves in response to temperature (in the laboratory and probably also in the field) and this therefore seems likely to result from either an increased advantage to adults of larger body size at lower temperature, or a reduction in the disadvantage of the larval growth needed to achieve large size (or indeed both). We now examine these two possibilities.

Thermal selection on *Drosophila* adult body size

To establish whether the size advantage to adults is indeed increased at lower temperature, it is necessary to manipulate body size and to examine the consequences for adult fitness at a range of experimental temperatures. It is important that only body size is manipulated, without interference from confounding variables, and these relatively straightforward experiments (e.g. using size-selected lines) have not been carried out.

Adult body size is altered by thermal evolution and also by rearing temperature, and both of these manipulations have been used recently to examine the fitness of the resulting adults at different temperatures. We established replicated laboratory thermal lines of *D. melanogaster* at 16.5 °C (after an initial year at 18 °C) and at 25 °C (Huey *et al.*, 1991). The lines were allowed to evolve in population cages and after 4 years they showed clear evidence of thermal evolution (Partridge *et al.*, 1994*a*), with body size (measured both as thorax length and wing area) greater in the 'cold' line flies when both sets of lines were reared either at 16.5 °C or at 25 °C (Fig. 3). Adult fitness components were examined at both rearing temperatures, and the results indicated that

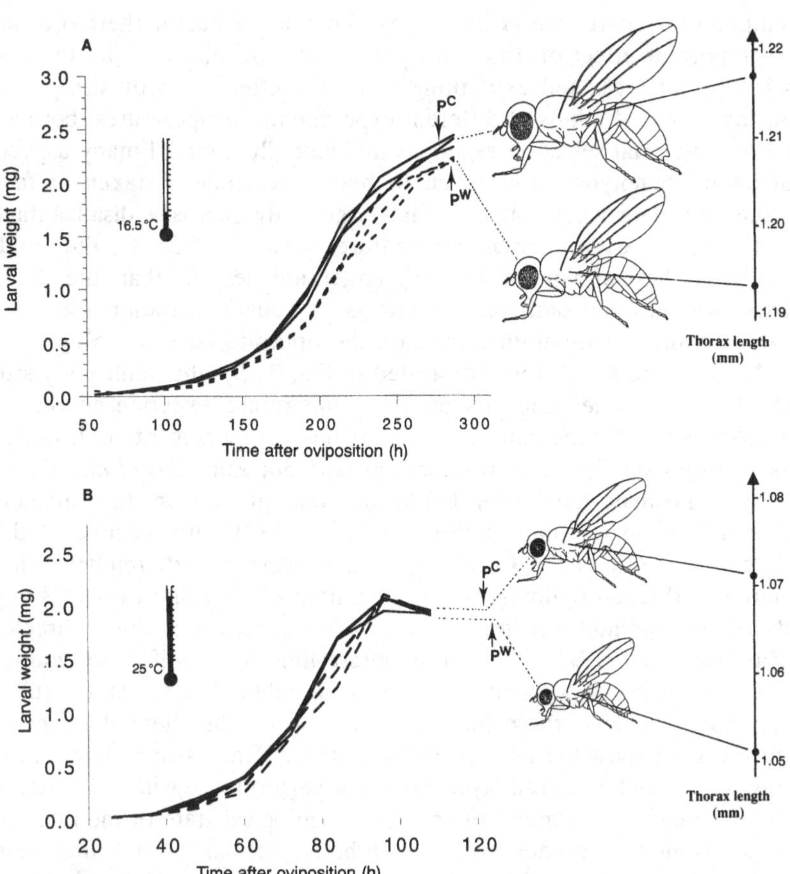

Fig. 3. Development of laboratory thermal selection lines of *Drosophila*. Larval growth, pupariation times and adult size for females from the 'Cold' (—) and 'Warm' (- - -) thermal selection lines, when reared (*a*) at 16.5 °C and (*b*) at 25 °C. See Fig. 2 for abbreviations and notations. P^C and P^W are mean pupariation times for the three Cold lines and the three Warm lines, respectively; the scale gives the mean thorax lengths for the resulting Cold line and Warm line female flies. Only females are shown; the males behaved similarly, but were smaller. (After Partridge *et al.*, 1994*a*,*b*.)

the flies were adapted to the temperature at which they had been evolving; the males and females lived for longer and the females were more fecund, relative to the flies from the other thermal selection regimen, when rearing and testing were at their own evolutionary temperature (Partridge *et al.*, 1995). Thermal selection therefore has an important effect on the adult part of the life history. The thermal selection lines can tell us nothing about the effect of body size *per se* on the fitness of adults at different experimental temperatures, because the two thermal selection regimens undoubtedly affected many aspects of adult phenotype, in addition to body size. Indeed, taken at face value, the data would suggest that large body size is a disadvantage at the higher experimental temperature because, at 25 °C, the 'cold' (16.5 °C) thermal lines were both larger and less fit than the 25 °C lines, whereas the data on the effects of genetic variation for body size at a single temperature indicate the opposite (see above).

As discussed above (and illustrated in Fig. 3*a*,*b*), the adult body size of *D. melanogaster* depends on the temperature experienced during development, and the consequences for one component of adult fitness have been examined recently (see pp. 205–38; Zamudio *et al.*, 1995). Success in male territorial behaviour was greater in the smaller, 25 °C-reared males than in those reared at 18 °C, irrespective of the temperature (18 or 27 °C) at which they were tested, implying that there is a thermal optimum for development (25 °C) that overrides any effects of thermal acclimation or of body size for this character (Zamudio *et al.*, 1995). It would be interesting to know if these findings extend to other aspects of fitness such as adult survival and fertility but, none the less, these flies, like those from the thermal selection lines, cannot show the influence of size *per se*. Body size is determined only by the epidermal cell layer, but adult performance will undoubtedly be influenced by the morphology and physiological state of the internal tissues (muscle, nervous system, fat body, genitalia, etc.) that may respond in different and complex ways to temperature during development and evolution.

Experiments directly relating body size to adult success at different experimental temperatures have yet to be performed. One aspect of adult size variation, however, is of interest for understanding the relationship between laboratory and field selection on body size, and between the evolutionary and developmental responses of body size to temperature. A change in insect body size can be achieved by a change in the number of epidermal cells and/or a change in cell size. In *Drosophila* it is relatively easy to measure these two factors in the adult wing blade because each epidermal cell produces a single hair, or 'trichome' (Dobzhansky, 1929). Cell density, and hence cell size, varies over the wing blade, but appears to be correlated between different regions (Partridge *et al.*, 1994*a*). Cell density in one area is therefore an index for the whole wing and can also be used, in conjunction with the wing area, to give an index of total number of cells in the wing blade. It has been shown repeatedly that the developmental response of body size to temperature, as measured in the wing blade, is achieved by a change in cell size, with little or no effect on cell number (Alpatov, 1930; Robertson, 1959; Delcour & Lints, 1966; Masry & Robertson, 1979; Partridge *et al.*, 1994*a*). The size response of the laboratory lines to selection appears also to be achieved entirely by a change in cell size (Partridge *et al.*, 1994*a*). Under various conditions, there is a good correlation between size changes in the wing area and elsewhere on the body of the fly (Wilkinson *et al.*, 1990; Cowley & Atchley, 1990) but it remains to be seen if their cellular basis is the same, as seems to be the case for interspecific size differences in Hawaiian *Drosophila* (Stevenson *et al.*, 1995).

The similarity between the developmental and laboratory evolutionary increases in cell size at low temperature suggest that the former may be a case of adaptive phenotypic plasticity (Schmalhausen, 1949; Bradshaw, 1965; Gomulkiewicz & Kirkpatrick, 1992). Plasticity of body size is known to be heritable and to respond to artificial selection (Scheiner & Lyman, 1989, 1991; Scheiner *et al.*, 1991), so one way to test whether it is indeed adaptive would be to select for increased and decreased plasticity, and then examine the consequences for adult fitness at different temperatures.

Geographical clines in body size have been found repeatedly in *Drosophila* (see Introduction). In a recent study of *D. melanogaster* populations collected in 20 comparable sites along the east coast of Australia, we have demonstrated a genetic size cline (James *et al.*, 1995). Flies from the more southerly populations (corresponding to cooler conditions in the southern hemisphere) were larger in thorax length and wing area when all populations were reared at standard temperature and low density in the laboratory (Fig. 4). Unlike the effects of laboratory thermal evolution, however, the clinal geographic variation in wing size appears to result mainly from a change in cell number, with only a small effect on cell size (James *et al.*, 1995). The additional changes in cell number associated with clinal variation may indicate a different mechanism of size change from that occurring in laboratory thermal evolution, perhaps because of selective agents in addition to temperature. Alternatively, selection in cool conditions has been going on for much longer in the field, so an initial increase in cell size may have been followed by a response in cell number. Extended observations on laboratory thermal lines could reveal if this latter sequence of events can occur.

The contribution of cell size to evolutionary change in body size poses some interesting problems because, at least in some species, there appears to be stabilising selection on cell size. In many organisms,

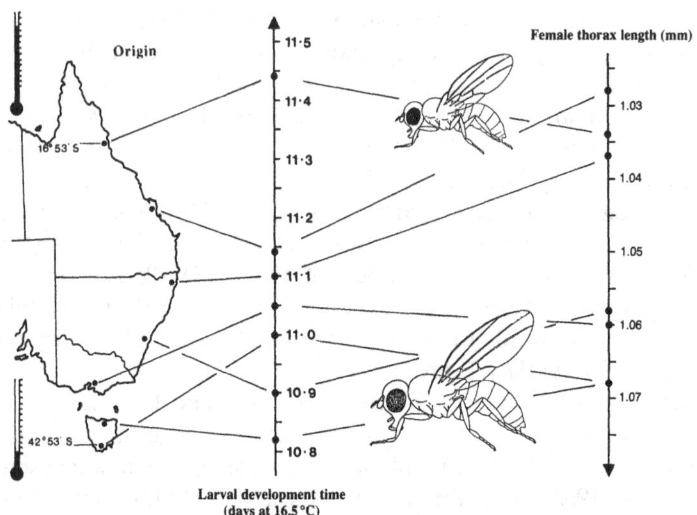

Fig. 4. Clinal variation in body size and development rate in *D. melanogaster*. Lines link the sites of collection (down the eastern coast of Australia) of geographical strains of *D. melanogaster* to their mean larval development times and mean female adult thorax lengths, when all strains were reared at low density and at 16.5 °C in the laboratory. There is clear evidence of a genetic cline, with a general decrease in development time and an increase in size with increasing latitude. For clarity, only seven of the strains in their cline are shown, but the others behaved similarly. (After James & Partridge 1995; James *et al.*, 1995)

including *Drosophila*, changes in genome size, including in ploidy level, result in correlated changes in cell size (e.g. Dobzhansky, 1929; Held, 1979; Nurse, 1985), presumably because both entry to the cell cycle and termination of growth in differentiated cells are regulated by ratio genome to cytoplasm. Long-established polyploids, however, show an evolutionary reversion of cell size to the ancestral value in taxonomically diverse organisms (Nurse, 1985). Furthermore, functional considerations lead to the conclusion that stabilising selection on cell size would occur, at least in some tissues (McNeill Alexander, 1995). Adult fly epidermis may be unusual in that its main function, the secretion of the overlying cuticle, does not occur in larval stages and, in many regions, such as the wing, epidermal cells die after the adult cuticle is secreted. This pattern of activity may somehow relieve the adult epidermis from stabilising selection on cell size (and the larval epidermis is a bizarre tissue, in that growth occurs solely by increase in cell size).

It remains to be shown whether large cell size *per se* is adaptive at low temperature and, if so, why. If not, why should the evolutionary and developmental thermal responses in body size involve cell size, whereas the response to direct selection for increased body size occurs through an increase in cell number, with no change in cell size (L. Partridge, K. Fowler, R. Langelin & V. French, unpublished results)?

Thermal selection on *Drosophila* larval growth

The second broad category of explanation for the thermal evolution of larger adult body size in cool conditions is that the disadvantages of extra larval growth are somehow reduced at low temperature. There is no evidence from *Drosophila* that a protracted larval period is less of a handicap in the cold. The extended larval development of the 'large' size-selected lines was associated with increased larval mortality at 25 °C, but also at 18 °C and at 29 °C, with survival relative to controls reduced at both the low and the high temperatures (Partridge & Fowler, 1993).

The study of populations which have evolved in cool conditions has revealed an important and intriguing alteration to larval growth, which suggests that the larval performance of the 'large' size-selected lines may not be relevant to the consequences for larvae of thermal evolution. Development from egg to adult occurs more rapidly (at all rearing temperatures investigated) in the laboratory 'cold' thermal lines than in the 'warm' lines, and this is achieved by a decrease in the duration of the larval period (Anderson, 1966; Huey *et al.*, 1991; Partridge *et al.*, 1994*b*; James & Partridge, 1995). Similarly, the more southerly field-collected populations of Australian *Drosophila* show the more rapid larval development (James & Partridge, 1995; James *et al.*, 1995). The selection lines and geographical populations with lower temperature thermal history therefore both give rise to larger adults but despite this they have more rapid larval development (Figs. 3 and 4), somehow converting larval food supply to 'adult size' more rapidly. Direct examination of larval growth in the laboratory thermal lines (Fig. 3) confirms that the 'cold' line larvae do gain weight more rapidly, regardless of rearing temperature (Partridge *et al.*, 1994*b*).

There seem to have been no studies of growth rate in relation to latitudinal size clines in other insects but the copepod crustacean *Scottolana canadensis*, like *Drosophila*, shows an increase with latitude in both adult size and larval growth rate (Lonsdale & Levinton, 1985). This stands in contrast to the direct correlation between body size and duration of larval development found in the *Drosophila* lines size-selected at a single temperature (Fig. 2; Partridge & Fowler, 1993) and it raises two important issues. What are the mechanisms responsible for the accelerated growth associated with a low temperature thermal history, and why do the higher temperature populations not also show this phenotype, as it would appear to be adaptive at any temperature if rapid larval development confers a selective advantage?

Despite the general features of growth and size control being better understood in insects (including *Drosophila*) than in any other animals, we do not yet understand how the developmental mechanisms may be modified by thermal evolution. As the completion of the last larval instar is somehow triggered by the larva reaching critical weight (see Developmental control of insect growth and size), we attempted to determine critical weights for the laboratory thermal lines by starving immature larvae of known weight. Surprisingly, each thermal line had the higher critical weight when rearing was at its own evolutionary temperature (Partridge *et al.*, 1994*b*). The differences in final larval and adult size between the thermal lines are consistent across rearing temperatures, however, so they cannot be caused by these differences in critical weight at which the hormonal changes are initiated.

More rapid growth to a larger adult size must result from some combination of a more rapid feeding rate, an increased rate of food absorption, a more efficient assimilation of food into larval body tissue and a more efficient conversion of larval into adult tissue at metamorphosis. Larval feeding rates in *Drosophila* can be estimated from the rate of cephalopharyngeal retraction (Bakker, 1961), but this has yet to be examined in lines differing in thermal history. It is difficult to measure directly the food (yeast) eaten and absorbed by *Drosophila* larvae feeding *ad lib*, so instead we have investigated the growth achieved on a limited amount of food. Single eggs were placed on agar with a fixed weight of yeast, allowed to develop and their resulting adult size measured. For both the laboratory thermal lines (Neat *et al.*, 1995) and the Australian geographic strains (A. James, S. Robinson & L. Partridge, unpublished results), an evolutionary history of low temperature was associated with development of a larger adult from a given food supply, at levels of yeast provision where body size was reduced, indicating that the larva had consumed all the food available (Fig. 5). Evolution at lower temperature therefore leads to more efficient growth, in that more of the food consumed is allocated to increasing the size of the resulting adult.

One way of maximising adult size in the lines with the colder thermal history would be to decrease that part of the larval metabolism that is associated with costly non-growth activities such as movement, detox-

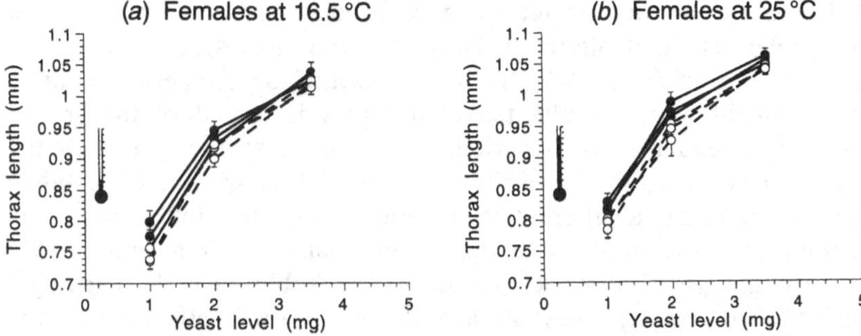

Fig. 5. Growth efficiency of thermal selection lines in *D. melanogaster*. Mean thorax lengths (with 95% confidence limits) of female flies produced by individual hatchling larvae from the 'Cold' (—●—) and 'Warm' (- - ○ - -) thermal selection lines, provided with fixed weights of yeast and reared (*a*) at 16.5 °C and (*b*) at 25 °C. Only females are shown; the males behaved similarly, but were smaller at each food level and temperature. (After Neat *et al.*, 1995.)

ification or somatic maintenance (the breakdown and resynthesis of proteins), and this might be detectable as a decrease in overall metabolic rate. The mass-specific metabolic rates of individual third instar larvae, measured at 18 or 25 °C indeed appeared to be lower in the 'cold' thermal selection lines, particularly when comparing the smaller (younger) larvae (D. Berrigan, J. McCabe & L. Partridge, unpublished results). Six of the Australian geographical populations were chosen to represent the two extremes and the centre of the clinal variation, but their metabolic rates showed no consistent relationship with latitude of their collection site. The significance of these data is not clear but, at least in the thermal selection lines, the overall metabolic rate appears to have evolved to lower levels in the cooler conditions.

The increase in adult size in the low temperature selection lines may occur partly through more efficient conversion of 'larval weight' into 'adult size' at metamorphosis, as the removal of immature larvae from the food showed that those from the 'cold' lines gave rise to the larger adults, for a given larval weight (Neat *et al.*, 1995). This may reflect a change in the balance between development of the temporary larval tissue, of the imaginal disc epidermis (which determines adult size) and of the internal adult tissues. A similar change in balance is also indicated in the size-selected lines, as starved larvae from the 'large' lines produced larger adults for a given larval weight (L. Partridge, R. Langelan, K. Fowler & V. French, unpublished results).

It is not clear why greater growth efficiency (and perhaps a lowered metabolic rate and altered balance between tissues) evolves at low temperature. In *Drosophila*, the direct effect of low temperature during development with an unlimited food supply is to reduce the growth rate, to extend the larval period, and to increase the weight of the mature larva and the size of the resulting adult (Figs. 3 and 4). When growth efficiency at different temperatures was estimated by restricted feeding of larvae from the laboratory thermal lines (Neat *et al.*, 1995) and the geographical stocks (A. James, S. Robinson & L. Partridge, unpublished results), they all had reduced growth efficiency at the lower temperature, making a smaller adult from a given weight of yeast (Fig. 5). Larvae from the laboratory thermal lines showed evidence of adaptation to their selection regimens, with each set of lines having the higher rates of survival when testing was at the temperature at which they had been evolving (Partridge *et al.*, 1994*b*). These results, however, cannot demonstrate a relationship between growth efficiency *per se* and larval fitness at different temperatures. As explained for size and adult fitness (see Thermal selection on *Drosophila* adult body size), the thermal selection lines undoubtedly differ in many respects, confounding any analysis of the causal relationships between subsets of the traits.

Some caution may be required in interpreting the restricted feeding experiments in *Drosophila* as larvae starved before maturity may expend energy wandering in search of food, and the extent to which they do this could vary with temperature or with thermal history. Growth efficiency, however, has been studied in conditions of *ad lib* feeding in many other species of insect by measuring the increase in body weight and the food ingested and assimilated and, here too, growth efficiency often decreases with low rearing temperature. For example, in the studies of ten species of lepidopteran larvae reviewed by Schroeder & Lawson (1992), size was greatest at low temperature in eight species, whereas growth efficiency (in terms of food assimilated) was highest at the higher temperatures in four of these species (but at lower temperatures in only one). Unfortunately, growth efficiencies were not compared between different latitudinal populations of these species, or the copepod *Scottolana canadensis* (Lonsdale & Levinton, 1985).

It seems that low temperature reduces growth efficiency, increasing the amount of food and length of time required to produce the (larger) adult, but the *Drosophila* results indicate that, over evolutionary time, there may be compensation by a change in the allocation between imaginal disc growth and other forms of metabolism such that adult size is maintained (even increased) while minimising the penalties of protracted development. The finding of the opposite effects of evolutionary and rearing temperature on the efficiency of conversion of food into adult size suggests that these two sources of size variation may operate through separate mechanisms. Certainly it has not so far proved possible to formulate a mechanism that could explain both findings. The suggestion that the developmental response of the body size to temperature may depend upon different rates of change in anabolism and catabolism in response to temperature change (von Bertalanffy, 1960) is logically flawed (pp. 183–204). An alternative suggestion is that some of the total catabolism may represent a cost of growth, and that the amount of catabolic cost per unit of growth for some reason decreases with growth rate (Wieser, 1994). Such a relationship would fit the facts. Growth rate is increased at higher environmental temperature, and growth is more efficient then. Low temperature line larvae grow both faster and more efficiently than do those from high temperature lines; however, there is no obvious theoretical reason why the cost of growth should vary inversely with growth rate. Another suggestion is the existence of a trade-off between eventual size and growth rate (Berrigan & Charnov, 1994). Such a relationship would fit the facts about development, but not about evolution at different temperatures, because in the latter case more rapid growth to larger size is the characteristic of low-temperature populations, the opposite to the relationship predicted by the proposed trade-off. Similar comments apply to the model of Perrin (1995), on the basis of differential responses of rates of food acquisition and metabolism to changing temperature.

If evolution at low temperature leads to a change in nutrient allocation to larval and imaginal disc growth, reducing development time and increasing final adult size, then the question remains: why is this strategy not followed also at higher temperatures? One factor that could mediate thermal evolution of growth efficiency and hence body size is the impact of competition. There has been no systematic investigation of the dynamics of *Drosophila* field populations in relation to latitude but it has been suggested that, because of reduced seasonality, tropical populations of *Drosophila* spend more time near or at their ecological carrying capacity (David & Capy, 1982). This is likely to give higher levels of competition, at least for some life-history stages. There may also be more direct effects of temperature, however. Laboratory culture of *D. melanogaster* thermal lines in population cages results in populations permanently at carrying capacity but, even under these conditions, where seasonal effects, predators and parasitoids are absent, as the temperature drops below 25 °C, population densities decline (Davis *et al.*, 1995). The impact of competition on larvae developing at lower temperatures may therefore be reduced, in both laboratory and field.

The evolution of *Drosophila* life history in response to variation in larval density is quite well studied. High larval densities lead to the evolution of reduced larval development time and a smaller adult (Roper *et al.*, in press) and of more rapid rates of larval feeding (Joshi & Mueller, 1989), which are associated with less efficient use of food in growth (Mueller, 1991). Reciprocally, higher larval feeding rates have been shown to lead to increased competitive ability (Burnet *et al.*, 1977). Also, a direct environmental effect of high larval density is to increase larval metabolic rate (D. Berrigan, J. McCabe & L. Partridge, unpublished results), which may result from higher larval feeding rates and perhaps additional costs such as an increased need for detoxification of ingested larval waste products.

A possible scenario, then, for thermal evolution in *Drosophila* is that low temperature reduces the impact of larval competition, and hence the need for the metabolically-costly activities associated with high competitiveness. This allows the larva to evolve a pattern of increased allocation of nutrients to growth, which in turn increases growth rate and allows the production of a larger adult, even in the face of the reduced growth efficiency that is a direct environmental effect of low temperature. The increased growth efficiency and larger adult body size of the low temperature thermal lines and southerly Australian strains are therefore consistent with an evolutionary history of low larval competition.

This argument could be tested by examining the impact of larval competition in population cages kept at different temperatures. For example, the number and size of enclosing adults could be compared when larval cultures are taken from cages and either allowed to develop undisturbed, or split up into subcultures so that the larvae are reared under relaxed competition. Measurements of metabolic rates and feeding rates of larvae taken from the cages could also be compared with those of larvae reared at known densities, to get an index of the perceived level of competition in cages at different temperatures.

Discussion

Neither the developmental nor the evolutionary responses of ectotherm body size to temperature are understood in terms either of their consequences for pre-adult or adult fitness. A picture is emerging for *Drosophila* of evolution at lower temperatures, in which a change in nutrient allocation during larval development results in more rapid growth and the production of a larger adult. What is far less clear is why this higher growth efficiency is restricted to evolution at low temperature, because we would suppose it to be advantageous at any temperature. The results therefore suggest that the high growth efficiency is achieved by a trade-off with other activities that are less beneficial at lower than at higher evolutionary temperatures, although there is little direct evidence for this at present. From the data on thermal evolution in *Drosophila* we have argued for a trade-off between growth efficiency and competitiveness, and that altering this balance in favour of the latter leads to higher larval fitness as temperature increases. This prediction could be tested by artificial selection on growth efficiency and examination of the consequences for larval survival at different temperatures and larval densities. Similarly, selection for increased larval competitiveness would be predicted to be especially beneficial at higher temperatures. If levels of larval competition do prove to be of importance in the thermal evolution of body size in *Drosophila*, it will be important to investigate their role in other organisms where thermal evolution has produced similar clines in body size, as in the copepod *Scottolana canadensis* (Lonsdale & Levinton, 1985), and also to determine if they have any bearing on the widespread developmental response of body size to temperature.

Body size may evolve in response to temperature not primarily because of the effects of thermal change on the physiology of individual organisms, but instead because of the impact of temperature on population parameters such as mortality and reproductive rate, which in turn determine the way in which natural selection acts on the life history. The operation of natural selection on age and size at maturity is well understood (Roff, 1992; Stearns, 1992), and depends on a number of ecological factors that could have a bearing on thermal evolution of body size (see pp. 183–204; also Atkinson, 1994). The basic picture, as illustrated by *Drosophila*, is that increased duration of development leads to increased juvenile mortality, but also to increased size and fitness of the resulting adult. The most important single factor affecting the optimal balance is the rate of juvenile mortality; as this increases, the optimum shifts towards early maturation at small size. In contrast, if adults have relatively high mortality, then the optimum shifts towards late maturation at large size. Atkinson (pp. 183–204; 1994) and Sibley & Atkinson (1994) discuss effects of temperature on pre-adult mortality rates; more data on this point would be valuable. Juvenile mortality could be increased at lower temperatures through the risk of starvation as a consequence of reduced efficiency of use of nutrients at lower temperatures. Such an impact would seem likely to be greater on juveniles that on adults, and certainly is in *Drosophila* (Partridge *et al.*, 1994*b*, 1995), but this would predict the evolution of decreased size at maturity at lower temperatures, the opposite response to that generally observed.

Variation in the size of the population could be another important consideration; if most reproduction occurs in populations that are expanding in numbers at the time, there is selection for rapid development, because the laws of compound interest apply, and the age of first breeding and hence the number of generations achieved by a genotype becomes relatively important (Cole, 1954; Lewontin, 1965). Clearly populations do not expand in numbers indefinitely, and this kind of consideration is particularly likely to be important in multivoltine organisms (ones with more than one generation per breeding season) in seasonal environments, where reproduction occurs only under favourable conditions. It is not clear if temperature might affect selection on body size through this route. The intensity of selection for early reproduction as a result of population expansion depends on the change in population size per generation, and this will not obviously change with temperature, which may simply change generation time. Temperature, however, can have an influence through the number of generations that can be accommodated in one breeding season.

In seasonal breeders, there is a selection for the whole of the time suitable for breeding to be used for that purpose. As the length of physiological time (day degrees) available for breeding increases with decreasing latitude and altitude, other things (particularly growth rate) being equal, there will initially be selection for increased size at maturity, to maximise adult fitness. This will produce a size cline in reverse direction to those so far described here, and this may explain recent data on the water strider *Aquarius remigis* (Blanckenhorn & Fairbairn, 1995). As season length increases further, selection for increased size and reproductive potential of adults may be reduced in favour of the completion of an extra generation per season, leading to the evolution of a 'saw-tooth' latitudinal size cline, with increases in adult size with decreasing latitude, alternating with sharp decreases associated with the insertion of an extra generation within the breeding season (Roff, 1980). Such clines have been documented in a number of cricket species, generally associated with changes from a uni- to a bivoltine life history with decreasing latitude (Mousseau & Roff, 1989; Roff, 1992). In the continuous parts of these clines, the direction of size change with latitude and the (presumed) direct relationship between size and development time, is opposite to that observed in the other thermal clines, in insects such as *Drosophila*, where higher latitudes and altitudes are associated with genetic changes enabling rapid development to a large body size. The two kinds of cline must have different underlying mechanisms, and it will be important to discover the reasons for the difference. Whatever these may be, the kinds of theoretical considerations that can explain the occurrence of the saw tooth clines appear to be irrelevant for the evolution of rapid development and larger body size at lower temperatures.

The response of many temperate ectotherms to rearing temperature may be linked to an important environmental factor that reliably indicates season: photoperiod. For example, studies of the growth and development of several North European butterfly species (Nylin *et al.*, 1989; Nylin, 1992) have yielded examples of variation in growth rate with photoperiod, and of constant growth rate regardless of rearing temperature at some photoperiods (but not at others). It is argued that this plasticity of larval development, which tends to maintain adult body size, has evolved as an adaptation fitting the insect's life history to its seasonal environment (Nylin *et al.*, 1989; Nylin, 1992), and that populations at different latitudes will evolve different and appropriate responses (Nylin *et al.*, 1995). The mechanisms and the full consequences (in terms of the fitness of the larvae and the resulting adults developed under particular temperature/photoperiod regimens) have yet to be explored.

In *Drosophila*, thermal evolution in cool conditions leads to a faster larval growth rate, a heavier larva, earlier pupation and a larger fly (Figs. 3 and 4). From the food limitation experiments, we have argued that this is achieved by an increase in the efficiency of food utilisation for growth during this time interval (see pp. 272–7). Experimental studies of imaginal disc development indicate, however, that adult cell number is intrinsically controlled (see pp. 272–7). Adult cell number is not limited by the extent to which the imaginal discs can grow before hormonal changes lead to the completion of larval development (except when the late larva is starved or underfed, when both cell number and cell size are reduced (Held, 1979)). If this conclusion is correct, an increased cell number, as occurs in laboratory size-selection (L. Partridge, K. Fowler, R. Langelin & V. French, unpublished results) and in the latitudinal size cline (James *et al.*, 1995), must result from changes in the disc cell interactions that control proliferation and establish morphological patterns. Some of the molecular signals mediating these interactions (e.g. products of the *wingless* and *decapen-*

taplegic genes (Cohen, 1993; Blair, 1995) are homologues of 'growth factors' initially known for their effects on mammalian cell proliferation. Developmental studies may indicate eventually how these signals, or the proliferative response to them, may be modulated to alter cell number. Almost nothing is known at present of the control of imaginal disc cell size, however, so there are few clues to the mechanism by which this component of adult size can be changed in thermal evolution, and also as a developmental response to reduced rearing temperature.

Drosophila's many practical advantages make it the ectotherm of choice for work on thermal evolution. It shows regular and repeatable geographical clines, rapidly undergoes thermal evolution in the laboratory and can be readily selected for changes in life history or morphology. The development, physiology and fitness components of the various genotypes can be studied in controlled conditions, and the molecular basis of developmental size control is (in principle) accessible. For a general understanding of thermal evolution of body size, however, many of these studies should be extended to other organisms, partly to test the generality of some of the *Drosophila* findings (e.g. the responses of cell number and cell size; the relationship between growth efficiency, development time and size increase; the possible role of competition). Other, larger insects may be more suitable for exploring evolutionary and developmental effects of temperature in changing the hormonal mechanisms that control moulting and metamorphosis. Also, it will be important to explore the relationship between the selection pressures and mechanisms that produce 'regular' size clines, and those operating in critically seasonal conditions to produce different patterns of response to temperature and other environmental variables.

Acknowledgements

We thank David Atkinson, Al Bennett, David Berrigan, Ray Huey, Ian Johnston and Jennie McCabe for their comments on the manuscript, and the NERC for financial support.

References

Alpatov, W.W. (1929). Biometrical studies on variation and races of the honey bee (*Apis mellifera*). *Quarterly Review of Biology* **4**, 1–58.

Alpatov, W.W. (1930). Phenotypical variation in body and cell size of *Drosophila melanogaster. Biology Bulletin* **58**, 85–103.

Anderson, W.W. (1966). Genetic divergence in M. Vetukhiv's experimental populations of *Drosophila pseudoobscura. Genetical Research, Cambridge* **7**, 255–66.

Anderson, W.W. (1973). Genetic divergence in body size among experimental populations of *Drosophila pseudoobscura* kept at different temperatures. *Evolution* **27**, 278–84.

Atkinson, D. (1994). Temperature and organism size: a biological law for ectotherms? *Advances in Ecological Research* **25**, 1–58.

Bakker, K. (1959). Feeding period, growth and pupation in larvae of *Drosophila melanogaster. Entomology Experimental and Applied*. **2** 171–86.

Bakker, K. (1961). An analysis of factors which determine success in competition for food among larvae of *Drosophila melanogaster. Archive Neerlandaises de Zoologie* **14** 200–81.

Bergmann, C. (1847). Uber die verhältnisse der Wärmeökonomie der Thiere zu ihrer Grösse. *Göttinger Studien* **1**, 595–708.

Berrigan, D. & Charnov, E.L. (1994). Reaction norms for age and size at maturity in response to temperature: a puzzle for life historians. *Oikos* **70**, 474–8.

Berven, K.A. (1982). The genetic basis of altitudinal variation in the wood frog *Rana sylvatica*. I. An experimental analysis of life history traits. *Evolution* **36**, 962–83.

Blair, S. (1995). Compartments and appendage development in *Drosophila. BioEssays* **17**, 299–309.

Blanckenhorn, W.U. & Fairbairn, D.J. (1995). Life-history adaptation along a latitudinal cline in the waterstrider *Aquarius remigis* (Heteroptera; Gerridae). *Journal of Evolutionary Biology* **8**, 21–41.

Bonner, J.T. (1965). *Size and Cycle*. New Jersey: Princeton University Press.

Bradshaw, A.D. (1965). Evolutionary significance of phenotypic plasticity in plants. *Advances in Genetics* **13**, 115–55.

Bryant, E.H. (1977). Morphometric adaptation of the housefly, *Musca domestica* L., in the United States. *Evolution* **31**, 580–96.

Bryant, P.J. (1987). Experimental and genetic analysis of growth and cell proliferation in *Drosophila* imaginal discs. In *Genetic Regulation of Development* (ed. W.F. Loomis), pp. 339–72. New York: Alan R. Liss.

Bryant, P.J. & Levinson, P. (1985). Intrinsic growth control in the imaginal primordia of *Drosophila*, and the autonomous action of a lethal mutation causing overgrowth. *Developmental Biology* **107**, 355–63.

Bryant, P.J. & Simpson, P. (1984). Intrinsic and extrinsic control of growth in developing organs. *Quarterly Review of Biology* **59**, 387–415.

Bulliere, D. & Bulliere, F. (1985). Regeneration. In *Comprehensive Insect Physiology, Biochemistry and Pharmacology* (ed. G.A. Kerkut & L.I. Gilbert). Oxford, UK: Pergamon Press.

Burnet, B., Sewell, D. & Bos, M. (1977). Genetic analysis of larval feeding behaviour in *Drosophila melanogaster. Genetical Research, Cambridge* **30**, 149–61.

Calder, W.A. (1984). *Size, Function and Life History*. Boston, MA: Harvard University Press.

Cohen, S.M. (1993). Imaginal disc development. In *The Development of* Drosophila melanogaster (ed. M. Bate & A. Martinez Arias). New York: Cold Spring Harbor Press.

Cole, L.C. (1954). The population consequences of life history phenomena. *Quarterly Review of Biology* **29**, 103–37.

Cowley, D.E. & Atchley, W.R. (1990). Development and quantitative genetics of correlation structure among body parts of *Drosophila melanogaster. American Naturalist* **135**, 242–68.

Coyne, J.A. & Beecham, E. (1987). Heritability of two morphological characters within and among natural populations of *Drosophila melanogaster. Genetics* **117**, 727–37.

Damuth, J. (1987). Population density and body size in mammals. *Nature* **290**, 699–700.

David, J.R. & Bocquet, C. (1975). Similarities and differences in latitudinal adaptation of two *Drosophila* sibling species. *Nature* **257**, 588–90.

David, J.R. & Capy, P. (1982). Genetics and origin of a *Drosophila melanogaster* population recently introduced to the Seychelles. *Genetical Research, Cambridge,* **40**, 295–303.

Davis, A.J., Jenkinson, L.S., Lawton, J.H., Shorrocks, B. & Wood, S. (1995). Global-warming, population dynamics and community structure in a model insect assemblage. In *Insects in a Changing Environment* (ed. N. Stork & R. Harrington.) 17th Royal Entomological Symposium, 1993.

Delcour, J. & Lints, F.A. (1966). Environmental and genetic variations on wing size, cell size, and cell division rate in *Drosophila melanogaster. Genetica* **37**, 543–56.

Dobzhansky, T. (1929). The influence of the quantity and quality of chromosomal material on the size of the cells in *Drosophila melanogaster. Archiv für Entwicklungsmechanik der Organismen* **115**, 363–79.

Endler, J.A. (1977). *Geographic Variation, Speciation and Clines*. Princeton Mongraphs in Population Biology No. 10. Princeton: Princeton University Press.

French, V. (1989). The control of growth and size during development. In *The Physiology of Human Growth*, (ed. J. Tanner & M. Preece, pp. 11–28). Society for the Study of Human Biology Symposium 29.

Gomulkiewicz, R. & Kirkpatrick, M. (1992). Quantitative genetics and the evolution of reaction norms. *Evolution* **46**, 390–411.

Held, L.I. (1979). Pattern as a function of cell number and cell size on the second-leg basitarsus of *Drosophila. Wilhelm Roux's Archives* **187**, 105–27.

Huey, R.B., Partridge, L. & Fowler, K. (1991). Thermal sensitivity of *Drosophila melanogaster* responds rapidly to laboratory natural selection. *Evolution* **45**, 751–6.

Imasheva, A.G., Bubli, O.A. & Lazebny, O.E. (1994). Variation in wing length in Eurasian natural populations of *Drosophila melanogaster*. *Heredity* **72**, 508–14.

James, A. & Partridge, L. (1995). Thermal evolution of rate of larval development in *Drosophila melanogaster* in laboratory and field populations. *Journal of Evolutionary Biology* **8**, 315–30.

James, A.C., Azevedo, R. & Partridge, L. (1995). Cellular basis and developmental timing in a size cline of *Drosophila melanogaster*. *Genetics* **140**, 659–66.

Joshi, A. & Mueller, L.D. (1989). Evolution of higher feeding rate in *Drosophila* due to density-dependent natural selection. *Evolution* **42**, 1090–3.

Lemeunier, F., David, J.R., Tsacas, L. & Ashburner, M. (1986). The *melanogaster* species group. In *Genetics and Biology of Drosophila*. vol. 3e. (ed. M. Ashburner, H.L. Carson & J.M. Thompson), pp. 147–256. New York: Academic Press.

Lewontin, R.C. (1965). Selection for colonizing ability. In *The Genetics of Colonizing Species* (ed. H.G. Baker & G.L. Stebbins), pp. 77–94. New York: Academic Press.

Long, A.D. & Singh, R.S. (1995). Molecules versus morphology: the detection of selection acting on morphological characters along a cline in *Drosophila melanogaster*. *Heredity* **74**, 569–81.

Lonsdale, D.J. & Levinton. J.S. (1985). Latitudinal differentiation in copepod growth: an adaptation to temperature. *Ecology* **66**, 1397–407.

McNeill Alexander, R. (1995). Big flies have bigger cells. *Nature* **375**, 20.

Masry, A.M. & Robertson, F.W. (1979). Cell size and number in the *Drosophila* wing. III. The influence of temperature differences during development. *Egyptian Journal of Genetics and Cytology* **8**, 71–9.

Mayr, E. (1963). *Animal Species and Evolution*. Cambridge, MA: Harvard University Press.

Mousseau, T.A. & Roff, D.A. (1989). Adaptation to seasonality in a cricket: patterns of phenotypic and genotypic variation in body size and diapause expression along a cline in season length. *Evolution* **43**, 1483–96.

Mueller, L.D. (1991). Ecological determinants of life history evolution. *Philosophical Transactions of the Royal Society of London B* **332**, 25–30.

Neat, F., Fowler, K., French, V. & Partridge, L. (1995). Thermal evolution of growth efficiency in *Drosophila melanogaster*. *Proceedings of the Royal Society of London B*. **260**, 73–8.

Nijhout, H.F. (1994). *Insect Hormones*. Princeton, N.J.: Princeton University Press.

Nurse, P. M. (1985). The genetic control of cell volume. In *The Evolution of Genome Size* (ed. T. Cavalier-Smith, pp. 185–96. London: Wiley.

Nylin, S. (1992). Seasonal plasticity in life-history traits: growth and development in *Polygonia c-album* (Lepidoptera: Nymphalidae) *Biological Journal of the Linnaean Society* **47**, 301–23.

Nylin, S., Wickman, P.O. & Wiklund. C. (1989). Seasonal plasticity in growth and development of the speckled wood butterfly, *Pararge aegeria* (Satyrinae). *Biological Journal of the Linnaean Society* **38**, 155–71.

Nylin, S., Wickman, P.O. & Wiklund, C. (1995). Life-cycle regulation and life history plasticity in the speckled wood butterfly: are reaction norms predictable? *Biological Journal of the Linnaean Society* **55**, 143–57.

Partridge, L. Barrie, B., Barton, N.H., Fowler, K. & French, V. (1995). Rapid laboratory evolution of adult life history traits in *Drosophila melanogaster* in response to temperature. *Evolution* **49**, 538–44.

Partridge, L., Barrie, B., Fowler, K. & French, V. (1994*a*). Evolution and development of body size and cell size in *Drosophila melanogaster* in response to temperature. *Evolution* **48**, 1269–76.

Partridge, L., Barrie, B., Fowler, K. & French, V. (1994*b*). Thermal evolution of pre-adult life history traits in *Drosophila melanogaster*. *Journal of Evolutionary Biology* **7**, 645–63.

Partridge, L. & Fowler, K. (1993). Direct and correlated responses to selection on thorax length in *Drosophila melanogaster*. *Evolution* **47**, 213–26.

Perrin, N. (1995). About Berrigan and Charnov's life-history puzzle. *Oikos* **73**, 137–9.

Peters, R.H. (1983). *The Ecological Implications of Body Size*. Cambridge: Cambridge University Press.

Prevosti, A. (1955). Geographical variability in quantitative traits in populations of *Drosophila subobscura*. *Cold Spring Harbor Symposium on Quantitative Biology* **20**, 294–9.

Prout, T. & Barker, J.S.F. (1989). Ecological aspects of the heritability of body size in *Drosophila buzzatii*. *Genetics* **123**, 803–13.

Ralls, K. & Harvey, P. (1985). Geographic variation in size and sexual dimorphism of North American weasels. *Biological Journal of the Linnaean Society* **25**, 119–67.

Ray, C. (1960). The application of Bergmann's and Allen's rules to the poikilotherms. *Journal of Morphology* **106**, 85–108.

Robertson, F.W. (1959). Studies in quantitative inheritance. XII. Cell size and number in relation to genetic and environmental variation of body size in *Drosophila*. *Genetics* **44**, 869–96.

Robertson, F.W. (1963). The ecological genetics of growth in *Drosophila*. 6. The genetic correlation between the duration of the larval period and body size in relation to larval diet. *Genetical Research, Cambridge* **4**, 74–92.

Robertson, F.W. (1987). Variation of body size within and between wild populations of *Drosophila buzzatii. Genetica* **72**, 111–25.

Robertson, F.W. & Reeve, E.C.R. (1952). Studies in quantitative inheritance. I. The effects of selection of wing and thorax length in *Drosophila melanogaster. Journal of Genetics* **50**, 414–48.

Roff, D.A. (1980). Optimizing development time in a seasonal environment: the 'ups and downs' of clinal variation. *Oecologia* **45**, 202–8.

Roff, D.A. (1992). *The Evolution of Life Histories*. London: Chapman and Hall.

Roper, C., Pignatelli, P. & Partridge, L. Evolutionary responses of *Drosophila melanogaster* life history to differences in larval density. *Journal of Evolutionary Biology* (in press).

Santos, M., Fowler, K. & Partridge, L. (1992). On the use of tester stocks to predict the competitive ability of genotypes. *Heredity* **69**, 489–95.

Santos, M., Fowler, K. & Partridge, L. (1994). Gene–environment interaction for body size and larval density in *Drosophila melanogaster*: an investigation of effects on development time, thorax length and adult sex ratio. *Heredity* **72**, 515–21.

Scheiner, S.M., Caplan, R.L. & Lyman, R.F. (1991). The genetics of phenotypic plasticity. III. Genetic correlations and fluctuating asymmetries. *Journal of Evolutionary Biology* **4**, 51–68.

Scheiner, S.M. & Lyman, R.F. (1989). The genetics of phenotypic plasticity. I. Heritability. *Journal of Evolutionary Biology* **2**, 95–107.

Scheiner, S.M. & Lyman, R.F. (1991). The genetics of phenotypic plasticity. II. Response to selection. *Journal of Evolutionary Biology* **4**, 23–50.

Schmalhausen, I.I. (1949). *The Factors of Evolution*. Philadelphia: Blakiston.

Schmidt-Neilsen, K. (1984). *Scaling: Why is Animal Size so Important?* Cambridge: Cambridge University Press.

Schroeder, L. & Lawson, J. (1992). Temperature effects on the growth and dry matter budgets of *Malacosoma americanum. Journal of Insect Physiology* **38**, 743–9.

Sehnal, F. (1985). Growth and life cycles. In *Comprehensive Insect Physiology, Biochemistry and Pharmacology* (ed. G.A. Kerkut & L. I. Gilbert). Oxford, UK: Pergamon Press.

Sehnal, F. & Bryant, P. J. (1993). Delayed pupariation in *Drosophila* imaginal disc overgrowth mutants is associated with reduced ecdysteroid titers. *Journal of Insect Physiology* **39**, 1051–9.

Sibly, R.M. & Atkinson, D. (1994). How rearing temperature affects optimal adult size in ectotherms. *Functional Ecology* **8**, 486–93.

Simpson, P., Berreur, P. & Berreur-Bonnenfant, J. (1980). The initiation of pupariation in *Drosophila*: dependence on growth of the imaginal discs. *Journal of Embryology and Experimental Morphology* **57**, 155–65.

Stalker, H.D. & Carson, H.L. (1947). Morphological variation in natural populations of *Drosophila robusta* Sturtevant. *Evolution* **1**, 237–48.

Stalker, H.D. & Carson, H.L. (1948). An altitudinal transect of *Drosophila robusta* Sturtevant. *Evolution* **2**, 295–305.

Stalker, H.D. & Carson, H.L. (1949). Seasonal variation in the morphology of *Drosophila robusta* Sturtevant. *Evolution* **3**, 330–43.

Stearns, S.C. (1992). *The Evolution of Life Histories*. Oxford: Oxford University Press.

Stevenson, R.D. (1985). Body size and limits to daily range of body temperatures in terrestrial ectotherms. *American Naturalist* **125**, 102–17.

Stevenson, R.D., Hill, M.F. & Bryant, P.J. (1995). Organ and cell allometry in Hawaiian *Drosophila*: how to make a fly big. *Proceedings of the Royal Society of London B* **259**, 105–10.

Tantawy, A. O. (1964). Studies on natural populations of *Drosophila*. III. Morphological and genetic differences of wing length in *Drosophila melanogaster* and *D. simulans* in relation to season. *Evolution* **18**, 560–70.

Von Bertalanffy, L. (1960). Principles and theory of growth. In *Fundamental Aspects of Normal and Malignant Growth* (ed. W.N. Nowinski), pp. 137–259. Amsterdam: Elsevier.

Wieser, W. (1994). Cost of growth in cells and organisms: general rules and comparative aspects. *Biological Reviews* **68**, 1–33.

Wilkinson, G.S., Fowler, K. & Partridge, L. (1990). Resistance of genetic correlation structure to directional selection in *Drosophila melanogaster*. *Evolution* **44**, 1990–2003.

Zamudio, K.R., Huey, R.B. & Crill, W.D. (1995). Bigger isn't always better: body size, developmental and parental temperature and male territorial success in *Drosophila melanogaster*. *Animal Behaviour* **49**, 671–77.

Zwann, B.J. (1993). *Genetical and environmental aspects of ageing in* Drosophila melanogaster, *an evolutionary perspective*. Ph.D. thesis, University of Groningen.

J.E.P.W. BICUDO

Physiological correlates of daily torpor in hummingbirds

Introduction

Hypometabolism is a widespread physiological feature among animals (Hochachka & Guppy, 1987). It can be regarded as an important strategy to overcome fluctuations of environmental parameters, such as cold ambient temperatures, lack of food or water, hypoxia, etc. Temperature, of course, is one of the most important environmental variables directly affecting the energy metabolism of animals, and some of the mechanisms that allow animals to adapt to different ambient temperatures have been extensively investigated over the years (Precht *et al.*, 1973). Endothermy combined with a high body temperature has to a great extent enabled some animals to be fairly independent of environmental temperature fluctuations, either diurnal or seasonal, and has evolved in different Phyla. Insects among the invertebrates, and some fish, some reptiles, birds and mammals among the vertebrates are well known examples of endothermic animals (McNab, 1983). This chapter will focus on some questions involving the temperature relationships as well as the energy metabolism of some birds and mammals that are capable of undergoing a deep metabolic depression within a 24-h cycle, commonly referred to as daily torpor (as opposed to hibernation, referred to as prolonged torpor).

Daily torpor in birds and mammals may be viewed as a well-regulated hypometabolic state which may last up to several hours within a 24-h cycle. It is expressed by a significant reduction of metabolism and body temperature (T_B). There are two important requirements for endothermic animals to meet the above definition. First, to undergo a cycle of metabolic depression within 24 h, animals must have high thermal conductances (C) to have low thermal inertias; and second, they must be able to tolerate low body temperatures because of low ambient temperatures. Thus, only a few and very specialised groups of endothermic animals undergo daily torpor, and can be used to investigate its mechanisms. Mammals such as bats and some shrews, and birds such as hummingbirds, are the most representative examples of endotherms that undergo torpor according to the definition presented above. They share a very important trait, i.e. a small body size. They have, during periods of activity, some of the highest mass specific metabolic rates that have been reported for vertebrates. For such animals, daily torpor seems to be an important adaptive strategy when the food supply is no longer readily available, either during the day (bats) or during the night (hummingbirds).

Hummingbirds are well known endothermic animals which undergo daily torpor and will be examined in greater detail throughout this chapter, mainly because they approach their limits in terms of performance or environmental adaptation. A few simple calculations will serve to demonstrate why they undergo daily torpor. Hummingbirds mainly feed on nectar, rich in carbohydrates and, when energy demands are high, they consume more than three times their body mass in fluid per day (Beuchat *et al.*, 1990). During the night, however, they are unable to see, and therefore cannot feed. If, for instance, a hypothetical 5 g hummingbird maintained a constant high body temperature (40 °C) during the night and were for this purpose to rely on its energy stores accumulated during the day, e.g. 0.04 g of fat in the liver (equivalent to 20% of the total volume of the liver, in contrast to 2% in mice (Zerbinatti, 1994)), the energy reserves would only last for approximately 4 h. This figure is based on a resting oxygen consumption ($\dot{V}_{O_2}$) during normothermia of about 20.8 ml O_2 h^{-1}, and assumes that 1 g fat consumes 2000 ml of oxygen. Obviously, the bird would not be able to survive until the next day. Therefore, a reduction of body temperature and metabolism is essential to survival in hummingbirds. If, on the other hand, the same hummingbird entered torpor with T_B staying around 15 °C and a $\dot{V}_{O_2}$ of 0.9 ml O_2 h^{-1} (20 times lower than in normothermia) was maintained throughout the entire period, disregarding at this point the costs of entering and arousing from torpor, and assuming the same values for fat content and oxygen consumption, the total time it could remain in torpor would be approximately 80 h. Of course, these are estimates which do not take into account fluctuations that may occur in both T_B and $\dot{V}_{O_2}$ during torpor, but they clearly show the importance of torpor as an indispensable strategy for energy conservation, and therefore for survival, in small endotherms like hummingbirds. In contrast, humans, for example, would not face such problems because of being large and thus having a resting mass specific oxygen consumption ($\dot{V}_{O_2}$/Mb) roughly 17 times lower than normothermic hummingbirds at rest, allows them to survive relatively longer periods of fasting (many days) without ever having actively to undergo hypothermia and torpor.

Although it has been known for a while why small endotherms undergo daily torpor, little is known about the physiological mechanisms that govern this intriguing biological process. What are the stimuli for torpor in these animals, for how long they can stay in such a state, and how arousal is triggered and accomplished are still puzzling questions. To make the picture even more complex, data obtained by Calder (1994) show that hummingbird torpor in nature does not appear to be a regular event. According to this author, its benefits come only to compensate for times when energy intake is unexpectedly low. This certainly poses fundamental questions regarding the ecology and evolution of hummingbirds, as the majority of the physiological studies to date have been carried out under laboratory conditions. The purpose of this chapter is therefore to discuss such questions, based on recent data reported in the literature, as well as possible directions for future research in the field.

Thermal conductance

To maintain a constant body temperature, an animal must satisfy the steady state condition in which the rate of metabolic heat production ($\dot{H}$) equals the rate of heat loss ($\dot{Q}$). To describe the heat loss, a simplified equation that relates heat loss ($\dot{Q}$) to temperature is presented below:

$$\dot{H} = \dot{Q} = C \times (T_B - T_A) \quad (1)$$

This equation simply states that the rate of metabolic heat production equals the heat loss, which in turn is proportional to the temperature difference between the body and ambient ($T_B - T_A$). The term C is the thermal conductance (heat flow per unit time per unit area per degree temperature difference) and has the units W (watts) m^{-2} °C^{-1}. For practical purposes conductance can also be expressed in terms of metabolism with the units ml O_2 h^{-1} °C^{-1} (i.e. 1 ml O_2 h^{-1} is roughly equal to 0.01 W). It is a measure of the heat flow from the animal to the surroundings. The term includes the flow of heat from deeper parts of the body to the skin surface and from the skin surface through the fur or feathers to the environment. When conductance is low, the insulation value is very high; in fact, insulation is the reciprocal value of conduction.

Body size and daily torpor

A considerable amount of material on thermal conductance in mammals and birds has been accumulated and discussed by Herreid & Kessel (1967). This was followed by several more recent compilations on thermal conductance relative to body size in, for example, eutherian mammals (Bradley & Deavers, 1980; Stone & Purvis, 1992), where mass specific conductance (C/Mb) is proportional to body mass $(\mathrm{Mb})^{-0.426}$. Thermal conductance also depends on body mass (Mb) in birds and Lasiewski *et al.* (1967) reported that C/Mb is proportional to $\mathrm{Mb}^{-0.508}$. A more complete allometric relationship applicable to both night and day mass specific conductances in birds has since been offered by Aschoff (1981), where C_{night}/Mb is proportional to $\mathrm{Mb}^{-0.583}$, and C_{day}/Mb is proportional to $\mathrm{Mb}^{-0.484}$. Thus, in general, for both mammals and birds, mass specific thermal conductances vary with $\mathrm{Mb}^{-0.50}$.

Heat loss from an animal takes place mainly from the outer surface, and larger animals have smaller surface-to-volume ratios (relative surface area). Mass specific conductance is proportional to $\mathrm{Mb}^{-0.50}$, and relative surface area to $\mathrm{Mb}^{-0.33}$, and so conductance decreases more rapidly with increasing body size than does body surface area (Schmidt-Nielsen, 1984); in other words, insulation is greater in the large animal. If, on the other hand, the mass exponent in the mass specific conductance expression (−0.50) is compared with that for mass specific heat production ($\dot{H}/\mathrm{Mb}$), i.e. $\mathrm{Mb}^{-0.25}$, it follows that conductance decreases more rapidly than does heat production (or metabolic rate). The fact that larger animals are better protected against heat loss, simply means that they will have a lower critical temperature below which heat production must be increased for them to remain in heat balance. A notable consequence of the mismatch between heat production and conductance is that the larger the animal, the more difficult it will be to transfer the metabolic heat produced. Thus, it is easy to understand why small birds and mammals may undergo daily torpor while this would be a physical impossibility for large ones. In fact, large mammals that undergo mild torpor do it only for prolonged periods of time, like hibernating bears.

Thermal conductances during normothermia and heterothermia

As T_A is lowered in hummingbirds, a linear increase of oxygen consumption ($\dot{V}_{O_2}$ or aerobic heat production) in the normothermic bird is found (Fig. 1). It is important to emphasise that because of their small body size the thermo-neutral zone of small endotherms is very narrow compared to that of large ones. The thermal conductance ($C = \dot{H} / (T_B - T_A)$), because $\dot{H}$ increases as T_A decreases, is normally stable in the whole range of ambient temperatures. Hummingbirds from which data are available reveal the same general pattern. The extrapolated body temperatures (when, for example in Fig. 1, the normothermic line encounters the *x*-axis) from different species are in general within the range of temperatures actually measured (Lasiewski, 1963; Lasiewski & Lasiewski, 1967; Lasiewski *et al.*, 1967; Wolf & Hainsworth, 1972; M. Berger, K. Johansen & J.E.P.W. Bicudo, unpublished results), with the exception of *Oreotrochilus estella*, in which the extrapolated T_B value (48.0 °C) deviates considerably from the directly measured T_B

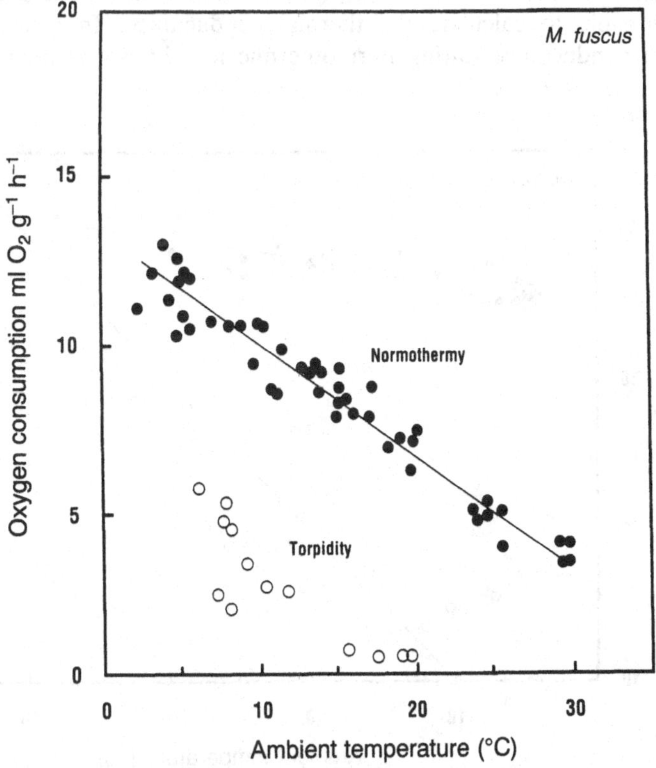

Fig. 1. Oxygen consumption of *Melanotrochilus fuscus* in relation to ambient temperature (T_A) in normothermia (●) and during torpor (○). (After Berger, M. Johansen & J.E.P.W. Bicudo, unpublished results.)

(Carpenter, 1974). With the change of T_B in response to T_A change such as seen in Fig. 2 upper curve, the thermal conductance is in fact not constant over the whole range of ambient temperatures. At lowered T_A, during torpor, the thermal conductance is somewhat increased (open symbols in Fig. 1 when T_A is lower than 15 °C).

Another empirical way of determining thermal conductance is by measuring the cooling constant while the birds are entering torpor. This, however, necessitates the assumption that a negligible heat production occurs during early entry into torpor. If so, the conductance will equal the specific heat multiplied by the cooling constant. With a specific heat of 0.83 cal g^{-1} °C^{-1}, which equals 0.17 ml O_2 g^{-1} °C^{-1}, and the measured cooling constant of $(\ln \Delta T_1 - \ln \Delta T_2)$ $(\Delta t)^{-1})$, where Δt is the difference between times t_1 and t_2, and T_1 and T_2 are equal to the difference between T_B and T_A (ΔT) at time t_1 and t_2, respectively, it is possible to calculate the thermal conductance. In general, the thermal conductance during normothermic metabolism is higher than

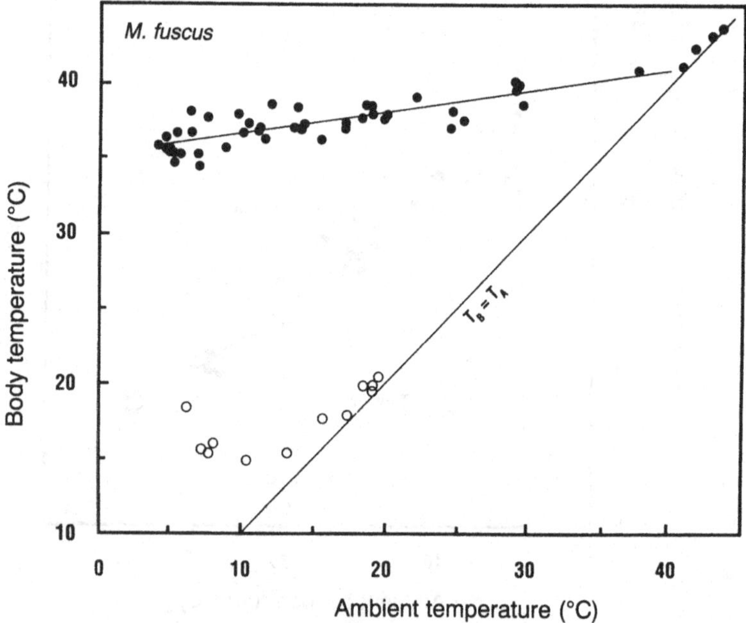

Fig. 2. Body temperature (T_B) plotted against ambient temperature (T_A) for *Melanotrochilus fuscus* in normothermia (●) and durıng topor (○). (After M. Berger, J. Johansen & J.E.P.W. Bicudo, unpublished results.)

that predictable from a multi-species comparison (based on the allometric equation, $C = 0.947\ Mb^{-0.583}$ (Mb in kg), given by Aschoff (1981)), whereas the values calculated from the cooling constants during entrance into torpor are, in general, lower than the values derived from data obtained during normothermia at rest. It is difficult to decide whether this difference (approximately 40%) is real, as the values of C derived from the cooling curves will inherently be elevated because of the remaining (although low) heat production during entry into torpor. If conductances calculated from metabolism in normothermic and torpid (hypothermic) birds are both compared, however, the values exactly match each other (Fig. 3). In manakins (*Manacus vitellinus*) thermal conductances measured in normothermic and hypothermic individuals were indistinguishable (Bartholomew *et al.*, 1983). The least

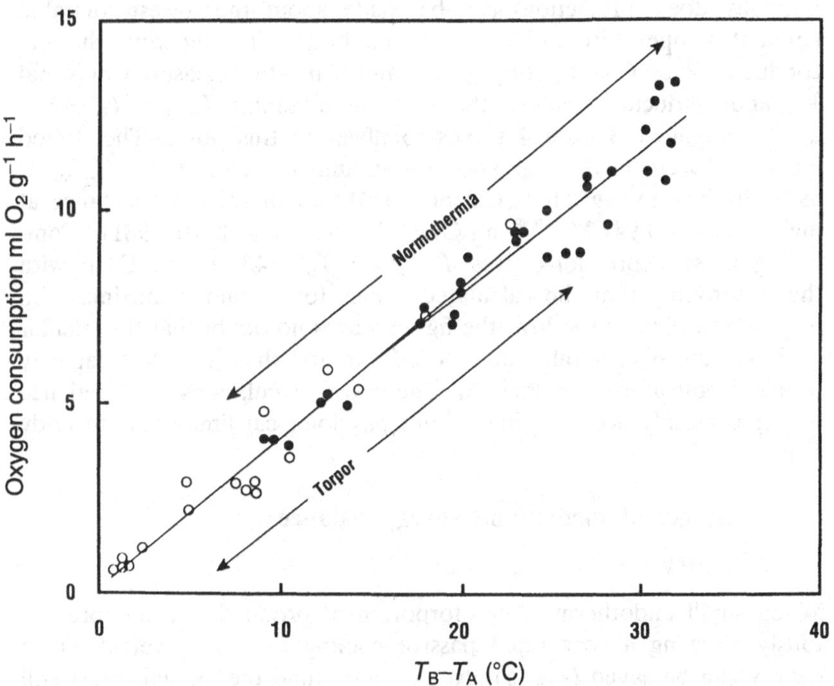

Fig. 3. Oxygen consumption in relation to the body to ambient temperature difference ($T_B - T_A$) during normothermia and torpor. The slopes of the regression lines show the conductance values which are equal in the two conditions. (After M. Berger, J. Johansen & J.E.P.W. Bicudo, unpublished results.)

square regression analysis for all 49 data points obtained for the black jacobin hummingbird, *Melanotrochilus fuscus* in normothermia and different stages of torpor, plotted in Fig. 3 is:

$$\dot{V}_{O_2} = 0.154 + 0.374(T_B - T_A)(r = 0.97);$$

for birds only in normothermia ($n = 31$) it is:

$$\dot{V}_{O_2} = 0.463 + 0.363(T_B - T_A)(r = 0.97);$$

and for birds in different stages of torpor ($n = 18$) it is:

$$\dot{V}_{O_2} = 0.174 + 0.387(T_B - T_A)(r = 0.98).$$ $\dot{V}_{O_2}$ values expressed as ml O_2 g^{-1} h^{-1}.

During activity, however, a much higher thermal conductance has been found (Berger & Hart, 1972). If it is assumed that the highest attainable non-flight $\dot{V}_{O_2}$ can be sustained for longer periods than it normally does, predictions can be made about the lowest tolerable ambient temperature in normothermic birds. Starting with thermal conductance as $C = \dot{V}_{O_2}/(T_B - T_A)$, and using the measured maximal $\dot{V}_{O_2}$ and conductance values, the following minimum $T_{Amin} = T_B - \dot{V}_{O_2\,max}/C$ is attained. Figure 4 serves to illustrate this point. The dotted lines are based on the regression for summit metabolism or $\dot{V}_{O_2\,max}$ as 39.8 $Mb^{-0.35}$ ml O_2 g^{-1} h^{-1} (Calder, 1974) and thermal conductance at night as $C = 0.947\ Mb^{-0.583}$ ml O_2 g^{-1} h^{-1} $°C^{-1}$ (Aschoff, 1981). Combining these expressions gives $T_{Amin} = = T_B - 42\ Mb^{0.233}$. Even with the uncertainty that the calculated values for sustained maximal $\dot{V}_{O_2}$ may be too high or too low, the figure leaves no doubt that the smaller birds are more vulnerable at reduced T_As and thus have less capacity to avoid obligatory hypothermia. The above calculations give credence to the generally accepted idea about physiological limitations to body size.

Stages of torpor and energy balance

Entry

When small endotherms enter torpor, heat production drops precipitously allowing a very rapid passive cooling to a T_B level at which energy can be saved (Fig. 5). At the same time the animals must still be able to retain integrity and alertness. The depth of torpor must depend on ambient climatic factors as well as on the individual energy status prior to torpor (Calder, 1994). The decrease in T_B during entry into torpor is in accordance with the theory of Newtonian cooling, where T_B is equal to T_A and a straight line during the initial cooling

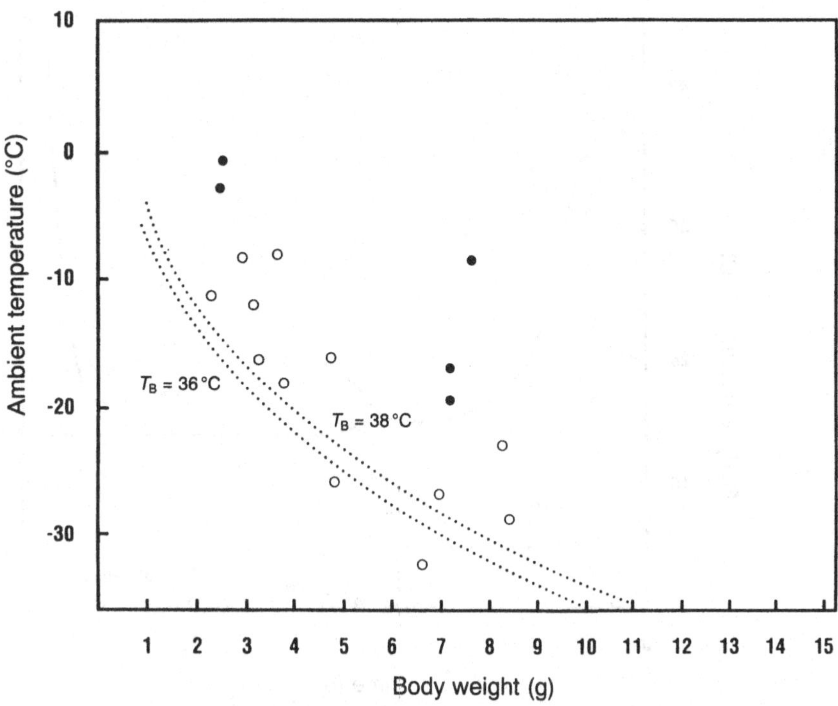

Fig. 4. Lowest tolerable ambient temperatures of normothermic hummingbird species ((●) *Calliphlox amethystina* (Mb = 2.6 g) and *Melanotrochilus fuscus* (Mb = 7.7 g); (○) other Brazilian hummingbird species) calculated form peak metabolism and thermal conductance. The dotted lines are limits for assumed body temperatures (T_B) using regression equations for peak metabolism and conductance (after M. Berger, J. Johansen & J.E.P.W. Bicudo, unpublished results.)

phase is observed (see torpid bird in Fig. 2, between T_A of 20 and 15 °C). This period is characterised by a very low heat production so that the slope of this part of the cooling curve must be determined mainly by insulative properties of the animal. This phase normally ends with a change in the rate of cooling and this inflection on the curve should be regarded as a set point for initiation of a reduced rate of cooling and must, as such, signal that a regulatory process has been engaged which involves an increased heat production (Fig. 1; lower curve).

It has been argued that normal Q_{10}s of 2–3 (Q_{10} is the change of metabolic rate over a 10 °C temperature range) cannot explain the

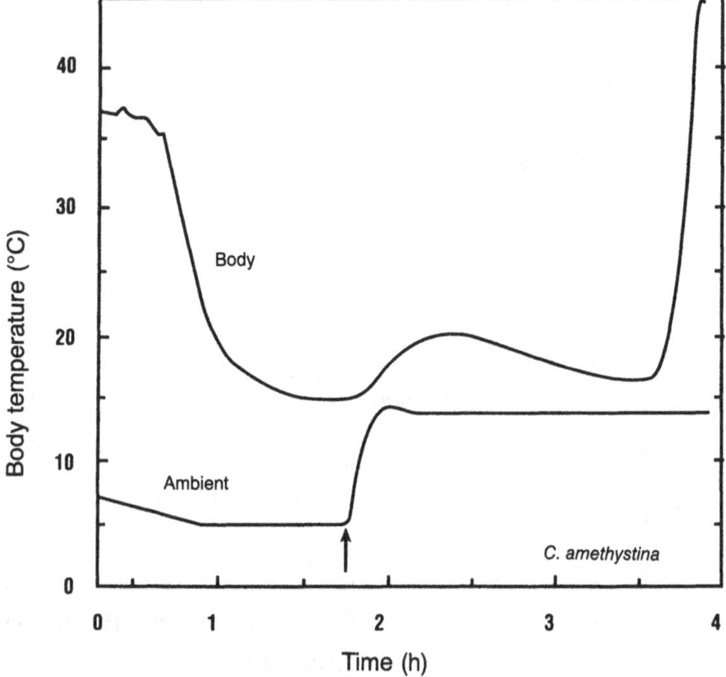

Fig. 5. Time course of body temperature (T_B) in a specimen of *Calliphlox amethystina* (amethyst woodstar hummingbird) entering spontaneous torpor when kept in darkness at low ambient temperature (After M. Berger, J. Johansen & J.E.P.W. Bicudo, unpublished results.)

substantial reduction of energy metabolism during torpor in mammalian hibernators (Malan, 1986). Q_{10}s between 3 and 4 have been found in hummingbirds (Lasiewski, 1963), which are considerably greater than the Q_{10} of 2–3 that characterises biological reactions in general. According to Geiser (1988), daily mammalian heterotherms, in contrast to hibernators, appear to be unable to reduce energy metabolism during torpor beyond a Q_{10} of 2–3. Birds that show daily torpor may differ from mammals in this respect because they show a relatively steep increase in Q_{10} at small body masses. One must consider, however, difficulties in obtaining good basal metabolic rate (BMR) values in the very small hummingbirds that have extremely narrow thermo-neutral zones; high Q_{10}s in birds at very small body masses may therefore be attributed to the inability to obtain true values of BMR under laboratory conditions.

Maintenance

In hummingbirds, the oxygen consumption at the range of ambient temperatures between 15 and 24 °C is very low, approximately 20 times lower than in normothermia (Fig. 1), and reasonably constant because $T_B - T_A$ difference is very small (Fig. 2), in general less than 2 °C. The heart rate is similarly greatly reduced to 0.5–1 beats s^{-1} (normothermic rates at rest are about 10–15 beats s^{-1}). If, however, T_A is lower than about 15 °C, neotropical hummingbirds usually maintain their T_B at 14–16 °C irrespective of further reductions in ambient temperature. Interestingly, the high altitude Andean species *Oreotrochilus estella* is able to regulate its body temperature to values as low as 6–9 °C, when ambient temperature may be as low as −2 °C (Carpenter, 1974).

Arousal

The process of arousal can supposedly be triggered by many factors, external as well as internal. Among the former, light, increased ambient temperature and acoustic signals seem evident. The increase in heat production needed for arousal must come from muscular activity. Heat production by shivering during arousal is localised mainly to the pectoral and supracoracoideus muscles (flight muscles), which represent one-third of the body mass in hummingbirds (Bicudo & Zerbinatti, 1995). Nevertheless, non-shivering thermogenesis could well take place in hummingbirds, considering that the extraordinary demands for hovering flight select for skeletal muscle fibre types (Salt, 1963) with high rates of substrate utilisation (Suarez *et al.*, 1986), and predispose the use of the flight muscles in hummingbirds as thermogenic organs. Although birds have been reported to have multi-ocular adipose tissue, with some similarities to mammalian brown adipose tissue (Saarela *et al.*, 1989), it lacks the key biochemical features (presence of uncoupling protein) and thus is functionally white adipose tissue. From the reported morphological and biochemical evidence, one must reject the hypothesis that birds have brown adipose tissue and look elsewhere for the cold-induced increased heat production (Block, 1994). A major difference in cold defence between mammals and birds is the lack of a specialised thermogenic organ in birds; however, the high oxidative capacity of the flight muscles may preclude the need for such a thermogenic organ. It is quite clear that the presence of a highly oxidative muscle phenotype for locomotion, as found in hummingbirds and also in tunas, predisposes the use of the muscular tissue as a furnace. This is further supported by the work of Barré *et al.* (1985) who demonstrated that cold exposure induces the development of nonshivering thermogenesis of muscular origin in ducklings.

Whether or not rewarming occurs by shivering or non-shivering thermogenesis and despite rapid blood circulation and consequent convective distribution of heat within the body, important temperature gradients may occur between core and surface during arousal in hummingbirds. After arousal from torpor has started, the rate of heat production increases until normothermia is restored. During torpor, when T_A increases to a level at which arousal is instigated, a very rapid rise in oxygen uptake followed by a similar rate of increase in T_B occurs. When T_B increases relative to T_A, oxygen consumption increases at a faster rate than the increase in T_B. During the final stages of arousal, oxygen consumption rapidly increases to a peak, then decreases rapidly to the normothermic value. The peak oxygen consumption is approximately one and a half times higher than the final oxygen consumption at that body temperature (Lasiewski, 1963). The rate of increase in body temperature during arousal is clearly higher in the smallest species. In their quantitative general analysis, Heinrich & Bartholomew (1971) found that the following relationship was present:

$$\text{Warm-up rate } (°\text{C min}^{-1}) = 2.03 \text{ Mb (g)}^{-0.40} \quad (2)$$

Their analysis included data from hummingbirds, swifts, nighthawks, mouse birds and mammals. The rate for a majority of hummingbirds, however, is better approximated by a regression which also includes insect data, where:

$$\text{Warm-up rate } (°\text{C min}^{-1}) = 3.22 \text{ Mb}^{-0.51}$$

(Henrich & Bartholomew, 1971) (3)

Rates of warm-up in hummingbirds, as in insects, appears to depend on ambient temperatures at the start and termination of arousal. At higher T_As the rate of increase in T_B is evidently higher than at lower T_As.

The time needed for entrance into torpor is much longer than the arousal time. The limitations large birds and mammals have in order to undergo daily torpor have already been discussed. Although larger endotherms may require several hours to cool down, it is easier for them to warm up because of their low thermal conductances.

Metabolic cost of arousal

Arousal typically terminates in a very sharp peak value for oxygen uptake that promptly subsides to a lower value. These peak values occur before T_B has reached its highest value, and represent the highest possible metabolism during arousal. The total oxygen consumption from the beginning of arousal until oxygen consumption has levelled off at normothermia represents the cost of arousal. During this time, some of the generated heat is used to raise T_B, while some is lost to the environment because of the temperature gradient, just as in resting normothermic animals. The cost of arousal, based on oxygen uptake measurements, can only be approximate as long as no precise measurements by direct calorimetry of actual heat loss are available. From the amount of energy necessary to raise T_B back to normothermic values, the specific heat (Hs) of the whole body can be calculated. For Brazilian neotropical hummingbirds a value of Hs = 0.69 cal g^{-1} $°C^{-1}$ has been found (M. Berger, K. Johansen & J.E.P.W. Bicudo, unpublished results). This figure is lower than the value of 0.8 which is used in most discussions about heat transfer (Pearson, 1950). Despite the very high peak oxygen consumption during arousal, the combined whole oxygen consumption during arousal is very small, in hummingbirds approximately 7.4 ml O_2 g^{-1} h^{-1}. This corresponds to normothermic metabolism at T_A between 15 and 20 °C.

Energy saving by undergoing torpor

As pointed out earlier the biological significance of torpor is clearly a saving of energy. The energy budget of a hummingbird during a torpor period of 10 h includes the energy expenditure during entry and arousal from torpor. The lowest costs are obviously found at highest T_A. The magnitude of total cost when related to T_A is mainly affected by metabolism during deep torpor. This implies that the regulation of T_B at low T_A has the greatest influence. To maintain torpor, at T_A = 10 °C, *Melanotrochilus fuscus* needs twice the amount of energy it needs at T_A = 20 °C; at T_A = 5 °C, it needs three to four times that amount (Fig. 1). Compared with resting normothermic metabolism which increases in the whole range with lowered T_A in order to keep T_B constant, *M. fuscus* during a torpor period has a metabolic rate that is between 15 and 33% of normothermic values. At ambient temperatures between 15 and 20 °C savings in the order of 85% have been recorded. It becomes clear that torpor at these ambient temperatures presents no problem concerning energy reserves. To stay normothermic overnight at low T_A, however, as for example in breeding birds (Calder, 1974), the storage of appreciable amounts of reserves would be a necessity in hummingbirds at least in some seasons, despite their ability to minimise energy expenditure by undergoing torpor.

The objective and strategy for the torpid hummingbird must be to minimise energy expenditure. To do so, the difference ($T_B - T_A$) and therefore T_B should be kept as low as possible. On the other hand, to maintain the needed thermogenesis at low T_As, hummingbirds regulate their T_B in torpor by increasing T_B as T_A is lowered beyond a critical value (Fig. 2). Hummingbirds, in general, are not able to tolerate a T_B lower than 12 °C, so they must solve this compromise.

Torpor and food deprivation

Some studies have indicated that hummingbirds may also enter into torpor at periods in which they are not exposed to food deprivation. Carpenter & Hixon (1988) showed that rufous hummingbirds may become torpid during the pre-migratory period, apparently to enable them to build up a fat reserve more rapidly. Carpenter (1974) reported torpor in the Andean hillstar hummingbird, *Oreotrochilus estella*, during natural roosting conditions, and found a clear seasonal difference in the use of torpor, with both the number of incidences and the duration of torpor being greater during the winter. Similar observations have been made on Brazilian hummingbirds under semi-natural conditions, as for example in the versicoloured emerald hummingbird (*Amazilia versicolor*), which may undergo nightly torpor during periods of high food availability (indicated by the onset of torpor in an individual caught at normal roosting time) (Bech *et al.*, 1994).

In bats, however, the situation seems to be more straightforward. According to Morris *et al.* (1994), unfed bats rapidly become torpid under laboratory conditions, whereas fed bats showed elevated oxygen consumption, carbon dioxide production and body temperature indicative of generally higher metabolic rates and sustained homeothermy.

Another interesting finding by Bech *et al.* (1994) is that hummingbirds under semi-natural conditions are able to enter into more than one period of torpor per night, with some of these periods lasting for only a few hours. Short periods of torpor have been previously reported in hummingbirds (Hainsworth *et al.*, 1977; Hiebert, 1990), probably related to a well fed state prior to torpor, with torpor bouts occurring in some cases late in the night. The observation of multiple bouts of torpor made by Bech *et al.* (1994) seems unusual and raises fundamental questions about the energetics of torpor. The time required for both entry into and arousal from torpor increases markedly with body size in birds, as does the energy requirement for arousal (Lasiewski & Lawiewski, 1967; Calder & King, 1974; Bartholomew, 1981). The increased energy requirement for arousal with increasing body mass thus limits the energy-saving potential for torpor. Thus, multiple torpor bouts shown by some species of hummingbirds indicate that they may not utilise the full time required to enter into torpor, but may actually arouse from torpor even before their body temperature has reached its lowest level. Such a pattern seems to indicate that these very short periods of torpor may be of thermoregulatory significance for the birds, and that the cost of arousal would not counteract the reduction in metabolic rate resulting from the short-term fall in T_B. Clearly, the energy cost of undergoing multiple shallow torpor bouts during the night is much lower than maintaining normothermia and the ensuing high metabolic rates during the entire night period.

Unusually low respiratory quotients

Recently, Chaui-Berlinck & Bicudo (1995) have found that well fed swallow-tailed hummingbirds (*Eupetomena mcroura*), whether they go into torpor during the night or not, invariably show dramatic changes in their respiratory quotient (RQ = carbon dioxide production/oxygen consumption), which can reach values as low as 0.2–0.3 over a period of time that can last more than 6 h, while maintaining normothermic T_B and metabolic rate. This unusual pattern can be summarised as follows: immediately after the bird has had its last meal, RQ stays higher than 1, indicating that sugar is being converted into fat; later on (usually less than 30 min), RQ drops to values ranging from 0.7 to 0.8 for no longer than a couple of hours; and, finally, RQ drops precipitously to values between 0.2 and 0.3, remaining so for more than 6 h. It is not known why the bird undergoes such unusual metabolic shifts, but it does not appear to be related to a ventilatory retention of carbon dioxide as there are no significant changes in the ventilatory pattern during this period (J.G. Chaui-Berlinck, personal communication). Walsberg & Wolf (1995) have also reported a low RQ value (0.66) for house sparrows during their nocturnal fast. This value, however, is much less extreme than those found for the swallow-tailed hummingbird. Non-fed bats (Morris *et al.*, 1994) appear to undergo a similar drop in RQ (0.41) during torpor, but in their case this is attributed to episodic respiration rather than to unusual metabolic shifts. Hummingbirds store fat and seem to depend on this substrate to overcome periods of food deprivation, so changes in fat beta-oxidation might be a possible explanation for the observed pattern. At present, however, the biochemical pathway that could result in the low RQ observed in fasting hummingbirds as well as its adaptive implications are not known. There is a broad range of low RQ variation within and among individuals (Chaui-Berlinck & Bicudo, 1995). This could be pointing to a mixed metabolism resulting in non-stable and non-precise low RQ values. Further biochemical studies, as measurements of ketone body formation and concentration, amino acids turnover, lipid composition, etc. should be performed during these periods of low RQ in hummingbirds. These rather intriguing and extreme patterns emphasise the need to conduct more integrative studies which should take into account, the inextricable links among energetics, nutrition and osmotic regulation observed in hummingbirds (Beuchat *et al.*, 1990).

Conclusion

Although torpor may be regarded as an integrated process that depends on internal and external variables, the basic physics of heat transfer as well as straightforward surface-to-volume considerations may explain why small endotherms can undergo torpor within a 24 hour cycle, whereas for endotherms larger than a critical body size this would be a physical impossibility. A more detailed comparison of the scaling of metabolic heat production and thermal conductance in endotherms further substantiates this concept. Thus, body size *per se* plays a crucial role in determining whether an endothermic animal is able and needs to undergo daily torpor.

It is quite clear that torpor (single or multiple bouts) allows small endotherms to have considerable energy savings, otherwise their survival would be severely threatened. It is not so obvious, at least in the case of hummingbirds, why they may or may not undergo nightly torpor during periods of high food availability. The results so far obtained suggest that there may exist a triggering system for torpor that operates only when a certain threshold is reached. Food availability, and thus energy reserve, appears not to be the sole variable in such a system. Low RQ values observed during several hours in normothermic hummingbirds point towards profound metabolic changes that may allow hummingbirds to delay torpor somehow. It would be extremely useful therefore to define such an internal signalling system that seems to permit them to 'choose' to undergo torpor or not under certain conditions.

Finally, it is not yet settled how rewarming in hummingbirds is accomplished. In cold-acclimated Muscovy ducklings, Barré *et al.* (1986) demonstrated a large increase in the metabolic rate that could not be attributed to shivering activity. A potent calorigenic effect of glucagon in birds, similar in scope to noradrenaline-induced thermogenesis in mammals, has also been documented (Barré *et al.*, 1987). Glucagon results in release of free fatty acids in birds, and it is postulated that in response to the hormone, increases of cytoplasmic free fatty acids would result in partial uncoupling of skeletal muscle mitochondria with a concomitant increase of respiration and heat production. The relative large sizes of flight muscles and liver (Zerbinatti, 1994) of hummingbirds combined with an elevated mitochondrial volume density in these tissues may well support the existence of non-shivering thermogenesis in hummingbirds.

References

Aschoff, J. (1981). Thermal conductance in mammals and birds: its dependence on body size and circadian phase. *Comparative Biochemistry and Physiology* **69A**, 611–19.

Barré, H., Cohen-Adad, F., Duchamp, C. & Rouanet, J. (1986). Multiocular adipocytes from muscovy ducklings differentiated in response to cold acclimation. *Journal of Physiology* **375**, 27–38.

Barré, H., Cohen-Adad, F. & Rouanet, J. (1987). Two daily glucagon injections induce nonshivering thermogenesis in Muscovy ducklings. *American Journal of Physiology* **252**, E616–E620.

Barré, H., Geloen, A., Chatonnet, J., Dittmar, A. & Rouanet, J. (1985). Potentiated muscular thermogenesis in cold-acclimated ducklings. *American Journal of Physiology* **249**, R533–R538.

Bartholomew, G.A. (1981). A matter of size: an examination of endothermy in insects and terrestrial vertebrates. In *Insect Thermoregulation* (ed. B. Heinrich), pp. 45–78. New York: John Wiley.

Bartholomew, G.A., Vleck, C.M. & Bucher, T.L. (1983). Energy metabolism and nocturnal hypothermia in two tropical passerine frugivores, *Manacus vitellinus* and *Pipra mentalis*. *Physiological Zoology* **56**, 370–9.

Bech, C., Abe, A.S., Steffensen, J.F., Berger, M. & Bicudo, J.E. P.W. (1994). Multiple torpor bouts in hummingbirds. In *Integrative and Cellular Aspects of Autonomic Functions* (ed. K. Pleschka, R. Gertsberger & K.F. Pirerau), pp. 323–8. Paris: John Libbey Eurotext.

Berger, M. & Hart, J.S. (1972). Die Atmung beim Kolibri *Amazilia fimbriata* wahrend des Schwirrfluges bei verschiedenen Umbegungstemperaturen. *Journal of Comparative Physiology B* **81**, 363–80.

Beuchat, C.A., Calder, W.A. & Braun, E.J. (1990). The integration of osmoregulation and energy balance in hummingbirds. *Physiological Zoology* **63**, 1059–81.

Bicudo, J.E.P.W. & Zerbinatti, C.V. (1995). Physiological constraints in the aerobic performance of hummingbirds. *Brazilian Journal of Medical and Biological Research* **28**, 1139–45.

Block, B.A. (1994). Thermogenesis in muscle. *Annual Review of Physiology* **56**, 535–77.

Bradley, S.R. & Deavers, D.R. (1980). A re-examination of the relationship between thermal conductance and body weight in mammals. *Comparative Biochemistry and Physiology* **65A**, 465–76.

Calder, W.A. (1974). Microhabitat selection during nesting of hummingbirds in the Rocky Mountains. *Ecology* **54**, 127–34.

Calder, W.A. (1994). When do hummingbirds use torpor in nature? *Physiological Zoology* **67**, 1051–76.

Calder, W.A. & King, J.R. (1974). Thermal and caloric relations in birds. In *Avian Biology*, vol. IV (ed. D.S. Farner & J.R. Kind), pp. 259–413. New York: Academic Press.

Carpenter, F.L. (1974). Torpor in the Andean hummingbird: its ecological significance. *Science* **183**, 545–7.

Carpenter, F.L. & Hixon, M.A. (1988). A new function for torpor: fat conservation in a wild migrant hummingbird. *Condor* **90**, 373–8.

Chaui-Berlinck, J.G. & Bicudo, J.E.P.W. (1995). Unusual metabolic shifts in fasting hummingbirds. *Auk* **112**(3), 774–8.

Geiser, F. (1988). Reduction of metabolism during hibernation and daily torpor in mammals and birds: temperature effect or physiological inhibition? *Journal of Comparative Physiology B* **158**, 25–37.

Hainsworth, F.R., Collins, B.G. & Wolf, L.L. (1977). The function of torpor in hummingbirds. *Physiological Zoology* **50**, 215–22.

Heinrich, B. & Bartholomew, G.A. (1971). An analysis of pre-flight warm-up in the sphinx moth, *Manduca sexta. Journal of Experimental Biology* **55**, 223–39.

Herreid, C.F., II & Kessel, B. (1967). Thermal conductance in birds and mammals. *Comparative Biochemistry and Physiology* **21A**, 405–14.

Hiebert, S.M. (1990). Energy costs and temporal organization of torpor in the Rufous hummingbird (*Selasphorus rufus). Physiological Zoology* **63**, 1082–97.

Hochachka, P.W. & Guppy, M. (1987). *Metabolic Arrest and the Control of Biological Time*. Cambridge: Harvard University Press.

Lasiewski, R.C. (1963). Oxygen consumption of torpid, resting, active and flying hummingbirds. *Physiological Zoology* **36**, 122–40.

Lasiewski, R.C. & Laseiwski, R. (1967). Physiological responses of the Blue-throated and Rivoli's hummingbirds. *Auk* **84**, 34–48.

Lasiewski, R.C., Weathers, W.W. & Bernstein, M. (1967). Physiological responses of the Giant hummingbird, *Patagona gigas. Comparative Biochemistry and Physiology* **23A**, 797–813.

McNab, B.K. (1983). Energetics, body size, and the limits to endothermy. *Journal of Zoology, London* **199**, 1–29.

Malan, A. (1986). pH as a control factor in hibernation. In *Living in the Cold* (ed. H.C. Heller, X.J. Musacchia & L.C.H. Wang), pp. 61–70. New York: Elsevier.

Morris, S, Curtin, A.L. & Thompson, M.B. (1994). Heterothermy, torpor, respiratory gas exchange, water balance and the effect of feeding in Gould's long-eared bat *Nyctophilus gouldi*. *Journal of Experimental Biology* **197**, 309–35.

Pearson, O.P. (1950). The metabolism of hummingbirds. *Condor* **52**, 145–52.

Precht, H., Christophersen, J., Hensel, H. & Larcher, W. (1973). *Temperature and Life*. New York: Springer-Verlag.

Saarela, S., Hissa, R., Pyörnilä, Harjula, R., Ojanen, M. & Orell, M. (1989). Do birds possess brown adipose tissue? *Comparative Biochemistry and Physiology* **92A**, 219–28.

Salt, W.R. (1963). The composition of the pectoralis muscles of some passerine birds. *Canadian Journal of Zoology* **41**, 1185–90.

Schmidt-Nielsen, K. (1984). *Scaling: Why Is Animal Size So Important?* Cambridge: Cambridge University Press.

Stone, G.N. & Purvis, A. (1992). Warm-up rates during arousal from torpor in heterothermic mammals: physiological correlates and a comparison with heterothermic insects. *Journal of Comparative Physiology B* **162**, 284–95.

Suarez, R.K., Brown, G.S. & Hochachka, P.W. (1986). Metabolic sources of energy for hummingbird flight. *American Journal of Physiology* **225**, R699–R702.

Walsberg, G.E. & Wolf, B.O. (1995). Variation in the respiratory quotient of birds and implications for indirect calorimetry using measurements of carbon dioxide production. *Journal of Experimental Biology* **198**, 213–19.

Wolf, L.L. & Hainsworth, F.R. (1972). Environmental influence on regulated body temperature in torpid hummingbirds. *Comparative Biochemistry and Physiology* **41A**, 167–73.

Zerbinatti, C.V. (1994). *Adaptações estruturais e funcionais do músculo peitoral, coração e figado à demanda oxidativa do vôo estacionário em beija-flores*. Ph.D. dissertation, University of São Paulo, Brazil.

Z.A. EPPLEY

Development of thermoregulation in birds: physiology, interspecific variation and adaptation to climate

Introduction

Birds and mammals have independently evolved substantial capacities for metabolic heat production which they use to maintain a high body temperature (approximately 37 °C for eutherian mammals, approximately 40 °C for neognathous birds). These high body temperatures result from their high rates of cellular metabolism (Else & Hulbert, 1987). Birds and mammals can further augment heat production as needed to offset heat loss. Skeletal muscle is the primary site of supplementary thermogenesis in birds (Hohtola, 1982; Duchamp & Barre, 1993). The uniquely mammalian tissue, brown fat, is a major site of thermogenesis (Hayward & Lisson, 1992) in some circumstances, such as during the neonatal period, hibernation and cold-acclimation. Birds and mammals living in extreme thermal habitats have adjustments in metabolic rate and insulation to compensate for the thermal extremes (Marsh & Dawson, 1989). The degree to which these phenotypic differences are genetically fixed or are environmentally determined (examples include Lynch *et al.*, 1976; James, 1983) are not well characterised (Garland & Adolph, 1991). Metabolic rate and insulation show some phenotypic plasticity and may be adjusted during thermal acclimation (Marsh & Dawson, 1989). Birds and mammals normally maintain their body temperature within narrow limits, whether they live in temperate or thermally extreme habitats. Notable exceptions are when species employ torpor to avoid energetically unfavourable conditions and abandon thermoregulation, or regulate at a considerably lower set point.

Neonate birds also have high body temperatures (approximately 38 °C in many species). While adults are competent thermoregulators, they are not born so. Thermoregulation develops through stages of increasing competency. First, the young lack metabolic compensation for cold (ectothermy). Ectothermic young have intact central regulatory systems and behaviourally thermoregulate using parental heat sources (Evans *et al.*, 1994). Later they defend body temperature by increasing their metabolic rate (endothermy), but this cannot offset heat loss. Ultimately, the metabolic response is sufficient to maintain a high body temperature (homeothermy). The transition from ectothermy to endothermy is sudden, while the perfection of homeothermy occurs more slowly. The thermal range over which homeothermy is maintained increases with age (Fig. 1*a*); however, homeothermy is conditional. Sufficiently low temperatures can overwhelm the thermoregulatory capacity of any endotherm and cause hypothermia (Fig. 1).

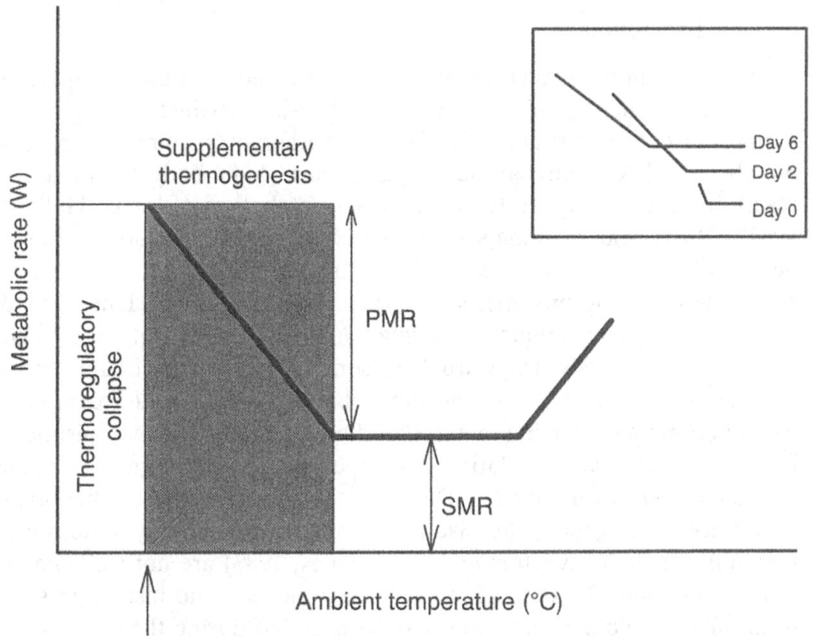

Fig. 1. Metabolic response of a homeothermic bird to a range of ambient temperatures. Standard metabolic rate (SMR), the metabolic rate of the resting animal, sets the minimum energetic costs for thermoregulation. Supplementary thermogenesis increases the metabolic rate to offset greater heat loss at cold temperatures. At temperatures where the rate of heat loss is greater than the maximum metabolic rate (peak metabolic rate, PMR), the animal will become hypothermic and undergo thermoregulatory collapse. (Inset). The thermoregulatory abilities of nestlings improve with age. For a given ambient temperature, older chicks may have greater costs than younger ones but work at a lower fraction of their maximum metabolic power. (After Eppley, 1991, 1994.)

Birds nest in diverse thermal habitats: grey gulls *Larus modestus* nest in the Atacama Desert in summer when temperatures of 50 °C are common (Howell *et al.*, 1974) whereas emperor penguins *Aptenodytes forsteri* breed in the depth of the Antarctic winter when temperatures may drop to −48 °C (LeMaho, 1977). Even mild temperatures (approximately 20 °C) can be an overwhelming cold challenge for the hatchlings of most species. Brooding by parents normally protects the young until they have achieved a basic degree of thermoregulation, and compensates for local thermal conditions (White & Kinney, 1974). During the period of thermal dependence, however, there are trade-offs between brooding and feeding opportunities (Ricklefs, 1988). Self-feeding young must stop feeding and seek brooding as they become hypothermic. Parents, whose young remain in the nest and are fed, may face a similar trade-off between the need to brood their young and the need to forage for them. Temperature extremes, storms or poor feeding conditions tip the balance between brooding and foraging, and young are exposed (for examples, see Nice, 1962: 175–6; Weeden, 1966; Kaiser *et al.*, 1990). At these times, chicks are dependent for their survival on their physiological abilities to thermoregulate. Thus, temperature is a potential selective agent on avian breeding biology acting on the young before they have perfected thermoregulation. Specific thermal and ecological conditions may favour young that develop thermal independence earlier or later.

Ecological variation in ontogenetic pattern

Birds show a spectrum of post-hatching developmental patterns: from altricial chicks that hatch naked with poorly developed sense and locomotor systems to the extremely precocial Megapods whose young are independent and can fly shortly after hatching (Nice, 1962). The terms altricial and precocial originally described the relative maturity of newborns, but they have also been used to describe other correlated aspects of postnatal development. Altricial young generally have short total development times, from conception to fledging, and high growth rates. Precocial young have longer development times and slower growth rates. Precocial species have rudimentary thermoregulatory abilities shortly after hatching and are thermally competent early, whereas altricial species are ectothermic for much of the nestling period. Altricial young, however, because of their shorter development time, may become thermally independent in fewer days after conception than more precocial species (Nice, 1962).

The terms, altricial and precocial, while useful, obscure the diversity in developmental patterns. For example, the downy young of galliform (chickens and pheasants) and anseriform (waterfowl) birds are precocial but gallinaceous hatchlings have more mature resting metabolic rates (Bucher, 1986), whereas ducklings have supplementary thermogenic systems that are more mature (Koskimies & Lahti, 1964). Semi-precocial young show a diverse and confusing mixture of traits. Semi-precocial gull species perfect thermoregulation earlier than the precocial galliforms (Wekstein & Zolman, 1971; Eppley, 1994), despite their being less precocial in locomotion. Within a semi-precocial chick, tissues show a diversity in their rates of growth and maturation, some more similar to altricial species while others are more similar to precocial species (Ricklefs, 1979*b*).

Ontogenetic pattern appears to be highly conserved among families of birds (Nice, 1962). Laboratory studies of hatchlings, however, have identified ecologically important variation in the cold tolerance of chicks among species within families of birds. Koskimies & Lahti (1964) examined the thermoregulatory ability of hatchlings in ten species of waterfowl whose thermal habitats differed. They measured metabolic rate and body temperature during long exposures to cold in mallard (*Anas platyrhynchus*), common teal (*A. crecca*), European widgeon (*A. penelope*), goldeneye (*Bucephala clangula*), tufted duck (*Aythya fuligula*), common poachard (*A. ferina*), common eider (*Somateria mollissima*), velvet scoter (*Melanitta fusca*), common merganser (*Mergus merganser*), and red-breasted merganser (*M. serrator*). Some species' young rapidly succumbed to hypothermia when exposed to low temperatures; others maintained normal body temperatures for hours. Homeothermic chicks would eventually suffer thermoregulatory collapse when their energy reserves were exhausted, but the time to collapse differed among species. The best thermoregulators were found among diving ducks and subarctic and arctic species. Differences among hatchlings in thermoregulatory ability were primarily the result of differences in their ability to increase their metabolic rate on cold exposure (supplementary thermogenesis), although body size and insulation also contributed. Koskimies & Lahti (1964) recognised that early metabolic maturation has survival value in cold climates, and releases parents from the need to brood frequently, but that the high energetic costs of metabolic maturity are a handicap when food is limiting. Metabolically immature young are more economical but are also more vulnerable to exposure. Chilled young may succumb to hypothermic coma, and even when the exposure is not lethal, growth is suspended and may resume at a lower rate (Webb, 1987).

Dawson and students (Dawson *et al.*, 1972, 1976; Dawson & Bennett, 1980, 1981) confirmed these findings over a finer phylogenetic scale.

They studied several species in the genus *Larus*, which contains more than 30 species and spans a latitudinal range from the arctic to the Antarctic. The standard metabolic rate (SMR) of hatchlings is higher for anseriforms than for gulls; however, gull hatchlings had much greater capacity for supplementary thermogenesis: 1.5–3.4 times SMR, compared with 1.5–1.7 for ducklings. Dawson *et al.* (1976) found that newly hatched gulls were capable of shivering but that the aerobic capacity of their muscles was low. They used field studies to assess the extent to which eggs and chicks are jeopardised by short-term disruption of parental attention. They found that temperatures at the breeding site exceed both the heat- and cold-tolerance of hatchlings on a daily basis, even in temperate climates. They reasoned that the hatchlings' thermoregulatory abilities were important primarily when parental attention was disrupted, such as during disturbance by predators or rivals.

Studies on the thermoregulatory abilities of hatchlings or the development of thermoregulation now number in the dozens. A disproportionate number of these studies involve extremely precocial species (examples include Booth, 1984; Eppley, 1984) or species nesting in extreme habitats (Bech *et al.*, 1991). Most of these studies address the physiological ecology of neonate birds, or the physiological mechanism for the development of thermoregulation. Few studies have assessed climatic adaptation in development, and those that do have limited their analysis to the thermoregulatory abilities of hatchlings. Previously mentioned studies of climatic adaptation in the thermoregulatory abilities of hatchlings (Koskimies & Lahti, 1964; Dawson *et al.*, 1972, 1976; Dawson & Bennett, 1980, 1981) ranked species but did not apply statistical analyses. More recently, Bech *et al.* (1991) found a positive correlation between the SMR of hatchling procellariiforms and their natal latitude, as did Klaassen (1994) for tern chicks. All these studies relied on interspecific allometry to generate predictions and detect climatic adaptation. Allometry has been challenged as a means of interspecific comparison because species are treated as statistically independent when, in fact, they are phylogenetically related (Felsenstein, 1985; Garland & Adolph, 1994; Garland & Carter, 1994). Comparative methods that account for phylogeny have not been applied to avian physiological development, with the exception of Konarzewski's (1995) analysis of hatchling SMR and postnatal growth rates, in which he found a positive correlation between SMR, growth rate and natal latitude for 36 species representing nine orders of birds. While previous studies have suggested that development shows climatic adaptation, the methods applied have been questioned and few species and points in development have been studied.

Mechanistic basis for the development of endothermy

The physiological basis for the development of endothermy and homeothermy has received much attention (Wekstein & Zolman, 1969, 1971; Untergasser & Hayward, 1972; Misson, 1977; Dunnington & Siegel, 1984; Tazawa *et al.*, 1988; Eppley, 1991; Choi *et al.*, 1993; Eppley, 1994). The broad consensus of these studies is that rapid and profound changes in the capacity for muscular thermogenesis are key to the development of endothermy (Steen & Gabrielsen, 1986; Grav *et al.*, 1988; Eppley, 1991). High SMR and good insulation also contribute to thermoregulatory performance and may develop earlier in species from cold climates (Koskimies & Lahti, 1964; Klaassen & Drent, 1991). The timing of endothermy and the muscle groups used for early thermogenesis differ among avian taxa. In passerines (Olson, 1994) and procellariiformes (Ricklefs *et al.*, 1980), the pectoral muscles are the principal site of thermogenesis, whereas the leg muscles are more important for heat production in the galliformes (Aulie & Grav, 1979), anseriiformes (Steen *et al.*, 1988) and charadriiformes (Eppley, 1991). In altricial redwinged blackbirds (*Agelaius phoeniceus*) (Olson, 1994), shivering by the pectoralis correlates with the onset of endothermy and precedes the development of shivering and locomotion by the leg muscles. In precocial birds, shivering occurs in both the pectoral and the leg muscles at the onset of endothermy (Eppley, 1991). The leg muscles of precocial birds are used both for shivering and locomotion; however, shivering develops suddenly and coincidently with endothermy while locomotor capacity develops more slowly (Z.A. Eppley, unpublished results). The greater mass and maturity of the leg muscles in precocial hatchlings may predispose their differential contribution to shivering thermogenesis (Ricklefs, 1983).

Several studies have examined one or a few biochemical factors in an attempt to identify the physiological changes responsible for the sudden development of muscular thermogenesis in young birds. These studies have found increases in myofibrillar ATPase (bank swallows *Riperia riperia*) (Marsh & Wickler, 1982), an index of the contractile speed of muscle, and cytochrome oxidase activities (common eiders); (Grav *et al.*, 1988) when endothermy develops, which suggest that increases in contractile speed and aerobic metabolism are involved. Increases in these factors at the onset of endothermy, however, have not been found in seven other species (Choi *et al.*, 1993; Eppley, 1991), indicating that these factors are not causal for all species. Probably, the maturation of several systems is involved in the development of muscular thermogenesis.

The tight correlation between structure and function in muscle permit ultrastructural studies to monitor the maturation of several systems. Ultrastructural changes in muscle during the development of endothermy have been examined twice. Once, briefly in common eiders (Grav *et al.*, 1988), in which qualitative increases in mitochondrial volume and cristae area were described in a thigh muscle (femorotibialis medius). A more thorough, stereological analysis of a chicken (*Gallus gallus*) thigh muscle (iliofibularis) across the threshold of endothermy shows a profound increase in the number of triads, structures central to excitation–contraction coupling, and intracellular lipid stores, needed to fuel sustained activity (Eppley & Russell, 1995). Significant increases in mitochondrial inner membrane surface area and maturity of the matrix (sites of aerobic metabolism) also occurred. Mitochondrial and myofibrillar proliferation were evident but were not significant across the threshold of endothermy.

Water content has been used as an index of muscle maturity (Ricklefs, 1979*b*; Ricklefs *et al.*, 1980) and is inversely related to myofibrillar content. While not a causal link, low water content is a convenient index and is supported by the stereological analysis of chicken muscle (Eppley & Russell, 1995). Major structural components (myofibrils, mitochondria and lipid) comprise 57% of the volume of ectothermic day 18 chick embryos whereas these components are 74% of the cell volume of endothermic 96-h chicks (Eppley & Russell, 1995). Myofibrillar volume accumulates slowly from 49% in day 18 embryo, to 60% in 96-h chicks to approximately 80% in adults. Muscle function in shivering thermogenesis depends on its capacity for activation and ability to generate sustainable metabolic power, not on its force production, *per se*.

Energetic consequences of the ontogenetic pattern

Altricial species have growth rates that are 3 to 4-fold higher than precocial species of the same adult mass (Ricklefs, 1973, 1979*a*). These differences are inherent and can not be abolished by manipulating diet (Ricklefs, 1983, 1985). Altricial and precocial chicks differ in the timing of their skeletal muscle maturation and their ability to produce heat by shivering. Growth rate and the skeletal muscle maturation are inversely related across ontogenetic patterns at the organismal level (Ricklefs, 1973), across muscles within a species (Ricklefs, 1979*b*) and within a muscle (Olson, 1992). Ricklefs proposed that rapid growth in skeletal muscle is incompatible with its function (Ricklefs *et al.*, 1994), because mature muscle contains a small number of myoblasts: myogenic stem cells that give rise to differentiated muscle cells. Ricklefs' hypothesis of a trade-off between muscle growth and function is true for individual muscle cells, but not for the muscle, as a whole. A skeletal muscle cell forms by the fusion of several myoblasts, the syncitium accumulates contractile proteins and individual nuclei permanently withdraw from the cell cycle (Olson, 1993). Birds have their full complement of muscle fibres at or shortly after birth (for an example, see Fredette & Landmesser, 1991). Further growth occurs by an increase in size of existing muscle fibres. Ricklefs suggests that functional muscle lacks the ability to grow rapidly because mature muscle cells have withdrawn from the cell cycle (Ricklefs, 1979*a*, *b*, 1983; Ricklefs *et al.*, 1994).

Studies on the cell biology of skeletal muscle show that the correlation between growth and function is not causal. Muscle can function soon after it is formed in the first quarter of embryonic development although it contracts slowly, produces low forces and lacks endurance (Landmesser & O'Donovan, 1984). Myoblasts proliferate to maintain their populations even when some differentiate as in growth and repair (Schultz & McCormick, 1994). The processes of new muscle fibre formation and existing fibre growth are not fundamentally different; in both, myoblasts proliferate and fuse with each other or with existing fibres. Mature muscle can form new fibres to repair damage (Kennedy *et al.*, 1988) and retains enough myoblasts to regrow through multiple cycles of regeneration (Schultz & McCormick, 1994). Increased activity, muscle damage and growth factors activate myoblast proliferation. Myoblast populations and their mitotic activity are regulated appropriate to the muscle's fibre type composition, metabolic activity and age (Schultz & McCormick, 1994). Ricklefs' trade-off between growth and function has heuristic value even if there is not a causal link. The link is the coordinate occurrence of slowed growth and high functional capacity late in the developmental program of skeletal muscle.

The development of thermoregulation affects the energy requirements of chicks, the time–energy budgets of the parents and the likelihood that chicks will survive interruptions in parental care, and thus may have profound ecological consequences for species. In the temperate nesting (34°N) western gull *Larus occidentalis*, homeothermic young that are brooded have energetic costs 3-fold greater than ectothermic young that are brooded. These costs increase to 6-fold greater when homeotothermic young are exposed to the minimum operative temperature at the nest site (Eppley, 1991). These costs are significant to the young, but are they significant in parental energy budgets? Parents provide energy to their chicks in the form of heat and food, thus the parental energy budget is where costs must be compared. One study on the costs of breeding allows this point to be addressed. Chappell *et al.* (1993) studied Adelie penguin (*Pygoscelis adeliae*) energetics in Antarctica. Breeding costs amounted to less than 10% of the annual energy budget. Most of the costs of breeding were incurred by parents early, during nest attendance, rather than later when they were feeding chicks with high growth or large thermoregulatory costs. This result is counter-intuitive because incubation is energetically inexpensive (approximate maintenance costs, 2.7 × SMR in Adelies) compared with the costs of foraging (8.2 × SMR). Nest attendance, however, precludes feeding and many bird species lose mass during this time (examples include orange-breasted sun birds *Nectaria nectaria* (Williams, 1993), Xantus' murrelets *Synthliboramphus hypoleucus* (Murray *et al.*, 1983), blue tits *Parus caeruleus* (Nur, 1984). Mass loss during incubation may be programmed (in least auklets *Aethia pusilla* Jones, 1994), and may be a hormonally-induced anorexia. High energetic costs accrue later in the breeding season when birds forage to replace the lost mass. In Adelie penguins, the cost of refeeding amounts to 52% of the cost of breeding for females and 67% of the costs of breeding for males (reflecting the costs incurred in territory establishment). Chick feeding accounted for 48% of the cost of breeding for females and 32% for males. This pattern is not universal. In other species (Wilson's storm petrel *Oceanities oceanicus*) chick feeding is the most expensive phase (Obst & Nagy, 1993).

The acquisition of thermoregulatory competency has large energetic consequences for young birds, but these costs are insignificant in the larger parental energy budget. What may be more significant is that the thermal independence of the young may release parents from the need to constantly attend the nest. In an extremely hot environment, Bartholomew & Dawson (1954) found that parental attention and chick thermoregulatory abilities were inversely related in sympatrically breeding pelicans, herons and gulls so that chicks with greater thermoregulatory abilities released their parents from the constant need to attend the nest. Even in temperate regions, brooding and chick thermoregulatory abilities showed a close relationship (Fig. 2). Likewise, the early development of thermoregulation may be advantageous in cold environments. The early development of thermoregulation has been cited for the preponderence of precocial species at high latitudes (Remmert, 1980), the lower mortality among precocial young compared with altricial young during arctic storms (Jehl & Hussell, 1966) and for differential survival among the young of sympatrically breeding penguin species in the Antarctic (Taylor, 1985). The possession of a particular developmental pattern does not bar species from nesting in

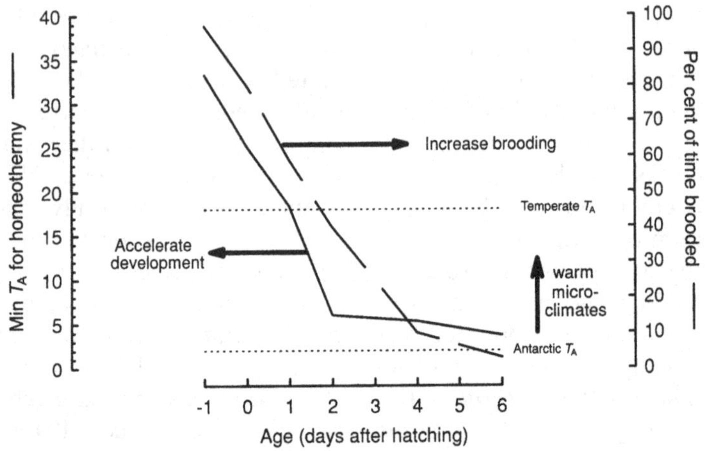

Fig. 2. Amount of time chicks are brooded (broken line) closely matches chick thermoregulatory abilities. Data are shown for western gulls on the Farallon Islands, California, where air temperatures average 14 °C (Eppley, 1996). The lowest temperature at which chicks could maintain homeothermy (homeothermic limit) (solid line) was calculated from laboratory values of standard (SMR) and peak (PMR) metabolic rate and thermal conductance. If western gulls were to invade the Antarctic nesting colonies of kelp gulls, where air temperatures average 2 °C, their chicks would be exposed to overwhelming low temperatures (dotted line) for the first week after hatching if brooding were not adjusted. The options for compensation for the colder conditions are for parents to choose warmer microclimates for their nest sites, for parents to brood their young longer or for chicks to mature faster.

specific climates. Both altricial and precocial species breed at high latitudes, and each shows the same mode of development as temperate members of their order. Thus, developmental pattern seems largely resistant to transformation and is highly conserved within orders regardless of the nesting climate (Nice, 1962; Ricklefs, 1983).

Climatic selection on ontogenetic pattern

How do avian species adapt to conserve reproduction when climatic change overcomes the breeding site or when they colonise a new thermal environment? A species that retains its temperature pattern of brooding and chick development in a colder climate will expose its young to potentially fatal low temperatures (Fig. 2). Thus, members of a temperate species must alter some aspect of their breeding biology to reproduce successfully in the colder climate.

Species may compensate in several ways (Fig. 2). Parents may simply brood their young longer until thermal independence is achieved at the lower temperature. Brooding behaviour may be plastic and easily adjusted; however, Johnson & Cowan's (1974) results suggest that some aspects of nest attendance are genetically determined. Female starlings (*Sturnus vulgaris*) attempting to extend their Northern breeding range did not increase their incubation bouts to compensate for colder conditions, and their eggs failed to hatch (Johnson & Cowan, 1974). Increasing the time devoted to nest attendance may profoundly impact parental energy budgets by limiting foraging, either directly – limiting foraging time, or indirectly – the hormonal state limiting the inclination to leave the nest. Species may avoid the cold by choosing warm microclimates for their nest sites. These may not be available, or there may be trade-offs involved in their use such as higher predation risk, greater competition for nest sites, or increased commuting distance to the foraging areas. Species may decrease the cooling rate of their young by building more insulating nests, by having bigger and better insulated young or by increasing brood size (Dunn, 1976). Alternatively, many aspects of the species' breeding biology may be conserved if the chicks' thermoregulatory development is accelerated. Young would then rapidly pass through the vulnerable period and, thus, would conserve parental time–energy budgets through most of the nestling period. Few studies have assessed the roles of these factors in compensation for climatic conditions; however, it has been suggested that behavioural change is the first adaptive response of organisms to a novel environment, and that behavioural plasticity may obviate the need for physiological evolution (Bartholomew, 1958; Slobodkin & Rapoport, 1974).

Colonisation of an extreme climate: a case study

If adjustment in postnatal development to climatic conditions occurs, it should be apparent among the Antarctic charadriiforms. The Antarctic is an extreme environment for nestlings and they will suffer selection by cold unless protected by their parents. The variation in postnatal ontogeny seen in the Charadriiformes is among the highest known (Nice, 1962) among monophyletic groups of birds. The Antarctic species represent different families (Chionidae, Stercoraridae, Laridae, Sternidae) the taxonomic level showing the greatest diversity in ontogenetic pattern, thus similarities across families are more likely to be convergent rather than be retained ancestral characteristics (pleisiomorphies). The relatively few colonisations of Antarctica (Fig. 3) by charadriiform birds make it easier to reconstruct evolutionary history.

The cold (≤1 °C) waters of the Southern Ocean create a cold maritime climate south of 50–60 °S latitude that is equivalent to polar climates found at much higher Northern latitudes. Members of four charadiiform families have independently colonised Antarctica (Fig. 3). These families radiated in within the last 30 million years (MY) (Warheit, 1992), after the Southern Ocean became cold (36 million years ago (MYA); Zachos *et al.*, 1992). The colonists invaded a cold Antarctica from more temperate regions, or invaded during warmer interludes and later experienced persistent cold. Five charadriiform species breed in Antarctica: kelp gulls (*Larus dominicanus*) and brown skuas (*Catharacta lonnbergi*) have Antarctic populations, whereas Antarctic terns (*Sterna vittata*), South Polar skuas (*Catharacta maccormicki*), and greater sheathbills (*Chionis alba*) nest exclusively in Antarctica. Eppley (1991, 1996) studied the development of thermoregulation in South Polar skuas, kelp gulls and greater sheathbills. Some comparative data are available on the growth and thermoregulation of Antarctic tern chicks (Kaiser *et al.*, 1990; Klaassen, 1994) and on the metabolic development of brown skuas (Eppley, 1991). Here, these studies are used to assess the roles of behavioural versus physiological compensation by these species for the rigorous climatic conditions. Has parental behaviour of the Antarctic species protected the young from selection by cold and completely compensated for the colder conditions or is there evidence of adjustment in the metabolic development of Antarctic young?

Interspecific comparisons have been the basis for inferring evolutionary adaptations to climate in avian ontogenetic patterns. Previous comparative studies suggested that metabolic development has responded to selection by climate, specifically, increased supplementary thermogenesis (Koskimies & Lahti, 1964; Dawson *et al.*, 1972, 1976; Dawson & Bennett, 1980, 1981) and greater SMR (Klaassen, 1994; Konarzewski, 1995) in hatchlings from cold climates and greater subsequent growth rates (Konarzewski, 1995). When phylogeny is ignored; the rates of Type 1 error (falsely rejecting the null hypothesis) are unacceptably high and power is low (Grafen, 1989; Martins & Garland, 1991).

Felsenstein's (1985) method of phylogenetic contrasts removes the phylogenetic component shared by sister taxa, yields acceptable Type 1 error rates and has relatively high power (Martins & Garland, 1991). Phylogenetic contrasts can be constructed for a set of species to be compared, if a phylogenetic tree is available with estimates of the branch lengths. Phylogenetic contrasts (Felsenstein, 1985) are calculated as the difference in a trait between sister taxa and show the unique evolution of the trait between the species. The contrasts are standardised by dividing by the square root of their branch lengths, an estimate of time since the evolutionary divergence between the taxa. This method allows analysis of the unique changes in traits between sister taxa since their divergence, and thus is well suited to compare the Antarctic charadriiforms to their non-Antarctic relatives. The phylogenetic contrasts can then be used in regression; however, the regression must pass through the origin because the contrasts are standardised.

Here, phylogenetic contrasts were used to determine whether the Antarctic charadriiforms have evolved greater maturity at hatching or have faster postnatal development than their relatives. Data on nesting latitude, adult female mass, egg mass, hatchling mass, hatchling SMR, hatchling PMR and chick growth rates for charadriiform birds were compiled from the literature. A phylogeny (Fig. 3) for the species to be compared was reconstructed from several sources (Schnell, 1970; Phillips, 1980; Furness, 1987; Sibley & Aylquist, 1991). Standardised phylogenetic contrasts of breeding latitude, log egg mass, log adult mass, log hatchling mass, log hatchling SMR, log hatchling peak metabolic rate (PMR) and log growth rate were calculated using the PDTREE program of Martins & Garland (1991). Linear regression of the standardised phylogenetic contrasts was used to determine whether the Antarctic species had diverged from their temperate relatives in the size and metabolic maturity of hatchlings and in their subsequent growth. Comparative data throughout development are lacking, so for postnatal development the Antarctic species were compared with a temperature member of their clade: kelp gull (no. 101, Fig. 3) and South Polar skua (no. 93, Fig. 3) with western gull (no. 106, Fig. 3); greater sheathbills (no. 66, Fig. 3) with lapwing (no. 13, Fig. 3). Western gulls are closely related to kelp gulls (same genus) and are more distantly related to South Polar skuas (different families) (Fig. 3). The Antarctic greater sheathbill and subAntarctic lesser sheathbill have no close temperate relatives (Fig. 3) and comparative data on metabolic development for the plover clade are scarce. Data for lapwings are from Visser & Ricklefs (1993), data for Antarctic terns are from Kaiser *et al.* (1990) and Klaassen (1994), and data for the other species are from Eppley (1991).

Behavioural compensation

Antarctic species compensated by brooding their young extensively. Chicks were brooded at least 50% of the time for 2 days after hatching in the temperate western gull, for 5 days in the South Polar skua and

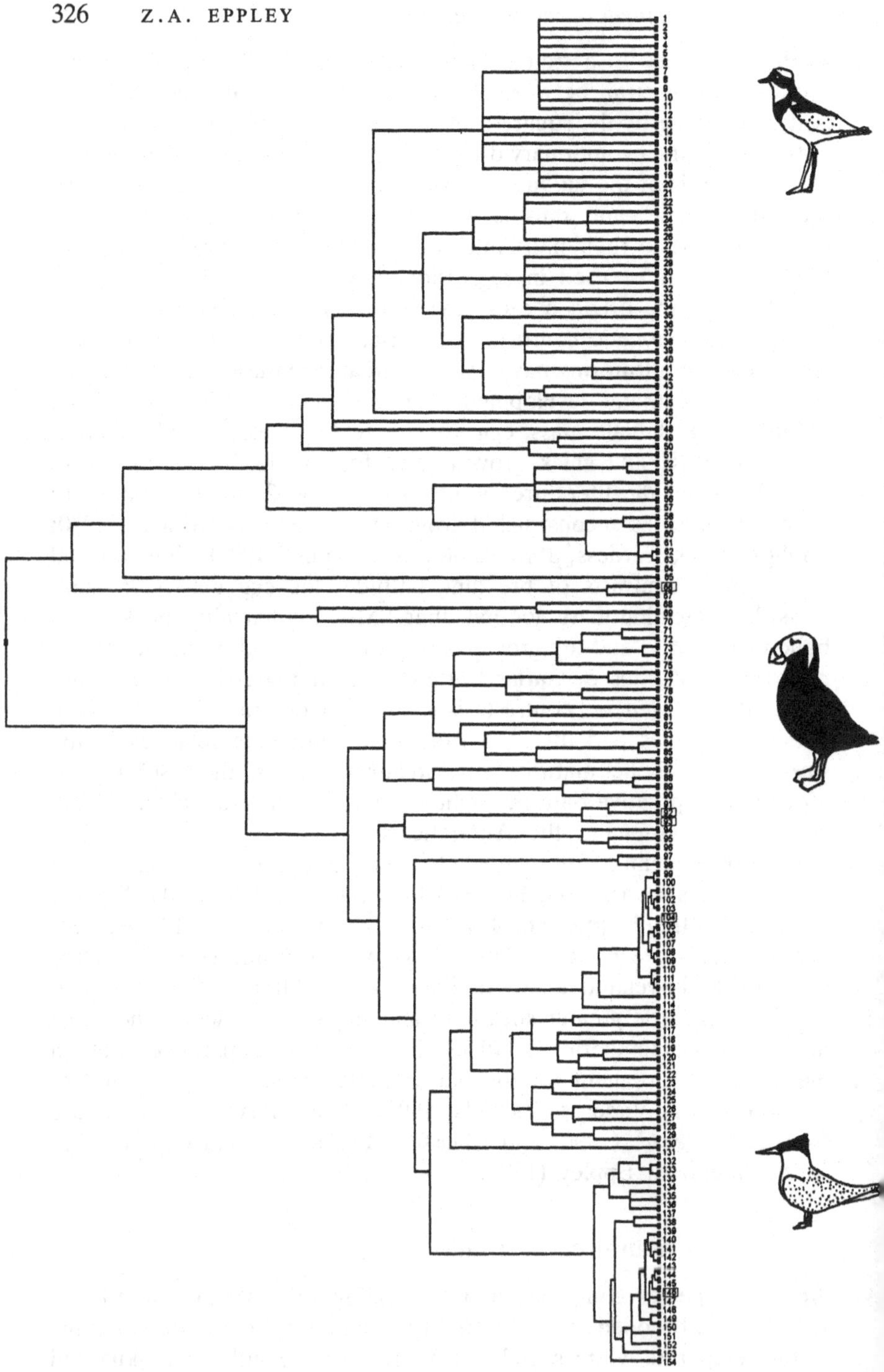
1 2 3 4 5 6 7 8 9 10 11 12 13 14 15 16 17 18 19 20 21 22 23 24 25 26 27 28 29 30 31 32 33 34 35 36 37 38 39 40 41 42 43 44 45 46 47 48 49 50 51 52 53 54 55 56 57 58 59 60 61 62 63 64 65 66 67 68 69 70 71 72 73 74 75 76 77 78 79 80 81 82 83 84 85 86 87 88 89 90 91 92 93 94 95 96 97 98 99 100 101 102 103 104 105 106 107 108 109 110 111 112 113 114 115 116 117 118 119 120 121 122 123 124 125 126 127 128 129 130 131 132 133 134 135 136 137 138 139 140 141 142 143 144 145 146 147 148 149 150 151 152 153 154

Fig. 3. Working hypothesis for the phylogenetic relationships among the plovers, gulls and allies used in constructing phylogenetic contrasts for the following comparative analyses (Eppley, 1996). The 67 species in the top side of the tree represents the 'plover' clade and includes lapwings, plovers, oystercatchers, avosets, stone curlews and sheathbills. The bottom side of the tree represents the 'gull' clade and includes pratincoles, alcids, skuas, skimmers, gulls and terns. The sandpipers, the third major clade of the charadriiformes, were excluded from the analysis because they have no Antarctic species. The tree is that of Sibley & Ahlquist (1991), with sister taxa added to the tree using other taxonomic or phylogenetic works (Schnell, 1970; Furness, 1987; Phillips, 1980). Taxa numbers enclosed in boxes nest in Antarctica. Charadriiform birds appear to have successfully colonised Antarctic at least four times, once by one species of greater sheathbills (66), once by the ancestor of skuas (92, 93), once by a population of kelp gulls (104), and once by one species of tern (146). Branch lengths represent time since divergence based on DNA hybridisation (Sibley & Ahlquist, 1991).

Taxa listed are: 1, *Vanellus armatus*; 2, *V. albiceps*; 3, *V. crassirostris*; 4, *V. senegallus*; 5 *V. coronatus*; 6, *V. lugubris*; 7, *V. melanopterus*; 8, *V. supercilliosis*; 9, *V. leucurus*; 10, *V. gregaris*; 11, *V. resplendens*; 12, *V. tectus*, 13, *V. vanellus*; 14, *V. chilensis*; 15, *V. malabaricus*; 16, *V. miles*; 17, *V. cinereus*; 18, *V. tricolor*; 19, *V. indicus*; 20, *V. spinosis*; 21, *Charadrius dubius*; 22, *C. melodus*; 23, *C. semipalmatus*; 24, *C. hiaticula*; 25, *C. placidus*; 26, *C. vociferus*; 27, *C. melanops*; 28, *C. modestus*; 29, *C. leschenaultii*; 30, *C. veredus*; 31, *C. asiaticus*; 32, *C. mongolus*; 33, *C. montanus*; 34, *C. cintus*; 35, *C. wilsonia*; 36, *C. falklandicus*; 37, *C. tricollaris*; 38, *C. collaris*; 39, *C. pecuarius*; 40, *C. marginatus*; 41, *C. alexandrinus*; 42, *C. ruficapillus*; 43, *C. bicinctus*; 44, *C. obscurus*; 45, *Peltohyas australis*; 46, *Anarhynchus frontalis*; 47, *Eudromias morinellus*; 48, *Oreopholis ruficollis*; 49, *Pluvialis apricaria*; 50, *P. dominica*; 51, *P. squatarola*; 52, *Haematopus unicolor*; 53, *H. ostralegus*; 54, *H. leucopodes*; 55, *H. ater*; 56, *H. fulignosis*; 57, *Cladorhynchus leucocephalus*; 58, *Himantopus himantopus*; 59, *H. mexicanus*; 60, *Recurvirostra americana*; 61, *R. andina*; 62, *R. avosetta*; 63, *R. avosetta*; 64, *R. novaehollandiae*; 65, *Burhinus oedicnemus*; 66, *Chionis alba*; 67, *C. minor*; 68, *Glareola nordmanni*; 69, *G. pratincola*; 70 *Dromas ardeola*; 71, *Uria aalge*; 72, *U. lomvia*; 73, *Alca torda*; 74, *A. torda*; 75, *Alle alle*; 76, *Cepphus grylle*; 77, *C. columba*; 78, *Synthliboramphus hypoleucus*; 79, *S. antiguus*; 80, *Brachyramphus brevirostris*; 81, *B. marmoratus*; 82, *Ptychoramphus aleutica*; 83, *Cyclorrhynchus psittacula*; 84, *Aethia cristatella*; 85, *A. pygmaea*; 86, *A. pusilla*; 87, *Cerorhinca monocerata*; 88, *Fratercula arctica*; 89, *F. corniculata*; 90, *F. cirrhata*; 91, *Catharacta skua*; 92, *C. lonnbergi*; 93, *C. maccormicki*; 94, *Stercorarius longicaudus*; 95, *S. pomarinus*; 96, *S. parasiticus*; 97 *Rynchops niger*; 98, *R. flavirostris*; 99, *Larus fuscus*; 100, *L. fuscus*; 101, *L. dominicanus*; 102, *L. dominicanus*; 103, *L. dominicanus*; 104, *L. dominicanus-Antarctic*; 105, *L. livens*; 106, *L. occidentalis*; 107, *L. argentatus*; 108, *L. argentatus*; 109, *L. argentatus*; 110, *L. glaucescens*; 111, *L. hyperboreus*; 112, *L. hyperboreus*; 113, *L. marinus*; 114, *L. ichthyaetus*; 115, *L. heermani*; 116, *L. hemprichii*; 117, *L. crassirostris*; 118, *L. audouinii*; 119, *L. delawarensis*; 120, *L. canus*; 121, *L. californicus*; 122, *L. atricilla*; 123, *L. pipxican*; 124, *L. ridibundus*; 125, *L. cirrocephalus*; 126, *L. novaehollandiae*; 127, *L. genei*; 128, *Rissa tridactyla*; 129, *R. brevirostris*; 130, *Creagrus furcatus*; 131, *Anous stolidus*; 132, *A. minutus*; 133, *A. tenuirostris*; 134, *Xena sabina*; 135, *Chlidonias leucopterus*; 136, *C. hybrida*; 137, *C. niger*; 138, *Sterna bergii*; 139, *S. maxima*; 140, *S. nilotica*; 141, *S. paradisaea*; 142, *S. bengalensis*; 143, *S. sandvicensis*; 144, *S. hirundo*; 145, *S. dougalli*; 146, *S. vitatta*; 147, *S. caspia*; 148, *S. repressa*; 149, *S. anaethetus*; 150, *S. fuscata*; 151, *S. albifrons*; 152, *S. balaenarum*; 153, *Gygis alba*; 154, *Pagophila eburnea*.

for 9 days in the Antarctic population of kelp gulls. Eppley (1991) was not able to observe brooding behaviour in greater sheathbills, but data are available for their sister species, *Chionis minor*, the lesser sheathbill, a subAntarctic species. Lesser sheathbill chicks were brooded at least 50% of the time for 3 weeks after hatching (Berger, 1981). Data for the lesser sheathbill were obtained at a subAnarctic colony (approximately 48°S) with milder temperatures and, thus, may underestimate brooding in the greater sheathbill breeding at approximately 65°S. Adjustments in brooding significantly affected parental time budgets; Antarctic kelp gulls invested 4.5-fold more time brooding their young than did temperate western gulls. Sheathbills, whose young are not thermally independent until approximately 40 days old (Berger, 1981), spend most of the nestling period tied to the nest by their chicks' need for constant brooding.

Antarctic species chose warmer (snow-free) nest sites. Nest scrapes provide shelter from the wind, and solar insolation warms the open nests of South Polar skuas and kelp gulls 9–15 °C over adjacent weather station temperatures (2 °C). Sheathbills nest in crevices that protect young and adults from the depradations by skuas during the long nestling period but block radiant heat gain. Operative temperatures in open nests are still 13–27 °C colder than in western gull nests in temperate colonies. The Antarctic species did not all compensate for the cold conditions by building well-insulated nests. Antarctic kelp gulls and sheathbills build well-insulated nests but South Polar skuas provide no nest insulation and many of their eggs are laid directly on gravel.

Behavioural compensation by parents did not fully protect the young from selection by cold during the colonisation of Antarctica. All three Antarctic species have young that are conscious and can call at extremely low body temperatures compared with their temperate relatives (Fig. 4). Their tolerance is surpassed by another Antarctic species, Wilson's storm petrel *Oceanites oceanicus*, whose young can call at body temperatures near 5 °C (Obst, 1986). The ability of chilled chicks to call to parents to elicit brooding is an important aspect of behavioural thermoregulation (Norton, 1973; Evans *et al.*, 1994) and has obvious survival implications. Despite evidence of selection on the prehomeothermic young of the Antarctic species, the development of thermoregulation was highly conserved (Fig. 5).

Physiology of hatchlings

Large body size confers a thermal advantage: there is more metabolically active tissue, more insulation and the reduced surface area/

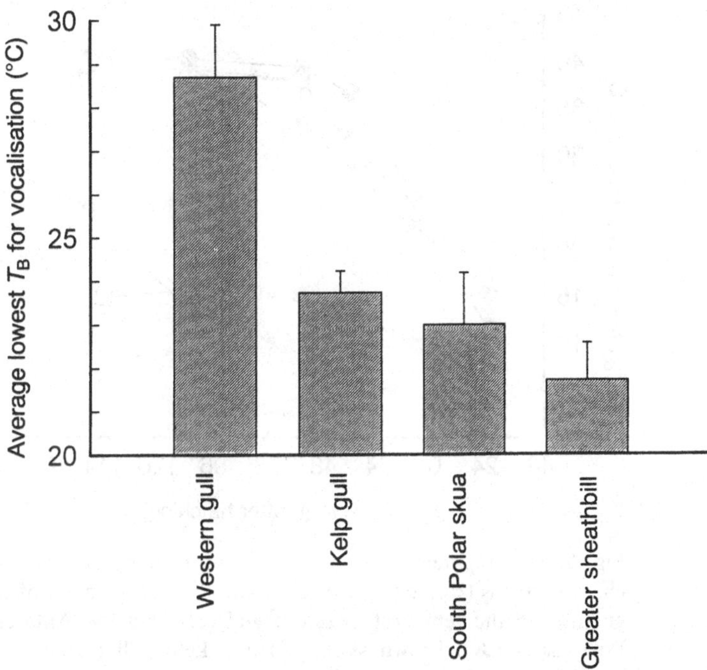

Fig. 4. The lowest body temperature at which chicks could call (x ± 95% CL). Chicks in metabolic chambers were exposed to 0 °C for 1 h. Core body temperature was obtained by an implanted thermocouple. The occurrence of chick vocalisations were recorded at 5 min intervals. (After Eppley, 1996.) Chicks call to their parents to elicit brooding when chilled. This ability has strong survival value and is uniformly lower in the Antarctic species, which suggests that these species have experienced selection for this enhanced ability.

volume helps retain heat. Antarctic hatchlings could gain a thermoregulatory advantage simply by being larger. A regression of phylogenetic contrasts of 154 taxa (145 species) of charadriiform birds was used to determine whether the Antarctic species lay disproportionately larger eggs (Fig. 6). Divergence in egg mass between the sister taxa is strongly correlated with divergence in adult mass ($r^2 = 0.78$, $p = 0.0001$), but is not related to divergence in the latitude of the nesting site ($r^2 = 0.0004$, ns). Sheathbills and Antarctic kelp gulls were larger and laid relatively smaller eggs than their temperate relatives, but their divergence was not usual for the order. In one case, the South Polar skua, the Antarctic species is slightly smaller than its subpolar relatives,

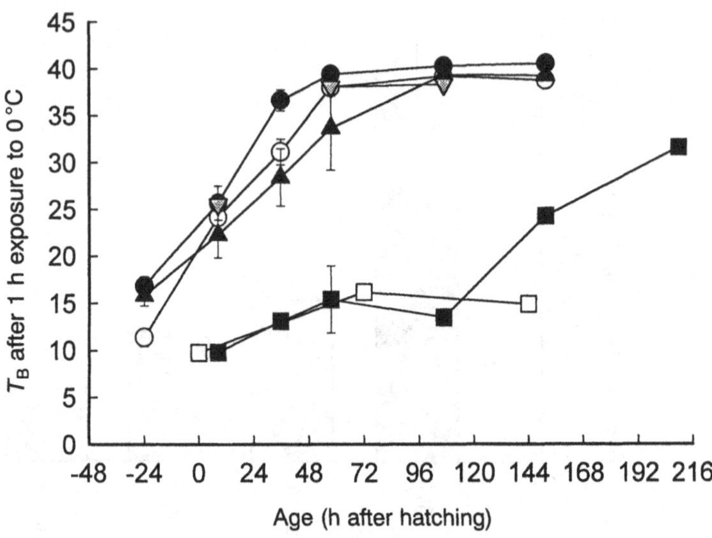

Fig. 5. Development of thermoregulation in temperate and Antarctic charadriiforms (Eppley, 1996). Compare the development of thermoregulation in the temperate western gull (○) with the Antarctic South Polar skua (▲), brown skua (▽) and kelp gull (●). Compare the development of thermoregulation in the temperate lapwing (□) with the Antarctic greater sheathbill (■). Two patterns are evident: rapid development for the temperate and Antarctic members of the gull clade, slower development for the temperate and Antarctic members of the plover clade. Data for the lapwing were estimated using standard (SMR) and peak (PMR) metabolic rate and thermal conductance from Visser & Ricklefs (1993) in the model of endothermic cooling (Eppley, 1994).

although egg size is conserved. The phylogenetic contrasts between Antarctic species and their temperate relatives are not remarkable compared with other phylogenetic contrasts in the analysis.

Greater metabolic maturity at hatching may also confer a thermal advantage by providing a greater SMR or greater supplementary thermogenesis. Previous studies (Klaassen & Drent, 1991) reported higher SMR for hatchlings from high latitudes. Data on the SMR of hatchlings are available for two species in the plover clade (lapwing *Vanellus vanellus* and greater sheathbill *Chionis alba*), and 23 species in the gull clade. A regression of phylogenetic contrasts for these 25 charadriiform species was used to determine whether Antarctic hatchlings had greater metabolic maturity than expected for their mass (Fig. 7). Divergence

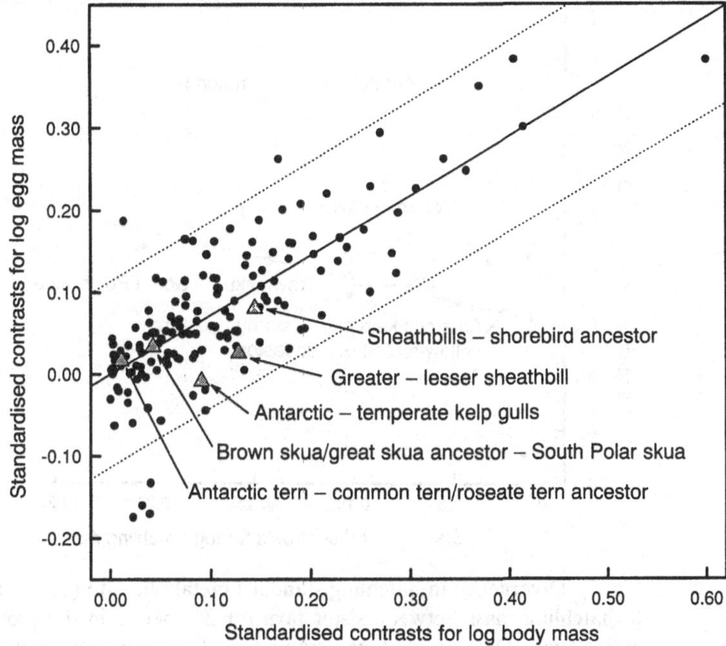

Fig. 6. Divergence in egg size relative to adult size between sister taxa for 154 species in the plover and gull clades. Data plotted are standardised phylogenetic contrasts of log egg mass and log adult body mass. Regression through the origin (95% confidence intervals are shown) is used because the phylogenetic contrasts are standardised (corrected for branch length differences) (Garland *et al.*, 1992). Divergence in adult body mass explains 78% of the divergence in egg mass, latitude of the breeding site explains $\ll$ 1% of the divergence in egg mass. Species included are listed in Fig. 3. (Sources are given in Eppley, 1991, 1996.)

in hatchling mass between sister taxa explained 52% of the divergence in hatchling SMR ($p = 0.0001$), while difference in natal latitude explained only 14% (ns). The Antarctic kelp gulls, greater sheathbills and Antarctic terns showed greater metabolic intensity than their temperate relatives, although only Antarctic terns were exceptional in their divergence. PMR, a measure of supplementary thermogenesis, was previously noted to be higher in gull hatchlings from higher latitudes (Dawson & Bennett, 1981). Comparative data on PMR are available for 15 species in the plover and gull clades. In this data set, neither

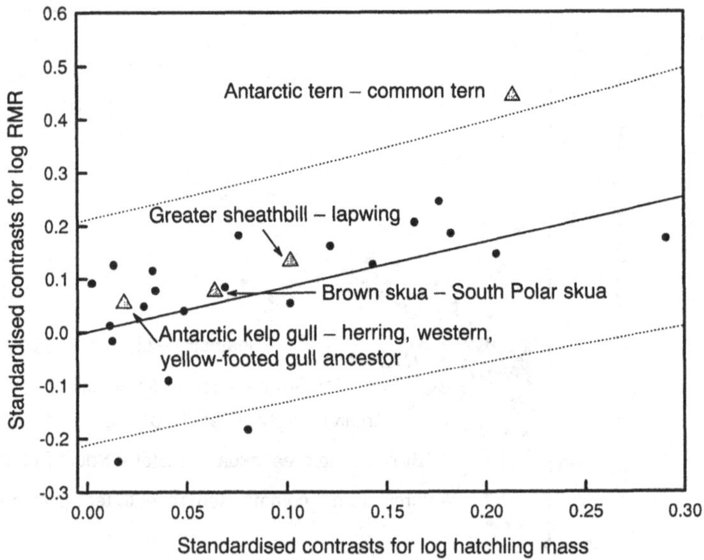

Fig. 7. Divergence in hatchling standard metabolic rate (SMR) relative to hatchling mass between sister taxa for 25 species in the plover and gull clades. Data plotted are standardised phylogenetic contrasts of log hatchling SMR and log hatchling body mass. Divergence in hatchling body mass explains 52% of the divergence in SMR, latitude of the natal site explains 14% of the divergence in SMR. Sources are given in Eppley (1991, 1996) and Ricklefs (1988). Species included are: *Vanellus vanellus* (13, Fig. 3), *Haematopus ostralegus* (53), *Chionis alba* (66), *Uria lomvia* (72), *Cepphus columba* (77), *Synthliboramphus hypoleucus* (78), *Fratercula arctica* (88), *Catharacta lonnbergi* (92), *C. maccormicki* (93), *Larus domincanus* (101), *L. livens* (105), *L. occidentalis* (106), *L. argentatus* (107), *L. glaucescens* (110), *L. delawarensis* (119), *L. atricilla* (122), *L. ridibundus* (124), *Rissa tridactyla* (128), *Anous stolidus* (131), *A. minutus* (132), *Sterna paradisaea* (2 populations, 141), *S. sandivicensis* (143), *S. hirundo* (144), *S. vitatta* (146), *S. fuscata* (150), *Gygis alba* (153).

divergence in hatchling mass (21%, ns), nor divergence in natal latitude (11%, ns) explains much of the variation in PMR. The Antarctic hatchlings are not exceptional in their divergence in PMR (Fig. 8).

Postnatal development

Young birds can more quickly pass through their thermally vulnerable period by growing rapidly. A regression of phylogenetic contrasts for

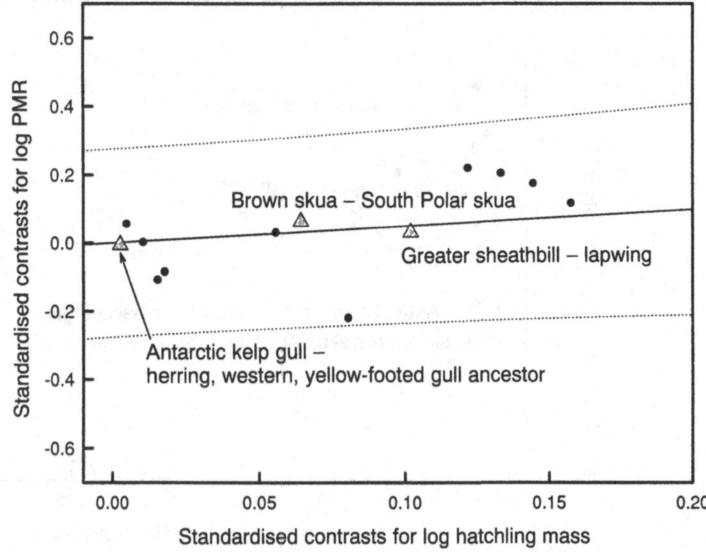

Fig. 8. Divergence in hatchling peak metabolic rate (PMR) relative to hatchling mass between sister taxa for 15 species in the plover and gull clades. Divergence in hatchling body mass explains 21% of the divergence in PMR, latitude of the natal site explains 11% of the divergence in PMR. Sources are given in Eppley (1991, 1996) and Ricklefs (1988). Species included are: *Vanellus vanellus* (13, Fig. 3), *Chionis alba* (66), *Uria lomvia* (72), *Synthiliboramphus hypoleucus* (78), *Fratercula arctica* (88), *Catharacta lonnbergi* (92), *C. maccormicki* (93), *Larus domincanus* (101), *L. livens* (105), *L. occidentalis* (106), *L. delawarensis* (119), *L. atricilla* (122), *L. ridibundus* (124), *Anous stolidus* (131), *S. fuscata* (150).

53 taxa of charadriiform birds in the plover/gull clades was used to determine if the Antarctic species grow more rapidly than expected for their size and clade. Divergence in body mass at the asymptote (Asy) of the growth curve explained 17% of the divergence in growth rate ($p = 0.002$) between sister taxa, while latitude explained >3% (ns) of the divergence. This finding differs from that of Konarzewski (1995), who found a significant latitudinal effect on growth using phylogenetic regression with a broader, although smaller data set. The Antarctic kelp gull and greater sheathbill grow more slowly than their more temperate relatives whereas the Antarctic tern and South Polar skua grow faster than their relatives that breed at lower latitudes (Fig. 9).

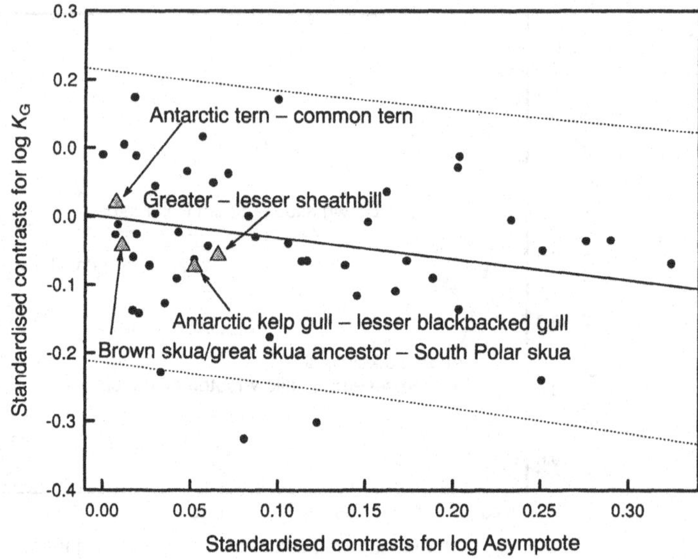

Fig. 9. Divergence in nestling growth relative to mass at the end of the growth period (asymptote mass) between sister taxa for 53 species in the plover and gull clades. Data plotted are standardised phylogenetic contrasts of K for Gompertz growth curves and log asymptote mass. Divergence in asymptote mass explains 17% of the divergence in growth rates, latitude of the natal site explains <3% of the divergence in growth rate. Sources are given in Eppley (1991, 1996) and Ricklefs (1973). Species used in the comparison are: *Vanellus vanellus* (2 populations, 13 Fig. 3), *Chardrius dubius* (21), *C. hiaticula* (2 populations, 24), *C. vociferous* (26), *Eudromias morinellus* (47), *Pluvialis apricaria* (49), *Haematopus ostralegus* (53), *H. palliatus* (54), *Chionis alba* (66), *C. minor* (67), *Cepphus columba* (77), *Brachyramphus marmoratus* (81), *Ptychoramphus aleuticus* (82), *Cyclorhynchus psittacula* (83), *Aethia cristatella* (84), *A. pusilla* (86), *Cerorhinca monocerata* (87), *Fratercula arctica* (88), *F. cirrhata* (90), *Catharacta skua* (91), *C. lonnbergi* (92), *C. maccormicki* (93), *Stercorarius longicauda* (94), *S. pomarinus* (95), *S. parasiticus* (96), *Larus fuscus* (100), *L. domincanus* (101), *L. livens* (105), *L. occidentalis* (106), *L. argentatus* (4 populations, 107), *L. glaucescens* (110), *L. marinus* (113), *L. modestus* (115), *L. delawarensis* (119), *L. canus* (120), *L. atricilla* (122), *L. ridibundus* (124), *Rissa tridactyla* (2 populations, 128), *Creagrus furcatus* (130), *Anous stolidus* (131), *A. minutus*, (132), *Sterna paradisaea*, (2 populations, 141), *S. sandvicensis* (143), *S. hirundo* (144), *S. vitatta*, (146), *S. lunata* (149), *S. fuscata* (150), *Gygis alba* (153).

While growth rate throughout the nestling period shows no clear pattern, the Antarctic species have faster early growth (Fig. 10), which is more relevant for the development of thermoregulation. The Antarctic species grow faster to 0.2 kg, a size more favourable for thermoregulation. Data on growth rates for plover species as large as sheathbills are not available. Therefore, Ricklefs' (1967) methods were used to calculate the time to grow to 0.2 kg for a hypothetical plover with the same asymptotic mass (0.62 kg). The growth constant, K_G, for this hypothetic species was calculated using Beintema & Visser's (1989) allometric regression.

In addition to their faster early growth, do the Antarctic species have accelerated metabolic development? When animals increase in

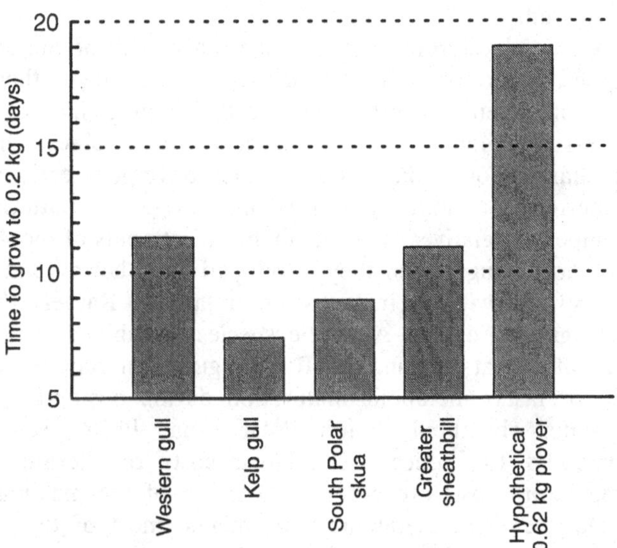

Fig. 10. Time required for hatchlings to grow to a size favourable for thermoregulation (0.2 kg). (After Eppley, 1996) The Antarctic kelp gull (adult mass 0.98 kg) (M. L. Chappell, unpublished results) and South Polar skuas (adult mass 1.188 kg) (Neilsen, 1984) grew more rapidly than the temperate western gull (adult female mass 1.011 kg) (Dunning, 1983). No growth curves have been published for temperate plover species as large as the greater sheathbill, therefore the time to grow to 0.2 kg for a hypothetical plover with an asymptote mass of 0.62 kg was calculated using Ricklefs (1967) methods with K_G obtained from the allometric regression for plovers (Beintema & Visser, 1989).

size but do not change their composition, metabolic rate is related to $mass^{0.67}$ (Heusner, 1984; Bucher, 1986). Metabolic maturation, independent of increases in mass, can be detected by changes in the ratio of $SMR/mass^{0.67}$ or $PMR/mass^{0.67}$. Metabolic development in the Antarctic gulls and skuas was compared with that of the temperate western gull, and the Antarctic sheathbill with the temperate lapwing (mass, SMR and PMR from Visser & Ricklefs, 1993). The gulls and skuas have higher mass-independent metabolic rates (both SMR and PMR) than do members of the plover clade (Fig. 11). The temperate and Antarctic gulls have similar metabolic development. Skuas show accelerated metabolic development of both SMR and PMR that diverges from that of gulls within a few days of hatching. Sheathbills have higher rates of resting mass-independent metabolism (SMR) than lapwings, but their development of PMR is similar. Metabolic maturation appears to be highly conserved.

The Antarctic charadriiform species share a variety of characteristics that compensate for cold: they brood their young longer; their young are tolerant of extreme hypothermia; and their young show rapid early growth, thus shortening their period of thermal vulnerability. The Antarctic charadriiforms did not show the convergent early development of thermoregulation, but showed an ontogenetic pattern similar to their temperate relatives. The metabolic adjustments of the Antarctic species did not change when they developed endothermy and homeothermy compared with their temperate relatives. Rather, the more rapid early growth of the Antarctic species, combined with greater metabolic maturity at hatching (SMR: kelp gull, Antarctic tern, greater sheathbill) or faster metabolic maturation during postnatal ontogeny (SMR: greater sheathbill, South Polar skua; PMR: South Polar skua) combined to reduce the relative costs of thermoregulation (thermoregulatory costs are a smaller fraction of the maximal power output). Despite these metabolic adjustments, most of the Antarctic young are more dependent on brooding than their temperate relatives, although they are better able to survive bouts of neglect resulting in hypothermia. Only in the South Polar skua do the metabolic adjustments in postnatal development substantially reduce the brooding requirements of the young. This metabolic adjustment may explain the ability of South Polar skuas to breed as far south as 78.5°S (Spellerberg, 1967), while the breeding range of the other Antarctic charadriiforms is limited to regions north of 68°S (Croxall *et al.*, 1984; Woehler, 1990).

Conservation of reproductive success has been achieved in different ways by the different Antarctic species. Kelp gulls primarily used

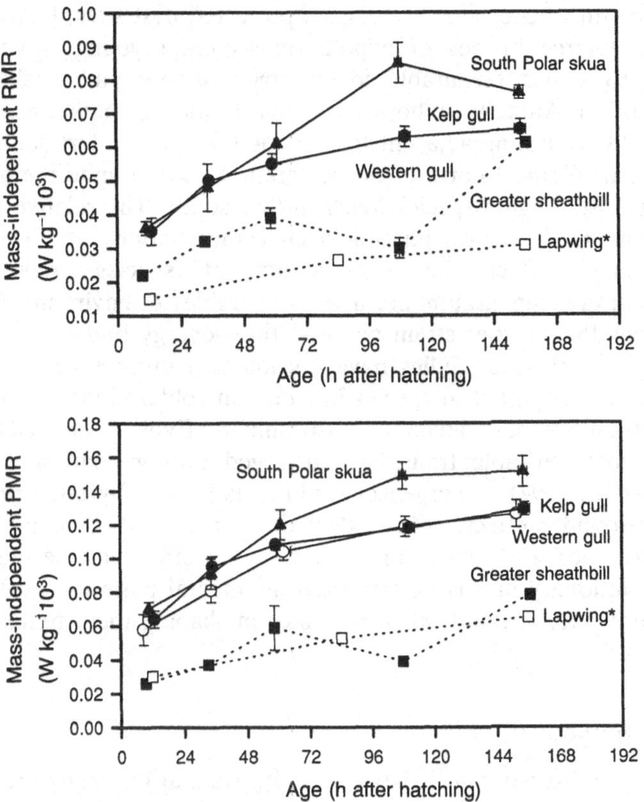

Fig. 11. Maturation of standard (SMR) and peak (PMR) metabolic rate for Antarctic and temperate charadriiform species. When organisms increase in size only, not in composition, metabolic rate scales as mass$^{0.67}$; therefore measurements were divided by this value to show maturation. Only skuas show accelerated metabolic development that diverges from that of gulls within a few days of hatching. Gull clade members have faster metabolic maturation than the plover clade members. Antarctic sheathbills have higher rates of mass-independent SMR than lapwings, but their development of PMR is similar. (After Eppley, 1996)

behavioural adjustments (longer brooding, thicker nests), slightly greater metabolic maturity of hatchlings and faster early growth. South Polar skuas relied less on behavioural compensation (brooding period is only slightly longer) and more on physiological adjustments in postnatal ontogeny (better insulation, more rapid metabolic maturation, faster early growth). Extensive brooding in protected nest sites, heavily-lined nests and large broods of hypothermia-tolerant young, which grow rapidly to a size favourable for thermoregulation, allow sheathbills to persist in Antarctica; however, their brooding requirements limit sheathbills to nesting adjacent to rich food sources (intertidal areas or in penguin colonies) north of the Antarctic Circle, despite the presence of their host penguin species much further south. The relative reliance on behavioural versus physiological compensations for cold may differentially affect the vulnerability of species' reproduction (behavioural compensators are more vulnerable) to environmental perturbations that further strain parental time–energy budgets.

This and previous studies have documented divergence in parental care and development in species breeding in colder climates, and have inferred evolutionary adaptation to climate. Evolutionary adaptation implies that heritable traits have diverged through natural selection (Leroi *et al.*, 1994). Divergences could equally arise by other processes such as founder effect, genetic drift or be the result of non-heritable processes such as learning or environmental effects on development. The evolutionary history of avian developmental patterns and the role of natural selection and other processes in shaping them remain to be discovered.

Summary

Adjustments to parental behaviour and postnatal ontogeny are critical to maintaining avian reproductive success in different thermal habitats. Adjustment of parental behaviour to compensate for climatic conditions is important and widespread but does not always protect young from thermal selection (Salzman, 1982; this chapter). The analyses made here using phylogenetic contrasts show that species have diverged in relative egg size, hatchling SMR and growth rates but that the Antarctic species are not exceptional for their order. Comparative studies of development among a few species suggest that relatively little change in the details of postnatal ontogeny occurs between sister taxa, rather, these changes occur deeper in the tree. Among the Antarctic charadriiforms, all retained the ancestral pattern for development of endothermy, yet diverged from their temperate relatives in certain physiological capacities related to survival in the cold: tolerance to hypothermia, faster early growth or metabolic development. Thus, there do not appear to be important changes in the timing of developmental events in response to selection by climate; however, minor adjustment in the physiological development of chicks can have profound effects on parental time–energy budgets and affect the likelihood that species are able to colonise more challenging climates or persist during climatic change.

Acknowledgements

I thank Ted Garland for the programs, discussions and advice on the phylogenetic analysis. Preparation of this manuscript was supported by NSF OPP9220775 to B.D. Sidell; DPP8702115 to A.F. Bennett supported Eppley's Antarctic studies.

References

Aulie, A. & Grav, H.J. (1979). Effect of cold acclimation of the oxidative capacity of skeletal muscles and liver in young bantam chicks. *Comparative Biochemistry and Physiology* **62A**, 335–8.

Bartholomew, G.A. (1958). The role of physiology in the distribution of terrestrial vertebrates. In *Zoogeography* (ed. C.L. Hubbs), pp. 81–95. Washington, DC: American Association for the Advancement of Science.

Bartholomew, G.A. & Dawson, W.R. (1954). Thermoregulation in young pelicans, herons and gulls. *Ecology* **35**, 466–72.

Bech, C., Memlum, F. & Halftorn, S. (1994). Thermoregulatory abilities in chicks of the Antarctic petrel (*Thalassoica antarctica*). *Polar Biology* **11**, 233–8.

Beintema, A.J. & Visser, G.H. (1989). Growth parameters in chicks of charadriiform birds. *Ardea* **77**, 169–80.

Berger, A.E. (1981). Time budgets, energy needs and kleptoparasitism in breeding lesser sheathbills (*Chionis minor*). *Condor* **83**, 106–12.

Booth, D.T. (1984). Thermoregulation in neonate mallee fowl *Leipoa ocellata*. *Physiological Zoology* **57**, 251–60.

Bucher, T.L. (1986). Ratios of hatchling and adult mass-independent metabolism: a physiological index to the altricial-precocial continuum. *Respiration Physiology* **56**, 465–83.

Chappell, M.A., James, D.N., Shoemaker, V.H., Bucher, T.L. & Mahoney, S.K. (1993). Reproductive effort in Adelie penguins. *Behavioral Ecology and Sociology* **33**, 173–82.

Choi, I.-H., Ricklefs, R.E. & Shea, R.E. (1993). Skeletal muscle growth, enzyme activities, and the development of thermogenesis: a comparison between altricial and precocial birds. *Physiological Zoology* **66**, 455–73.

Croxall, J.P., Evans, J.P. & Schreiber, R.W. (eds). (1984). *Status and Conservation of the World's Seabirds*. Cambridge: International Council for Bird Preservation.

Dawson, W.R. & Bennett, A.F. (1980). Metabolism and thermoregulation in hatchling western gulls (*Larus occidentalis livens*). *Condor* **82**, 103–5.

Dawson, W.R. & Bennett, A.F. (1981). Field and laboratory studies of the thermal relations of hatchling western gulls. *Physiological Zoology* **54**, 155–64.

Dawson, W.R., Bennett, A.F. & Hudson, J.W. (1976). Metabolism and thermoregulation in hatchling ring-billed gulls. *Condor* **78**, 49–60.

Dawson, W.R., Hudson, J.W. & Hill, R.W. (1972). Temperature regulation in newly hatched laughing gulls (*Larus atricilla*). *Condor* **74**, 177–184.

Duchamp, C. & Barre, H. (1993). Skeletal muscle as the major site of non-shivering thermogenesis in cold-acclimated ducklings. *American Journal of Physiology* **265** (*Regulatory, Integrative and Comparative Physiology* **34**), R1076–R1083.

Dunn, E.H. (1976). The relationship between brood size and age of effective homeothermy in nestling house wrens. *Wilson Bulletin* **88**, 478–82.

Dunning, J. (1983). *CRC Handbook of Avian Body Mass*. New York: CRC Press.

Dunnington, E.A. & Siegel, P.B. (1984). Thermoregulation of newly hatched chicks. *Poultry Science* **63**, 1303–13.

Else, P.L. & Hulbert, A.J. (1987). Evolution of mammalian endothermic metabolism: 'leaky' membranes as a source of heat. *American Journal of Physiology* **253** (*Regulatory, Integrative and Comparative Physiology* **22**) R1–R7.

Eppley, Z.A. (1984). The development of thermoregulatory abilities in Xantus' murrelet chicks *Synthilboramphus hypoleucus*. *Physiological Zoology* **57**, 307–17.

Eppley, Z.A. (1991). *The ontogeny of endothermy in charadriiform birds: functional bases, ecological adaptations and phylogenetic constraints*. PhD dissertation, University of California, Irvine, California.

Eppley, Z.A. (1994). A mathematical model of heat flux applied to developing endotherms. *Physiological Zoology* **67**, 829–54.

Eppley, Z.A, (1996). Charadriiform birds in Antarctica: behavioral, morphological and physiological adaptations conserving reproductive success. *Physiological Zoology* **69** (in press).

Eppley, Z.A. & Russell, B. (1995). Perinatal changes in avian muscle: implications from ultrastructure for the development of endothermy. *Journal of Morphology* **225**, 357–67.

Evans, R.M., Whitaker, A. & Wiebe, M.O. (1994). Development of vocal regulation of temperature by embryos in pipped eggs of ring-billed gulls. *Auk* **111**, 596–604.

Felsenstein, J. (1985). Phylogenies and the comparative method. *American Naturalist* **125**, 1–15.

Fredette, B.J. & Landmesser, L.T. (1991). Relationship of primary and secondary myogenesis to fiber type development in embryonic chick muscle. *Developmental Biology* **143**, 1–18.

Furness, R.W. (1987). *The Skuas.* Staffordshire: T & AD Poyser.

Garland, T., Jr. & Adolph, S.C. (1991). Physiological differentiation of vertebrate populations. *Annual Review of Ecology and Systematics* **22**, 193–228.

Garland, T., Jr. & Adolph, S.C. (1994). Why not do two-species comparative studies: limitations on inferring adaptation. *Physiological Zoology* **67**, 797–828.

Garland, T., Jr. & Carter, P.A. (1994). Evolutionary physiology. *Annual Review of Physiology* **56**, 579–621.

Garland, T., Jr., Harvey, P.H. & Ives, A.R. (1992). Procedures for the analysis of comparative data using phylogenetically independent contrasts. *Systematic Biology* **41**, 18–32.

Grafen, A. (1989). The phylogenetic regression. *Philosophical Transactions of the Royal Society of London, Series B* **326**, 119–57.

Grav, H.J., Borch-Iohnsen, B., Dahl, H.A., Gabrielsen, G.W. & Steen, J.B. (1988). Oxidative capacity of tissues contributing to thermogenesis in eider (*Somateria mollissima*) ducklings: changes associated with hatching. *Journal of Comparative Physiology* **158B**, 513–18.

Hayward, J.S. & Lisson, P.A. (1992). Evolution of brown fat: its absence in marsupials and monotremes. *Canadian Journal of Zoology* **70**, 171–9.

Heusner, A.A. (1984). Biological similitude: statistical and functional relationships in comparative physiology. *American Journal of Physiology* **246** (*Regulatory, Integrative and Comparative Physiology* **15**), R839–R845.

Hohtola, E. (1982). Shivering thermogenesis in birds. *Acta Universitatis Ouluensis Series A. Scientiae Rerum Naturalium 139, Biologica* **17**. Oulu, Finland: University of Oulu.

Howell, T.R., Araya, B. & Millie, W.R. (1974). Breeding biology of the gray gull, *Larus modestus*. *University of California Publications in Zoology* **104**, 1–57.

James, F.C. (1983). Environmental component of morphological differentiation in birds. *Science* **221**, 184–6.

Jehl, J.R., Jr. & Hussell, D.J.T. (1966). Effects of weather on reproductive success of birds at Churchill, Manitoba. *Arctic* **19**, 185–91.

Johnson, S.R. & Cowan, I.M. (1974). Thermal adaptation as a factor affecting colonizing success of introduced Sturnidae (Aves) in North America. *Canadian Journal of Zoology* **52**, 1559–76.

Jones, I.L. (1994). Mass changes of least auklets *Aethia pusilla* during the breeding season – evidence for programmed loss of mass. *Journal of Animal Ecology* **63**, 71–8.

Kaiser, M., Gebauer, A. & Peter, H.-U. (1990). Thermoregulation in the Antarctic tern *Sterna vittata* (Gmelin, 1789). *Geodatische und geophysikalische Verfoffentlichugene, Reiche I* **16**, 429–38.

Kennedy, J.M., Eisenberg, B.R., Reid, S.K., Sweeney, L.J. & Zak, R. (1988). Nascent fiber appearance in overloaded chicken slow tonic muscle. *American Journal of Anatomy* **181**, 203–15.

Klaassen, M. (1994). Growth and energetics of tern chicks from temperate and polar environments. *Auk* **111**, 525–44.

Klaassen, M. & Drent, R.H. (1991). An analysis of hatchling resting metabolism: in search of ecological correlates that explain deviations from allometric relations. *Condor* **93**, 612–29.

Konarzewski, M. (1995). Allocation of energy to growth and respiration in avian postnatal growth. *Ecology* **76**, 8–19.

Koskimies, J. & Lahti, L. (1964). Cold-hardiness of the newly hatched young in relation to ecology and distribution in ten species of European ducks. *Auk* **81**, 281–307.

Landmesser, L.T. & O'Donovan, M.J. (1984). Activation patterns of embryonic chick hind limb muscles recorded *in ovo* and in an isolated spinal cord preparation. *Journal of Physiology* **347**, 189–204.

LeMaho, Y. (1977). The emperor penguin: a strategy to live and breed in the cold. *American Scientist* **65**, 680–93.

Leroi, A.M., Rose, M.R. & Lauder, G.V. (1994). What does the comparative method reveal about adaptation? *American Naturalist* **143**, 381–402.

Lynch, G.R., Lynch, C.B., Dube, M. & Allen, C. (1976). Early cold exposure: effects on behavioral and physiological thermoregulation in the house mouse *Mus musculus*. *Physiological Zoology* **49**, 191–9.

Marsh, R.L. & Dawson, W.R. (1989). Avian adjustments to cold. In *Advances in Comparative and Environmental Physiology* (ed. L.C.H. Wang), vol. 4, pp. 205–53. Berlin: Springer-Verlag.

Marsh, R.L. & Wickler, S.J. (1982). The role of muscle development in the transition to endothermy in nestling bank swallows, *Riperia riperia*. *Journal of Comparative Physiology* **149B**, 99–105.

Martins, E.P. & Garland, T. Jr. (1991). Phylogenetic analyses of the correlated evolution of continuous traits: a simulation study. *Evolution* **45**, 534–57.

Misson, B.H. (1977). The relationship between age, mass, body temperature and metabolic rate in the neonatal fowl (*Gallus domesticus*). *Journal of Thermal Biology* **2**, 107–10.

Murray, K.G., Winnett-Murray, K., Eppley, Z.A., Hunt, G.L. Jr. & Schwartz, D.B. (1983). Breeding biology of Xantus' murrelet. *Condor* **85**, 12–21.

Neilsen, D.R. (1984). *Ecological and behavioral aspects of the sympatric breeding of the South Polar skua* (Catharacta maccormicki) *and the brown skua* (Catharacta lonnbergi) *near the Antarctic Peninsula*. M.S. thesis, University of Minnesota, Minneapolis, Minnesota.

Nice, M.M. (1962). Development of behavior in precocial birds. *Transactions of the Linnean Society of New York* **8**, 1–211.

Norton, D.W. (1973). *Ecological energetics of calidrine sandpipers breeding at Barrow, Alaska*. PhD thesis, University of Alaska, Fairbanks, Alaska.

Nur, N. (1984). The consequences of brood size for breeding blue tits: nestling weight, offspring survival and optimal brood size. *Journal of Animal Ecology* **53**, 497–517.

Obst, B.S. (1986). *The energetics of Wilson's storm-petrel* (Oceanites oceanicus) *breeding at Palmer Station, Antarctica*. Ph.D. dissertation, University of California, Los Angeles, California.

Obst, B.S. & Nagy, K.A. (1993). Stomach oil and the energy budget of Wilson's storm-petrel nestlings. *Condor* **95**, 792–805.

Olson, E.N. (1993). Regulation of muscle transcription by the MyoD family. *Circulation Research* **72**, 1–6.

Olson, J.M. (1992). Growth, the development of endothermy, and the allocation of energy in red-winged blackbirds (*Agelaius phoenicus*) during the nestling period. *Physiological Zoology* **65**, 124–52.

Olson, J.M. (1994). The ontogeny of shivering thermogenesis in the red-winged blackbird (*Agelaius phoenicus*). *Journal of Experimental Biology* **191**, 59–88.

Phillips, R.E. (1980). Behaviour and systematics of New Zealand plovers. *Emu* **80**, 177–97.

Remmert, H. (1980). *Arctic Animal Ecology*. Berlin: Springer-Verlag.

Ricklefs, R.E. (1967). A graphical method of fitting equations to growth curves. *Ecology* **48**, 978–83.

Ricklefs, R.E. (1973). Patterns of growth in birds. II. Growth rate and mode of development. *Ibis* **115**, 177–201.

Ricklefs, R.E. (1979*a*). Adaptation, constraint and compromise in avian postnatal development. *Biological Review* **54**, 269–90.

Ricklefs, R.E. (1979*b*). Patterns of growth in birds. V. A comparative study of development in the starling, common tern, and Japanese quail. *Auk* **96**, 10–30.

Ricklefs, R.E. (1983). Avian postnatal development. *Avian Biology* **7**, 1–83.

Ricklefs, R.E. (1985). Modification of growth and development of muscles of poultry. *Poultry Science* **64**, 1563–76.

Ricklefs, R.E. (1988). Adaptation to cold in bird chicks. In *Physiology of Cold Adaptation in Birds* (ed. C. Bech & R.E. Reinertsen), pp. 329–38. New York: Plenum Press.

Ricklefs, R.E., Shea, R.E. & Choi, I.-H. (1994). Inverse relationship between functional maturity and exponential growth rate of avian skeletal muscle: a constraint on evolutionary response. *Evolution* **48**, 1080–8.

Ricklefs, R.E., White, S.C. & Cullen, J. (1980). Energetics of post-natal growth in Leach's storm-petrel. *Auk* **97**, 566–75.

Salzman, A.G. (1982). The selective importance of heat stress in gull nest location. *Ecology* **63**, 742–51.

Schnell, G.D. (1970). A phentic study of the suborder Lari (*Aves*). I. Methods and results of principal components analyses. *Systematic Zoology* **19**, 35–57.

Schultz, E. & McCormick, K.M. (1994). Skeletal muscle satellite cells. *Review of Physiology, Biochemistry and Pharmacology* **123**, 213–57.

Sibley, C.G. & Alquist, J.E. (1991). *Phylogeny and Classification of Birds. A Study in Molecular Evolution*. New Haven, Connecticut: Yale University Press.

Slobodkin, L.B. & Rapoport, A. (1974). An optimal strategy of evolution. *Quarterly Review of Biology* **49**, 181–200.

Spellerberg, I.F. (1967). The distribution of the McCormick skua (*Catharacta maccormicki*). *Notornis* **14**, 201–7.

Steen, J.B. & Gabrielsen, G.W. (1986). Thermogenesis in newly hatched eider (*Somateria mollissima*) and long-tailed (*Clangula hyemalis*) ducklings and barnacle goose (*Branta leucopsis*) goslings. *Polar Research* **4**, 180–6.

Steen, J.B., Grav, H., Borch-Iohnsen, B. & Gabrielsen, G.W. (1988). Strategies of homeothermy in eider ducklings (*Somateria mollissima*). *In Physiology of Cold Adaptation in Birds* (ed. C. Bech & R.E. Reinertsen), pp. 361–70. New York: Plenum Press.

Taylor, J.R.E. (1985). Ontogeny of thermoregulation and energy metabolism in pygoscelid penguin chicks. *Journal of Comparative Physiology* **155B**, 615–27.

Tazawa, H., Wakayama, H., Turner, J.S. & Paganelli, C.V. (1988). Metabolic compensation for gradual cooling in developing chick embryos. *Comparative Biochemistry and Physiology* **89A**, 125–9.

Untergasser, G. & Hayward, S. (1972). Development of thermoregulation in ducklings. *Canadian Journal of Zoology* **50**, 1243–50.

Visser, G.H. & Ricklefs, R.E. (1993). Development of temperature regulation in shorebirds. *Physiological Zoology* **66**, 771–92.

Warheit, K.I. (1992). A review of the fossil seabirds from the Teriary of the North Pacific: plate tectonics, paleoceanography, and faunal change. *Paleobiology* **18**, 401–24.

Webb, D.R. (1987). Thermal tolerance of avian embryos: a review. *Condor* **89**, 874–98.

Weeden, J.S. (1966). Diurnal rhythm of attentiveness of incubating female tree sparrows (*Spizella arborea*) at northern latitudes. *Auk* **83**, 368–88.

Wekstein, D.R. & Zolman, J.F. (1969). Ontogeny of heat production in chicks. *Federation Proceedings* **28**, 1023–7.

Wekstein, D.R. & Zolman, J.F. (1971). Cold stress regulation in young chickens. *Poultry Science* **50**, 56–61.

White, F.N. & Kinney, J.L. (1974). Avian incubation: interactions among behavior, environment, net and eggs result in regulation of egg temperature. *Science* **186**, 107–15.

Williams, J.B. (1993). Energetics of incubation in free-living orange breasted sun birds in South Africa. *Condor* **95**, 115–26.

Woehler, E. (1990). The distribution of seabird biomass in the Australian Antarctic territory: implications for conservation. *Environmental Conservation* **17**, 256–62.

Zachos, J.C., Breza, J.R. & Wise, S.W. (1992). Early Oligocene ice-sheet expansion on Antarctica: stable isotope and sedimentological evidence from the Kergueken Plateau, Southern Indian Ocean. *Geology* **20**, 569–73.

J. RUBEN

Evolution of endothermy in mammals, birds and their ancestors

Introduction

Endothermy in extant vertebrates

Birds and mammals generally maintain aerobic metabolism at rates about 5 to 10-fold those of reptiles of equivalent size and body temperature. These high rates of endogenous heat production, or endothermy, along with the insulation afforded by feathers and fur, enable birds and mammals to maintain thermal homeostasis over a wide range of ambient temperatures. As a result, these taxa are able to thrive in environments with cold or highly variable thermal conditions and in nocturnal niches generally unavailable to ectothermic vertebrates. Furthermore, the increased aerobic capacity associated with endothermy allows them to sustain activity levels well beyond the capacity of ectotherms (Bennett, 1991). With some noteworthy exceptions, ectotherms, such as reptiles, typically rely on non-sustainable, anaerobic metabolism for all activities beyond relatively slow movements. Although capable of often spectacular bursts of intense exercise, ectotherms generally fatigue rapidly as a result of lactic acid accumulation. Alternatively, endotherms are able to sustain relatively high levels of activity for extended periods of time, enabling these animals to forage widely and to migrate over extensive distances. The physiological capacity of birds and bats to sustain long-distance powered flight is far beyond the capabilities of modern ectotherms (Ruben, 1991).

Some large-size fish (including some billfish, tunas and lamnid sharks) and a few snakes (e.g. *Python*) maintain somewhat greater than ambient core or deep body temperatures (Block, 1991; van Mierop & Barnard, 1978). Generally, elevated core temperatures occur in fish when heat from muscular exercise is conserved by the function of well-developed vascular countercurrent heat exchange systems; some billfish utilise highly specialised heater cells that help to elevate central nervous system temperatures only. Elevated temperatures in brooding female pythons result from enhanced heat production associated with powerful, spasmodic axial muscle contractions. Endothermy in fish and snakes, however, is not directly comparable to that in mammals and birds: without chronic skeletal muscle contractions, these animals cannot produce endogenous heat sufficient to maintain elevated deep-body temperatures. Moreover, core temperatures in these taxa generally do not exceed ambient temperature by more than about 10–20 °C. Clearly, the source and magnitude of caloric expenditure, as well as the stability and marked elevation of body temperature associated with avian and mammalian endothermy, is unique among the vertebrates and among the most significant evolutionary developments of the Metazoa.

Endothermy in extinct vertebrates

Unfortunately, deciphering the evolutionary history of tetrapod endothermy has not been straightforward. In the past few decades, many paleontologists have flailed away with largely futile efforts to demonstrate the endothermic status of various extinct Mesozoic Era reptilian and avian taxa. These have included, especially, dinosaurs (for reviews, see Farlow, 1990; Farlow *et al.*, 1995) and therapsids (the ancestors of mammals) (Carrier, 1987). Other candidates for possible endothermy have, from time to time, included pterosaurs (Padian, 1983), ichthyosaurs (Buffrenil & Mazin, 1990), rhynchosaurs (Triassic Era relatives of the living tuatara) (Carrier, 1987), early crocodilians (Carrier, 1987) and the earliest bird, *Archaeopteryx* (Regal, 1985).

Until very recently, endothermy has been virtually impossible to demonstrate clearly in extinct forms. Endothermy is almost exclusively an attribute of the 'soft anatomy', which leaves a poor, or usually non-existent, fossil record. Physiologically, endothermy is achieved through prodigious rates of cellular oxygen consumption: in the laboratory, mammalian resting oxygen consumption rates are typically about 6–10 times greater than those of reptiles of the same body mass and temperature; avian resting rates are greater still, up to 15 times reptilian rates. Field metabolic rates of mammals and birds often exceed those of equivalent-size ectotherms by 16 to 40-fold. To support such high oxygen consumption levels, endotherms possess profound structural and functional modifications to facilitate oxygen uptake, transport and delivery. Both mammals and birds have greatly expanded pulmonary capacity and ventilation rates, fully separated pulmonary and systemic circulatory systems and expanded cardiac output. They also have greatly increased blood volume and blood oxygen carrying capacities, as well as increased tissue mitochondrial density and enzymatic activities (Ruben, 1995). These key features of endothermic physiology are unknown to have ever been preserved in fossils: mammalian, avian or otherwise.

Consequently, most paleontologically-based conjecture concerning the possible presence of endothermy in extinct vertebrates has relied primarily on weakly-supported correlations of metabolic rate with a variety of far-flung criteria (including, but not limited to, features such as predator–prey ratios, upright posture, trackways, brain size, and paleogeographic distribution (for reviews, see Farlow, 1990; Farlow *et al.*, 1995). Close scrutiny has revealed that virtually all of these correlations are, at best, equivocal. Attempts have also been made to associate supposedly high overall growth rate in endotherms with hypothesised fast growth and endothermy in some dinosaurs. Growth rates, however, in a variety of extant endotherms and at least one dinosaur overlap with the American alligator (Ruben, 1995). More recently, fossilised bone oxygen isotope ratios were similarly purported to demonstrate endothermy in some dinosaurs (Barrick & Showers, 1994). The physiological literature is broadly at odds with major assumptions underlying these isotope-based studies and fossilised bone oxygen isotope ratios in dinosaurs are likely to reveal little, if anything, about dinosaur thermoregulatory or metabolic physiology (Ruben, 1995).

Perhaps most importantly, virtually all of the arguments used previously are based predominantly on apparent similarities to the mammalian or avian condition, without a clear functional correlation to endothermic processes themselves. Until very recently, no empirical studies were available that described an unambiguous, and exclusive, functional relationship to endothermy of a preservable morphological characteristic.

Selective basis for endothermy

Although endothermy almost surely developed independently in birds and mammals (Kemp, 1988; Ruben, 1995), the selective factor(s) responsible for the origin of avian and, especially, mammalian endothermy have historically been the subject of considerable speculation and debate among comparative physiologists. A broad range of selective factors resulting in the evolution of avian and mammalian endothermy have been proposed, but the most viable scenarios centre on selection for either: (i) increased aerobic capacity during exercise (Bennett & Ruben, 1979; Bennett, 1991); or (ii) enhanced thermoregulatory capacity (Crompton *et al.*, 1978; McNab, 1978; Bock, 1985).

On the basis of a relatively constant ratio of minimal : maximal oxygen consumption rates in a variety of active endo- and ectothermic tetrapods, the 'aerobic capacity' model suggests that the initial factor responsible for the evolution of high mammalian and avian resting rates of aerobiosis was selection for increased powers of sustainable, aerobically-supported physical activity. According to this scenario, elevated avian and mammalian aerobic metabolic rates were initially advantageous for the generation of higher levels of sustainable activity. The original selective advantages of elevated aerobic metabolic rates, therefore, had little to do with selection for increased resting metabolic rates or thermoregulatory powers. The continued selection for the expansion of aerobically-supported activity levels, combined with the described linkage of resting and active metabolic rates, led simultaneously to elevated rates of resting oxygen consumption. Eventually, resting metabolic rates in birds and mammals were sufficiently elevated to achieve endothermic homeothermy.

In the 'thermoregulatory' hypothesis (the most traditional), the evolution of high resting metabolic rate in mammals has been attributed to selection for endothermically-based thermoregulation in early (Mesozoic) mammals and their ancestors, the cynodont therapsid reptiles. Selection for a high, stable body temperature is hypothesised to have optimised rates of thermally-dependent processes and, thereby, to have facilitated 'niche expansion', via extended periods of diel trophic activity and/or ability to survive in cool environments.

The problems with interpreting evolutionary patterns of endothermy in tetrapods have been exacerbated because data from the paleontological and physiological literature are seldom integrated. Thus, for example, physiologists considering the evolution of tetrapod endothermy seldom consider that mammals probably evolved in climates so chronically mild that even the ancestors of extant lizards would have had little trouble maintaining constant thermal homeostasis. Similarly, paleontologists frequently assume incorrectly that the ectothermic nature of extant reptiles is automatically associated with a 'cold-blooded', sluggish lifestyle and a constricted pattern of activity physiology.

Fortunately, a variety of illuminating data relating to the physiology of endothermy have appeared within the past decade. Moreover, new information has recently been provided which promises to clarify the metabolic status of many extinct taxa, including dinosaurs, therapsids and early birds. Until recently, no empirical studies were available that described a preservable structure with an unambiguous and exclusive functional relationship to endothermy. This situation has changed with the discovery that the nasal respiratory turbinate bones in mammals and birds are tightly and causally linked to high ventilation rates and endothermy in these taxa (Hillenius, 1992). Evidence for the presence or absence of these structures is often preserved in fossils of long-extinct species, and promises exciting new insight into the chronology and selective factors associated with the evolution of endothermy. Here, I review our current understanding of pertinent aspects of the physiology and paleontology of birds, mammals and their ancestors, with a view toward synthesis of the most likely scenario for the origin of endothermy in these taxa.

Respiratory turbinates and the fossil evidence for the metabolic status of the therapsids, early mammals, dinosaurs and early birds

Respiratory turbinates

There is a strong causal association with endothermy for the respiratory turbinates of almost all extant mammals: the maxilloturbinate bones (Fig. 1). These thin but highly complex structures in the anterior nasal passages counteract the desiccating effects associated with high ventilation rates by facilitating a countercurrent exchange of respiratory heat and water between respired air and the moist lining of the turbinals. As cool external air is inhaled, it absorbs heat and moisture from the turbinal linings. This hydration prevents desiccation of the lungs and also cools the turbinal epithelia and creates a thermal gradient along the turbinates. During exhalation, this process is reversed: warm air from the lungs, now fully saturated with water vapour, is cooled as it passes over the turbinates in the opposite direction. The exhaled air becomes supersaturated as a result of this cooling, and 'excess' water vapour condenses on the turbinal surfaces, where it can be reclaimed and recycled rather than lost to the environment. Over time, a substantial amount of water and heat can thus be conserved (Fig. 2) (Hillenius, 1992).

This respiratory water recovery mechanism is well documented in a series of extant mammalian species from both xeric and mesic environments: when use of the maxilloturbinates was experimentally precluded, respiratory water loss was significantly above normal in all cases (Hillenius, 1992). Furthermore, complex respiratory turbinates occur in almost all birds (Bang, 1971) but are completely absent in all extant reptiles, including species from arid habitats (Hillenius, 1992). These structures appear to have a fundamental association with high pulmonary ventilation rates, and thus with the high levels of aerobic

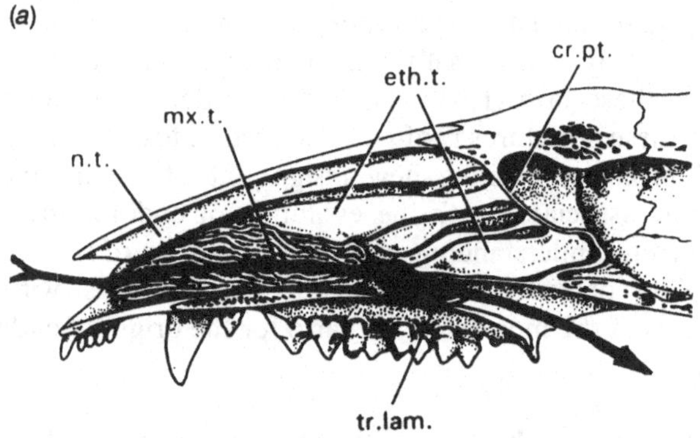

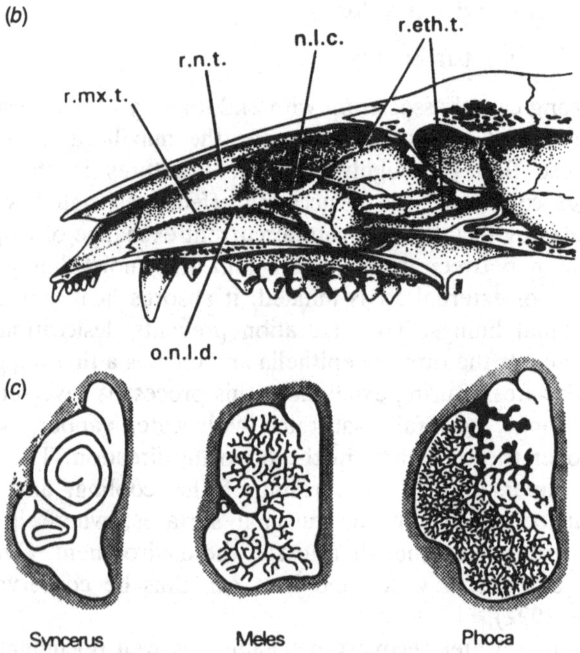

Fig. 1. Nasal turbinates of most mammals. The arrow describes the path of air flow through the nasal region into the oral cavity. (*a*) Right sagittal section of the skull of the opossum, *Didelphis*. (*b*) Similar, but with turbinates removed to reveal turbinate attachment ridges. (*c*) Cross-sections through the anterior nasal turbinals of several mammalian taxa. Not to scale. (After Hillenius, 1994.) Abbreviations: cr.pt., cribiform plate; eth.t., ethmoturbinals (olfactory); mx.t., maxilloturbinal, or respiratory turbinate; n.l.c., nasolacrimal canal; n.t., nasoturbinal (olfactory); o.n.l.d., opening of nasolacrimal canal; r.eth.t., ridges for olfactory ethmoturbinals; r.mx.t., ridge for maxilloturbinal; r.n.t., ridge for nasoturbinal (olfactory).

metabolism that support endothermy itself. In both birds and mammals, endothermy is supported by elevated lung ventilation rates. Even at rest, ventilation rates of birds and mammals in the laboratory are at least 3.5–5 times greater than those of similar-sized reptiles of equal body temperature (Bennett, 1973). Ratios of these differences in the field must be markedly higher, probably approaching 17 to 30-fold. In the absence of respiratory turbinates, continuously high rates of oxidative metabolism and endothermy might well be unsustainable insofar as respiratory water loss is likely to exceed tolerable levels, even in many species of non-desert environments (Fig. 2). Respiratory water loss is

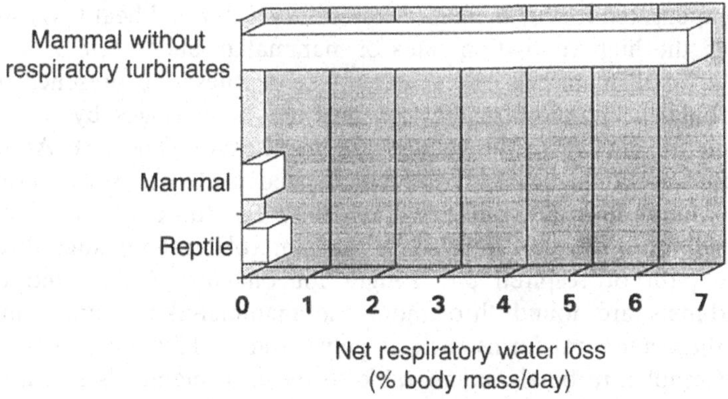

Fig. 2. Daily net respiratory evaporative water loss rates (= metabolic water production + respiratory evaporative water loss) for a normal, field-active 1 kg reptile, a 1 kg mammal, and probable net respiratory evaporative water loss for field-active mammal lacking the use of respiratory turbinates (i.e. with a reptile-like nasal anatomy and reptile-like net respiratory water loss O_2^{-1} consumed). Without the water-conserving function of the respiratory turbinates, daily water flux rates in mammals and birds would be out of balance by about 40–75%. Based on field metabolic and water flux rates for lizards and eutherian mammals (regressions provided by Nagy (1987) and Nagy & Peterson (1988) for doubly-labelled water (DLW; D_2O^{18} field studies) and observed rates of net respiratory evaporative water loss in temperate lizards and intact and experimentally-altered mammals (Hillenius, 1992). For thermoregulating lizards (T_B = 37 °C), net respiratory water loss (at ambient temperature = 15 °C) approximates 1.5 mg H_2O cc O_2^{-1} consumed; for intact mammals, net respiratory water loss cc O_2^{-1} consumed is negligible or slightly positive at ambient temperatures = 15 °C (see Hillenius, 1992).

of much greater potential significance to endotherms and the need for its reduction more immediate.

Mammalian maxilloturbinals have no homologues among extant ectotherms (Wittmer, 1992). In extant reptiles, one to three simple nasal 'conchae' may be present, but these are exclusively olfactory in function. Like the mammalian naso- and ethmoturbinals, these are typically located in the posterodorsal, olfactory portion of the nasal cavity. There are no structures in the reptilian nasal cavity specifically designed for recovery of respiratory water vapour, nor are they needed. Reptilian lung ventilation rates are sufficiently low that pulmonary water loss probably seldom creates a significant site of water loss, even for desert species (Hillenius, 1992). It is, therefore, likely that maxilloturbinates compensated for the increased respiratory water and heat loss associated with the high ventilation rates of mammalian endothermy.

In fossil mammals and mammal-like reptiles, the presence of nasal turbinates are most readily revealed via bony ridges by which these structures attach to the walls of the nasal cavity (Fig. 1*b*). Attachment ridges for olfactory turbinals are located posterodorsally, away from the main flow of respiratory air, whereas those of the respiratory maxilloturbinals are situated in the anterolateral portion, directly in the path of respired air. Ridges for olfactory (naso- and ethmo-) turbinals are found throughout the mammal-like reptiles, including earlier, ancestral forms such as pelycoasaurs (Hillenius, 1994).

Complex turbinals, comparable to those of mammals are also found in all birds (Bang, 1971), and it is likely that they share a similar function as well. The extent and complexity of the nasal cavity of birds varies widely with the shape of the bill, but in general, the avian nasal passage is elongate with three cartilaginous, or sometimes ossified, turbinates in succession (Fig. 3). The anterior turbinal is often relatively simple but the others, particularly the middle turbinal, are often more highly developed into prominent scrolls with multiple turns. Sensory (olfactory) epithelium is restricted to the posterior turbinal. Like mammalian olfactory turbinals, this structure is situated outside the main respiratory air stream, often in a separate olfactory chamber. Embryological and anatomical studies indicate that only the posterior turbinal is homologous to those of reptiles; the anterior and middle turbinals are avian neomorphs.

The anterior and middle turbinals of birds, like the respiratory turbinates of mammals, are situated directly in the respiratory passage, and are covered primarily with respiratory epithelium. The position of these turbinals leaves them well positioned to modify bulk respired air. Substantial data suggest that these turbinates function as well as,

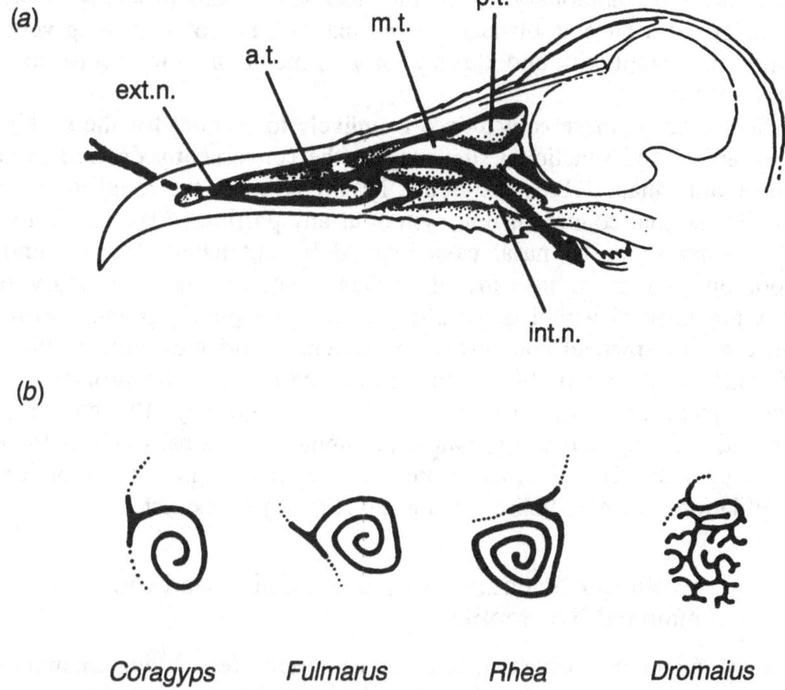

Fig. 3. Nasal turbinates of the gull (*Larus*). The arrow describes the path of air flow through the nasal cavity. Anterior and middle turbinates are respiratory in function; posterior turbinates are olfactory. (*a*) Right sagittal section through the nasal cavity of a gull, *Larus*. (*b*) Cross-sections through the middle (respiratory) turbinals of a vulture (*Coragyps*), a fulmar (*Fulmarus*), a rhea (*Rhea*) and an emu (*Dromaius*). Not to scale. (After Hillenius, 1994.) Abbreviations: ext.n., nostrils; int.n., internal nares or choanae; a.t., m.t., anterior and middle (respiratory) turbinates; p.t., posterior (olfactory) turbinates.

or superior to, mammalian respiratory turbinals for the recovery of water vapour contained in exhaled air. Consequently, these structures in birds probably represent an adaptation to high lung ventilation rates and endothermy, fully analogous to the respiratory turbinals of mammals.

To summarise, physiological data imply that independent selection for endothermy in birds, mammals and/or their ancestors was, by necessity, tightly associated with the convergent evolution of respiratory tubinates in these taxa. In the absence of these structures, unacceptably high rates of pulmonary water and heat loss would probably always have posed a chronic obstacle to the maintenance of bulk lung ventilation consistent with endothermy, or with metabolic rates approaching endothermy.

Significantly, mere coincidence is unlikely to account for the striking anatomical and functional similarity of the convergently derived avian and mammalian turbinate systems. Maintenance of the requisite alternating thermal countercurrent system at any portion of the respiratory tree other than the nasal cavity would be untenable. Intermittently cool and warm countercurrent exchange sites in the secondary or primary bronchi would necessarily preclude deep-body homeothermy; an efficient tracheal countercurrent system would inevitably result in chronic oscillation of brain temperature because of the proximity of the carotid circulation to the trachea. Consequently, the confirmed absence of respiratory turbinates, or similar structures, is likely to be strongly indicative of ectothermic, or near ectothermic, rates of lung ventilation and metabolism in any taxa, living or extinct.

Evolution of endothermy in mammals and the mammal-like reptiles

Extant mammals, which include prototherians (egg-laying mammals) and therians (marsupial and placental mammals), probably last shared a common ancestor about 160 million years ago (My) (Rowe, 1992). At any particular mass, all extant mammals maintain generally similar temperature-corrected metabolic rates, and are alike in a wide range of other endothermy-related physiological processes and anatomical structures (e.g. hair, non-nucleated red blood cells, lung structure, blood oxygen carrying capacity, diaphragm, sweat glands, etc.). Consequently, it is most parsimonious to assume the groups' common Mesozoic Era ancestor had achieved a similar, endothermic grade of aerobic physiology (Bennett & Ruben, 1986).

Interpretation of more ancient mammalian or pre-mammalian physiological status rests with linkage of fossilised morphology metabolic rate. Mammalia probably evolved from cynodont therapsids, or 'mammal-like' reptiles, by sometime late in the Triassic Period (approximately 200 My) (Rowe, 1992) and, based on the presence of a secondary palate and parasagittal limb stride in Triassic Era cynodonts, it has long been suggested that at least near-endothermy had been achieved in late therapsids (Brink, 1956). In fact, rudimentary anterolateral rostral ridges for support of respiratory (maxillo-) turbinate bones are now known to first appear in some late Paleozoic Era therocephalian therapsids, e.g. *Glanosuchus*, a wolf-like pristerognathid therocephalian (Hillenius, 1994).

Accordingly, initial phases in the evolution of 'mammalian' oxygen consumption rates may have begun as early as the Late Permian Period (250 My), some 40–50 million years prior to the origin of the Mammalia. Lower-Middle Triassic cynodont therapsids (e.g. the galesaurid *Thrinaxodon* and the traversodontid *Massetognathus*), as well as the earliest mammals (e.g. the Late Triassic *Morganucodon* and the Early Jurassic *Docodon*), appear to have had maxilloturbinate development comparable to that of extant mammals, presumably with a similar capacity for respiratory water recovery. This is the first compelling evidence that lung ventilation rates and, by extension, metabolic rates of the earliest mammals (and at least some mammal-like reptiles) approached, or even equalled, those of extant mammals (Hillenius, 1994).

It is difficult to overstate the importance of finding incipient maxilloturbinates in therocephalians for our understanding of the evolution of endothermy. Thermoregulatory-based models for the origin of mammalian endothermy are largely falsified if the earliest, incremental steps toward mammalian metabolic rate occurred in large, Late Permian therapsids. Many therocephalians appear to have been active, dog-, bear- or lion-like, medium to large (20–100 kg) carnivores which inhabited regions with subtropical to tropical climates. They were sufficiently large so that, whether hairy or scaly, their thermal conductance was probably similar to that of large mammals (McNab & Auffenberg, 1976). Accordingly, they were probably inertial homeotherms and thermoregulatory-related expenditure of metabolic energy was unlikely to have been necessary (McNab, 1978).

Metabolic rates in dinosaurs and early birds

Birds are accomplished endotherms and many (especially the songbirds) maintain the highest body temperatures and metabolic rates of any tetrapods. The fossil record indicates that modern bird orders were fairly well defined by about 60 My (Feduccia, 1995), therefore avian endothermy is likely to have been fully developed by about Late Cretaceous–Early Tertiary Periods.

As the birds are a monophyletic group, no extant endothermic sister taxon exists from which pre-Cenozoic Era inheritance of endothermy can be logically inferred (in contrast to the monotreme, marsupial and eutherian taxa of the mammals). The earliest known bird is the famous *Archaeopteryx lithographica*, a late Jurassic (145 My) archaeornithine with (at least) complete wing and tail plumage and a striking superficial skeletal similarity to some carnivorous dinosaurs (Ostrom, 1976). Recently described circumstantial evidence hints that *Archaeopteryx* might well have been an ectotherm (Ruben, 1991; Chinsamy *et al.*, 1994). Nevertheless *Archaeopteryx* is conventially assumed to have been endothermic because its well-developed flight plumage is thought to have been reflective of ambient radiation, sufficient to preclude effective ectothermic thermoregulation (Bock, 1985). The presence of even a fully developed set of flight and contour feathers in Mesozoic Era avian ancestors of extant birds need not necessarily signal the presence of endothermy, or even an approach to it. Like modern reptiles, some living birds utilise behavioural thermoregulation to absorb ambient heat across feathered skin. During nocturnal periods of low ambient temperatures, body temperature in the roadrunner (*Geococcyx californianus*) declines by about 4 °C. After sunrise, the roadrunner exposes poorly feathered parts of its body to solar radiation and warms ectothermically to normal body temperature. Additionally, a number of other fully-feathered extant birds can readily absorb and use incident radiant solar energy. An ecotothermic *Archaeopteryx*, which is thought to have lived in a warm, sunny climate, might easily have had a similar behavioural thermoregulatory capacity. A fully feathered *Archaeopteryx*, whether ectothermic or endothermic, could easily have achieved homeothermy. The appearance of well-developed plumage so early in avian history might well have had to do strictly with the evolution of powered flight, rather than with any particular thermoregulatory pattern (Ruben, 1991; see also Parkes, 1966).

The apparent anatomical similarity between some Mesozoic archosaurs and *Archaeopteryx* is marked, and many workers accept a close phyletic relationship between the theropod dromaeosaurid dinosaurs (e.g. *Velociraptor*, *Deinonychus*) and primitive birds (Ostrom, 1990); others have proposed that birds, as well as dinosaurs, are direct descendants of the Triassic thecodont archosaurs (Feduccia, 1993). If Triassic thecodonts were endothermic, then bird-like metabolic status is likely to have predated the origin of birds, as well as dinosaurs. Alternatively, ectothermy in dromaeosaurid dinosaurs probably signals a non-endothermic status for thecodonts and early birds, whatever their ancestry, as dinosaurs were unlikely to have reassumed reptilian metabolic rates if they had been derived from endothermic ancestors. The presence or absence of endothermy in dinosaurs has been a topic of considerable discussion and sometimes acrimonious debate over the past 20 years. In the past, proponents have advanced a variety of elaborate scenarios for dinosaur endothermy on the basis of, for example, dinosaur posture (Bakker, 1971), predator-prey ratios (Bakker, 1980), trackways (Bakker, 1986) and fossil bone oxygen isotope ratios (Barrick & Showers, 1994). Certainly, many large dinosaurs were inertial homeotherms (Spotilla *et al.*, 1991) but few, if any, of these arguments provide unambiguous evidence for endothermy in dinosaurs (Bennett & Dalzell, 1973; Bennett & Ruben, 1986; Farlow, 1990; Farlow *et al.*, 1995; Ruben, 1995).

The confirmed presence, or absence, of respiratory turbinates is likely bellweather indicators of lung ventilation and metabolic rates in virtually all terrestrial taxa, living or extinct. Although complex respiratory and olfactory turbinates are virtually ubiquitous among extant birds, the presence or absence of these structures has remained poorly known in fossil birds or in their ancestors, the dinosaurs. The widespread occurrence of respiratory turbinates among living birds (Fig. 3) suggests that these structures may predate the last common ancestor of extant avian taxa, perhaps Upper Cretaceous Era in age (Feduccia, 1995), but it is uncertain how much older, or widespread, respiratory turbinates might be in the dinosaurian–avian lineage.

Several problems complicate the study of the evolutionary history of turbinates in birds and their ancestors. Although they are occasionally discovered in extinct taxa (e.g. olfactory turbinates in phytosaurs (Camp, 1930), very early mammals (Lillegraven & Krusat, 1991)), turbinates are highly fragile structures in living taxa, and are generally poorly preserved or absent in fossilised specimens. Furthermore, although they ossify or calcify in many extant taxa, these structures often remain cartilaginous in birds, which further decreases their chances for preservation. Nevertheless, my colleagues (J. Hillenius, A. Leitch & N. Geist) and I have determined that the presence of turbinates is inevitably associated with certain distinct structures and/or morphologies that function to support, or otherwise accommodate, turbinates. Just as bony attachment ridges, or scars, on the medial walls of the nasal cavity mark former attachment sites of maxilloturbinate structures in mammals, avian respiratory turbinates are frequently associated with the presence of an osseous ventromedial crest, or 'schwele', in the nasal cavity. Moreover, the ducts of the lateral nasal glands typically open just anterior to this crest, whose shape appears to assist in the vaporisation of exudate from these glands. A similar nasal crest is unknown from animals that lack respiratory turbinates including all existing reptiles. Other gross features of the nasal cavity, such as the shape of the maxillary palatine process, the generally expanded morphology of the nasal vestibular region and, in some cases, the location of the internal nares (= choanae) often correlate with the presence of the avian respiratory turbinal apparatus.

Some, if not all, of these features have been observed in a variety of whole and partial fossil specimens. Significantly, preliminary evaluation based on these criteria suggest that respiratory turbinates and, consequently, endothermy, were unlikely to have occurred in *Archaeopteryx* or in its presumed ancestors, the dromaeosaurian (maniraptoran) theropod dinosaurs (Fig. 4).

Additional evidence is provided by the recent application of computed axial tomography (CAT or CT) scans to paleontological specimens, which has greatly facilitated non-invasive study of the fine details of the nasal region in fossilised specimens, especially those which have

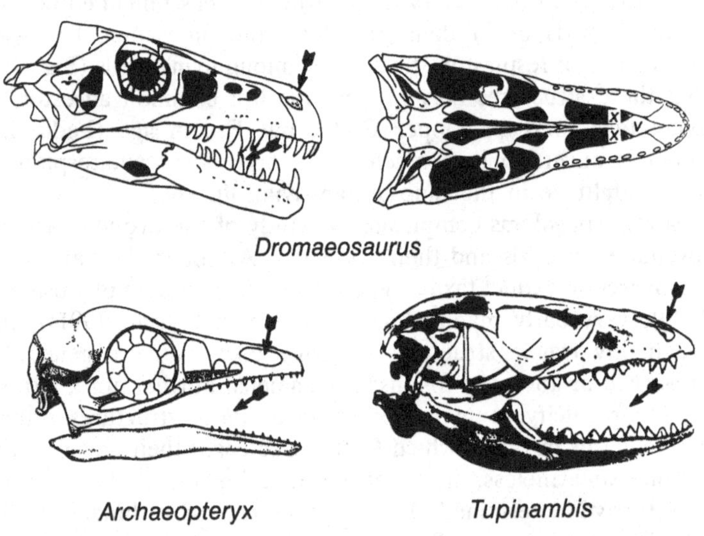

Fig. 4. Lateral and palatal view of the skulls of the dromaeosaurid theropod *Dromaeosaurus* and the earliest bird *Archaeopteryx* and the extant teid lizard, *Tupinambis*. The arrow describes the path of air flow through the nasal region into the oral cavity. For *Dromaeosaurus* and *Archaeopteryx*, the presumed air flow route is based on the location of the nostrils and the anteriorly displaced choanae (as indicated by the anterior placement of the vomer). In *Dromaeosaurus* and *Archaeopteryx*, the short, direct path of inhaled air into the oral cavity is highly reminiscent of the condition of the nasal region in many extant reptiles and almost certainly precluded sufficient space to have accommodated respiratory turbinates. Abbreviations: *V.*, vomer; *X*, internal nares. (*Dromaeosaurus* after Colbert & Russell, 1969; *Archaeopteryx* after Buhler, 1985; *Tupinambis* after Hildebrand, 1995.)

been 'incompletely' prepared. In some cases, CT scans of particularly well-preserved specimens have revealed delicate remnants of calcified, cartilaginous or lightly-calcified cartilaginous structures. In the small tyrannosaurid *Nanotyrannus* (Carnosauria: Tyrannosauridae), CT clearly demonstrates that in life this animal boasted particularly well-developed olfactory turbinates (and, probably, olfactory powers), but had no respiratory turbinates. This condition is reminiscent of the nasal region of many extant reptiles (e.g. *Alligator*) and is strong evidence for the ectothermic nature of this dinosaur (Fig. 5). The olfactory turbinates (not shown in Fig. 5) of *Nanotyrannus* were closely associated

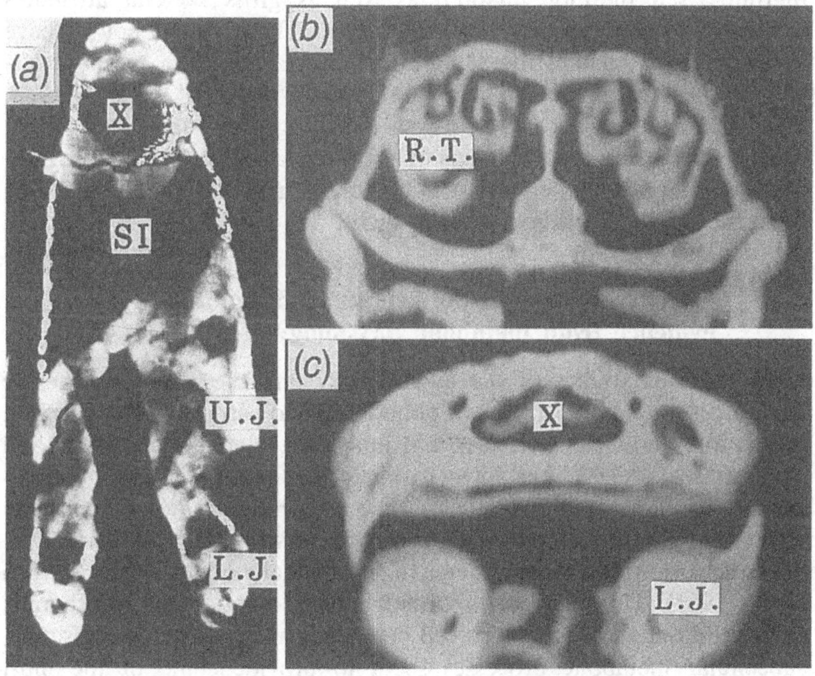

Fig. 5. Cross-sectional, computed tomography of the mid-snout region in: (*a*) tyrannosaurid dinosaur *Nanotyrannus*; (*b*) ostrich (*Struthio*); (*c*) crocodile (*Crocodylus*). Respiratory turbinates in the ostrich are housed in voluminous nasal passageways. The relatively narrow nasal passageway in *Nanotyrannus* resembles that in the crocodile and appears too small to have accommodated respiratory turbinates. Not to scale. Abbreviations: *SI.*, maxillary sinus; *R.T.*, respiratory turbinate; *U.J.*, *L.J.*, upper and lower jaws; *X*, nasal passage. (*Nanotyrannus* CT scan by courtesy of A. Leitch.)

with a well-preserved and substantial olfactory bulb that extended forward of the braincase as far as the animal's lacrimal bar. This structure represented almost half the total braincase endocranial volume.

On the basis of the general structure of its pectoral girdle and flight apparatus, I have suggested elsewhere that *Archaeopteryx* seems unlikely to have achieved an endothermic metabolic status (Ruben, 1991; see also Feduccia, 1985). Similarly, histological patterns of bone growth in early Cretaceous Era enantiornines indicate that these birds were likely to have maintained an ectothermic thermoregulatory pattern (Chinsamy *et al.*, 1994). Patterns of bone growth consistent with endothermic-based thermoregulation, as well as gross skeletal attributes indicative of the capacity for long-distance flight, seem not to have appeared until perhaps the Middle–Late Cretaceous Era, perhaps 50–60 My after the presumed mid–late Jurassic origin of birds (Ruben, 1991; Chinsamy *et al.*, 1994). These data, together with the apparent absence of respiratory turbinates in *Archaeopteryx*, are strongly suggestive that fully-developed avian endothermy may have appeared long after the origin of birds and avian flight.

Selective factors in the evolution of endothermy: evidence from metabolic physiology

Most previous studies of the evolutionary physiology of endothermy have focused on comparisons of pulmonary and cardiovascular function in amniote vertebrates. Hearts and lungs are relatively accessible and comparative observations have clarified the parallel origin of endothermy in birds and mammals; some experimental work has even yielded insight into selective factors associated with the origin of individual structures within these systems (e.g. mammalian diaphragm) (Ruben *et al.*, 1987). However, aerobic metabolism, whether at endothermic or ectothermic rates, occurs in cell organelles and, perhaps surprisingly, subcellular metabolic processes seem to provide some of the most revealing physiological data for understanding the evolution of endothermy.

The principal function of substrate metabolism is to sustain mitochondrial generation of ATP, the ubiquitous and readily usable form of energy for most kinds of cellular work. Metabolically-generated heat is usually a byproduct of biochemical 'inefficiencies' associated with the synthesis and utilisation of ATP. There is a consistent difference between the mitochondrial membrane surface area of a variety of tissues from endotherms and ectotherms that indicate cellular and even subcellular bases for the increased metabolic rate of endotherms.

In endotherms, expanded individual mitochondrial ATP output and/ or net mitochondrial membrane density provide the large sources of energy required to sustain endothermic metabolic rates (Else & Hulbert, 1987; Hulbert & Else, 1989). These observations do not, however, account for physiological processes in birds and mammals that necessitate the generation of elevated quantities of metabolic energy. A partial explanation might involve the nature of mammalian and avian plasma membranes and maintenance of their sodium–potassium ion gradients. These gradients, which are generated by Na^+–K^+-ATPase pumps (= sodium pumps), are necessary for the support of a variety of functions, including action potential generation and the active cellular uptake of organic molecules (e.g. amino acids, sugars, etc.) for cellular nutrition and protein synthesis (Clausen *et al.*, 1991). The plasma membranes of pigeon (*Columbia*) liver cells and laboratory rat liver and kidney cells, when compared with those of lizards, seem particularly 'leaky' to sodium (Na^+) and potassium (K^+) ions (Hulbert & Else, 1990). These leaky plasma membranes apparently require the generation of significantly increased levels of metabolic energy to maintain normal solute concentration gradients at the plasma membrane (i.e. relatively high [K^+] at the inner surface and relatively high [Na^+] at the outer surface) (Else & Hulbert, 1987).

This increased 'leakiness' in mammalian plasma membranes is correlated with marked increments (approximately 50–60%) in tissue protein and phospholipid concentrations. Compared with reptiles, there is also a qualitative alteration in plasma phospholipid composition, at least in the mammalian liver and kidney. Plasma membranes from each of these organs in mammals have significantly reduced total unsaturated fatty acids, but a 35% greater concentration of polyunsaturated fatty acids (= greater unsaturation index). In addition, certain long-chain polyunsaturated fatty acids, specifically arachidonic and docosahexanoic acids, 20- and 22-carbon phospholipids, respectively, are relatively more plentiful. Another polyunsaturated long-chain fatty acid, linoleic acid (18-carbon), occurs with less frequency in mammals than in reptiles (Hulbert & Else, 1989).

These changes in membrane phospholipid and protein composition seem causally linked to increased 'leakiness' and, ultimately, to the increased metabolic cost of plasma membrane ion-gradient maintenance in the cells of endotherms. Compared with reptiles, mammals exhibit increased metabolic sensitivity to thyroxin. In incubated mammalian liver slices, metabolic rate, mitochondrial membrane surface area, membrane polyunsaturated fatty acid composition and liver transmembrane Na^+–K^+ flux are all enhanced by increased thyroxin levels (Ismaeli-Begi, 1988; Brand *et al.*, 1991). Additionally, there is an inverse relationship between increasing body weight versus mass specific metabolic rate and cardiac tissue concentrations of docosahexanoic acid, one of the phospholipids that also occur in especially high concentrations in mammalian liver and kidney cells (Gudbarjarnason *et al.*, 1978).

It has been suggested that the evolution of these leaky, supposedly 'inefficient' plasma membranes in endotherms was linked expressly to the necessity for increased thermogenesis, rather than to fulfil expanded demands for plasma membrane work *per se* (Hulbert & Else, 1989). A similar, 'thermogenesis-dedicated' explanation has been offered for enhanced heat production associated with increased rates of futile proton (H^+) cycling at the apparently leaky inner mitochondrial membranes of birds and mammals (Brand *et al.*, 1991). As with mass-specific metabolism, mass-specific rates of mitochondrial proton leakage seem to decrease with increasing body size in mammals. Accordingly, variation in thermogenic proton leakiness has been proposed as accounting for mass-related differences in mammalian resting metabolic rates (i.e. Kleiber's 'mouse–elephant' metabolism curve) (Porter & Brand, 1993). This explanation fails to account for virtually identical mass-related changes in metabolic rates of almost all ectotherms (Hemmingsen, 1960), taxa in which it is hardly likely that any aspect of cellular metabolism is dedicated expressly to heat production.

If the ultimate function of greatly expanded plasma- and/or mitochondrial-membrane leakiness were simply to facilitate endogenous heat production in endotherms then, compared with reptiles, a greater proportion of total cellular metabolism in mammals and birds should be associated with these processes. Vertebrate tissues clearly specialised for thermogenesis dedicate high fractions of total cellular metabolism to membrane ion-gradient maintenance, e.g. mammalian brown adipose tissue (Horwitz, 1989; Himms-Hagen, 1990) and billfish brain 'heater' organs (Block, 1991). *In vitro*, however, incubation of ectotherm and endotherm liver and kidney slices demonstrates that similar fractions of total tissue oxygen consumption (approximately 25–35%) are devoted to the maintenance of plasma- and mitochondrial-ion homeostasis (Hulbert & Else, 1990; Brand *et al.*, 1991). There is no compelling reason to assume that either the mitochondrial or plasma 'leaky membrane' fractions of general resting metabolism have much more to do with a necessity for heat production in endotherms than they do in ectotherms. Certainly, there is a large increment in absolute rates of metabolism devoted to the maintenance of leaky membranes in endotherms; however, metabolic rates devoted to all cellular functions in endotherms have increased by about the same magnitude.

Overall, these observations suggest that the evolution of endothermy may have initially involved more substantive quantitative, rather than qualitative, changes in many aspects of cellular physiology. Symmetry of membrane- and non-membrane devoted expenditures of resting metabolic energy in endotherms parallels the equality of the ATP-generating capacity in many mitochondria of ecto- and endotherms. Perhaps the evolution of endothermy at the cellular and subcellular level was associated largely with the need to accommodate intensified rates of cellular functions, many of which were qualitatively similar to those of the ectothermic ancestors of birds and mammals. Concomitant requirements for increased production of ATP might have been fuelled, at least initially, by incremental addition of relatively unmodified mitochondria.

Rather than focusing on the energy consumption rate of any one particular aspect of cellular metabolism, perhaps it is more appropriate to ask why the total oxygen consumption rate in endotherms is so high. The physiological patterns described above indicate that the elevation of metabolic rate primarily for endothermic thermoregulation seems unlikely. Additionally, incipient endothermy in 'protomammals' (Late Permian (250 My) theriodont therapsids) appeared first in wolf-like inhabitants of equable, subtropical regions (see later). These animals were probably inertial homeotherms, and the presumed benefit of elevated endogenous heat production rates for thermoregulatory purposes was unlikely to have been worth the increased metabolic cost (McNab, 1978).

Endothermic metabolic rates could be driven by elevated rates of molecular turnover and synthesis. In mammals, protein turnover rates and metabolism scale equivalently with increased body mass (Munro & Downie, 1964) and, compared with reptiles, the 50% higher protein and phospholipid contents of mammalian tissues (see earlier) might signal increased levels of turnover and resynthesis of these compounds, necessitating higher rates of aerobic metabolism and ATP production. The leaky plasma membranes of endotherms are also consistent with this scenario: increased sodium pump activity is associated with enhanced plasma membrane active cotransport of a variety of molecules, including nutrients and amino acids essential for metabolism and molecular synthesis (Clausen *et al.*, 1991).

What parameters of mammalian and avian physiology might require higher rates of molecular turnover and synthesis? In laboratory rats, trout (*Oncorhynchus*) and carp (*Cyprinus*) high endurance, frequently used 'oxidative-type' skeletal muscle fibres analogous to those which predominate in most mammals and birds, are associated with markedly higher rates of protein turnover than are their high speed, but less often used, white-type muscle fibres, similar to those that predominate in many reptiles (Goldspink, 1972; Putnam *et al.*, 1980; Goldspink *et al.*, 1984; Lewis *et al.*, 1984; Garlick *et al.*, 1989; Butler, 1991; Houlihan, 1991). Increased frequency of exercise *per se* is also associated with accelerated protein turnover rates (Houlihan & Laurent, 1987). Perhaps the origin of avian and mammalian endothermy involved selection for increased levels of routine, sustainable activity supported by high concentrations of metabolically-expensive, high endurance, oxidative-type skeletal muscle fibres. Consequently, the elevation of ancestral resting metabolic rate might ultimately have been driven by enhanced requirements for ATP production, linked tightly to an increased demand for a variety of 'support and resupply' functions provided by the viscera. These enhanced visceral (= resting) functions probably include(d) synthesis of a variety of amino acids and proteins, urea or uric acid production and elimination, digestion and storage of nutrients, processing of lactic acid, etc.

In this context, it is particularly significant that in all classes of vertebrates, maximum rates of aerobiosis exceed resting rates by an average ratio of about 10 to 15, being less in sedentary species and greater in more active taxa (Bennett & Ruben, 1979). In addition, following a thorough review, Hayes & Garland (1995) conclude that analysis of a wide variety of intra- and interspecific data support the phenotypic linkage of resting and maximal rates of aerobiosis in vertebrates. Increased capacities or aerobic metabolism also seem mediated by genetic factors insofar as they have been shown to be heritable, persistent through time and variable among individuals (Bennett, 1991).

The aerobic capacity model for the origin of mammalian endothermy seems reinforced by these data: selection for increased aerobic capacity in Permian theriodont therapsids, many of which appear to have been active predators, would have yielded immediate benefits in terms of an expanded capacity for prey pursuit, predator avoidance, courtship and for the maintenance of increased territory size (Bennett & Ruben, 1979). Presumably, expansion of maximal capacity for aerobiosis was linked with increased resting metabolic rates in Permian therapsids, as it is in extant taxa. The continued selection for expanded aerobic capacity in Permian and Triassic therapsids eventually resulted in resting thermogenesis perhaps sufficient for endogenously-based thermoregulation in late cynodonts and/or early mammals. (Some of the preceding section is modified from Ruben, 1995.)

Integrating the physiological and fossil data

Given the antiquity of the earliest birds and mammals, many factors that influenced their origins will never be unearthed. Nevertheless, recent physiological and paleontological discoveries facilitate broad new insight into the evolution of endothermy in these groups.

Subcellular physiological data probably reveal far more than has been previously recognised about thermoregulatory versus aerobic capacity models for the origin of mammalian and avian endothermy. Overall rates of aerobiosis are high in both taxa, and in both approximately 65–75% of energy production is probably devoted to molecular synthesis and transport, nitrogen excretion, digestion, muscle contraction, etc.; the remaining 25–35% seems to be associated with futile shuttling of Na^+ and K^+ at 'leaky' plasma membranes. Consequently, given the high metabolic rates of birds and mammals, absolute amounts of energy/heat production devoted to heat-producing maintenance of leaky plasma membranes are far higher than in reptiles. This association has been cited as evidence for avian and mammalian endothermy having resulted from selection for enhanced endogenous heat production, and thermoregulatory capacity.

In particular cases, however, where elevated metabolism can be confidently ascribed to the need for greater heat production for thermoregulatory purposes (e.g. brown fat in young mammals, cephalic endothermy in some fish), a particularly high proportion of total cellular metabolism is clearly associated with heat production for its own sake, while relatively less is devoted to more conventional varieties of cellular work (Himms-Hagen, 1990; Block, 1991). Similarly, if avian and mammalian endothermy evolved in response to requirements for increased endogenous heat production, we might expect to find a relatively higher proportion of their metabolism devoted to futile Na^+–K^+ shuttling than in the ancestral condition. This is not the case. Modern reptiles, birds and mammals devote roughly similar fractions of total energy expenditure to leaky membrane-associated heat production. Thus, comparative metabolic differences in birds and mammals versus reptiles are primarily quantitative, not qualitative, in nature. Significantly, given the relative constancy of metabolic factorial scope in endo- and ectotherms, the aerobic capacity model is fully consistent with the qualitative similarity of metabolism in reptiles, birds and mammals.

The fossil evidence for the evolution of mammals from therapsids is also inconsistent with a thermoregulatory-based origin for mammalian endothermy. In the therapsid-mammal lineage, elevated lung ventilation rates and, by extension, accelerated routine metabolic rates, in large (>40 kg), Late Permian Period therocephalians from warm climates preceded the evolution of mammals by some 40 My. Subsequently, mammalian, or near-mammalian, metabolic rates may well have been attained in late cynodont (Triassic Period) therapsids. Given the probable inertially homeothermic status of these early, large therocephalians, it is difficult to attribute their incipient endothermy to selection for enhanced thermoregulatory capacity. Alternatively, the increasingly mammal-like structure of these Late Permian therapsids (Kemp, 1982) is fully consistent with their incipient endothermy having been associated with an elevated demand for cellular work and aerobically-based ATP synthesis, rather than for thermogenesis *per se*. Elevated tissue demand or oxygen was presumably facilitated by expansion of pulmonary-cardiovascular capacities and primarily by quantitative, rather than qualitative, alterations at cellular and tissue levels (e.g. increments in oxidative enzyme activity, increased numbers of mitochondria, etc.). It is possible, but as yet unproven, that attainment of endothermic metabolic status in late therapsids and/or early mammals was accompanied by the simultaneous development of homeothermic thermoregulatory patterns typical of most extant mammals. The existence of an extensive pelage (as yet unknown) in late therapsids and/or very early mammals would be consistent with at least near-capacity for fully developed endothermic homeothermy.

The chronological development of endothermy in the archosaurian-avian lineage is also becoming increasingly well understood. There is little, if any, compelling indirect evidence for elevated metabolic rates in any dinosaurs or early (Late Jurassic–Early Cretaceous Periods) birds. To the contrary, the apparent absence of respiratory turbinates in dromaeosaurid dinosaurs and in *Archaeopteryx* itself, provides the first direct evidence that endothermic metabolic rates had probably not been attained in either early birds or their immediate ancestors. This scenario for the ectothermic, or near-ectothermic, metabolic status of the earliest birds is also consistent with the presence of reptile-like annular growth rings in the long bones of some Early Cretaceous enantiornine birds (Chinsamy *et al.*, 1994). Alternatively, parsimony dictates that respiratory turbinates and endothermy were probably present in the Late Cretaceous ancestor of modern birds. Consequently, it is reasonable to suggest that not only were the earliest birds ectothermic, but that complete endothermy may not have been achieved until mid–late Cretaceous times, perhaps 50 My after the appearance of *Archaeopteryx*.

If the earliest birds were metabolically unlike their living descendants, could they nevertheless have flown? I have previously presented evidence that an ectothermic *Archaeopteryx* utilising the high power locomotor muscle typical of some extant reptiles would probably have been fully capable of short-distance, flapping flight (Ruben, 1991, 1993). It has been asserted recently, however, that *Archaeopteryx* was incapable of flight because: (i) it could not have generated sufficient muscle power to sustain flapping flight, supposedly because there is really no evidence for especially high-power skeletal muscle in reptiles (Speakman, 1993); (ii) *Archaeopteryx* lacked primary feather vane asymmetry consistent with powered flight (Speakman & Thomson, 1994); and (iii) *Archaeopteryx* lacked certain wrist-locking mechanisms present in all birds capable of powered flight (Vazquez, 1992). None of these assertions withstand close scrutiny. There is, in fact, reliable (although limited) data indicating the existence of particularly high-power locomotor muscle in some extant reptiles (Josephson, 1993: table 1, fig. 3; Swope *et al.*, 1993; James *et al.*, 1995; fig. 8). Consequently, an ectothermic *Archaeopteryx*, with 'all-out' metabolic and muscle power capacity equal to that of some extant reptiles, could probably have produced 50–60% more metabolic power than that required for level flight (Ruben, 1991, 1993). Moreover, even with the lower power, high endurance flight muscle typical of modern birds, the aerodynamically advanced design of *Archaeopteryx* (discussed below) probably gave it the capacity for a 'ground upward' powered take-off (Marden, 1994).

The conclusion that the magnitude of *Archaeopteryx*' primary feather asymmetry was inconsistent with flight was based on the comparison of vane asymmetry in primary feathers 4, 5 and 6 of the Berlin specimen, and 3 and 4 of the London specimen; but these measurements are compared with vane asymmetry in primaries 1 and 2, the most asymmetrical of living birds. Given that vane asymmetry diminishes as one proceeds inward on the wing of flying birds, and drops off dramatically after the fourth primary feather (Fig. 1, Speakman & Thomson, 1994), these comparisons are not legitimate. A. Feduccia and L. Martin (personal communication) and Feduccia (1996) measured vane asymmetry on the second primary of the Berlin specimen using a high quality silastic cast of the counterpart slab (where the feathers are best preserved). Their measurements (taken at a '25%' position on the flight feather, as in the measurements for extant birds) revealed an asymmetry index of 2.6 for the right wing and 2.8 for the left, thus falling within reported values for modern flying birds (see also Norberg,

1995). Finally, it has been claimed that as *Archaeopteryx* apparently lacked wrist-stabilising facets and ridges on the carpometacarpus and distal ulna (= trochlea carpalis), its distal wing would have been 'un-linked' and too unstable to accomplish flapping flight. Ostrom (1994) points out that, in fact, the shape and position of *Archaeopteryx*' semi-lunate carpal (which adjoined the first two metacarpals) ensured an automatic linkage of hand supination with wrist flexion, as well as linking pronation of the manus with wrist extension, very like that accomplished by the modern trochlea carpalis of birds.

These previous objections notwithstanding, *Archaeopteryx* clearly presents abundant evidence that it was indeed capable of powered flight. It was certainly aerodynamically advanced, with its forelimb proportions and arrangement, as well as the anatomy of its primary and secondary flight feathers, having been structurally similar to those of modern birds: its overall wing shape was similar to that of extant birds that fly easily through broken vegetation (e.g. woodcocks, woodpeckers). The flight feathers of *Archaeopteryx* exhibit a marked longitudinal feather curvature correlated only with flapping flight and the 'slotting' of its primary feathers (identical to that in many modern birds) is consistent with high-lift devices particularly important at, and shortly after, take-off, when forward speed is low. In addition, the vanes of *Archaeopteryx*' flight feathers were reinforced by a longitudinal furrow, virtually identical to a similar furrow that helps flight feathers to resist excessive dorsoventro flexion during wing-flapping in modern birds. It strains credulity to relate this combination of features in *Archaeopteryx* to any activity other than flapping flight (Norberg, 1985).

In retrospect, variation in the overall evolutionary patterns resulting in the evolution of modern birds and mammals is striking. In the therapsid-mammal lineage, the evolution of mammalian morphology, habit and endothermic metabolic physiology seem to have proceeded in a simultaneous, relatively seamless manner. Consequently, the latest therapsids and earliest mammals (Late Triassic–Early Jurassic Era) were probably not radically different in metabolic physiology and morphology, or in their general ecological niche from many of their extant descendants. This similarity suggests that selection for endothermy would have been sustained in the therapsid ancestors of mammals for many of the same criteria that benefit mammals today, including enhanced metabolic capacity to overcome the high net costs of terrestrial transport, thereby expanding the capacity for routine, and higher, levels of activity.

Given that the evolution of birds, unlike mammals, involved a radical departure in lifestyle from presumably flightless ancestors, perhaps it is to be expected that the development of the modern avian condition was not the synchronised, smooth sequence that seems to have characterised the evolution of mammals. Certainly, the fossil evidence for the evolution of the modern avian condition from archosaurian ancestors indicates a more abrupt, step-like sequence, with the aerodynamic plumage and capacity for powered flight appearing well before the evolution of fully-developed endothermy. Significantly, aspects of the anatomy and reproduction of modern birds also strongly suggest that flight preceded endothermy in birds (Randolph, 1994). If avian endothermy did not originate in association with the capacity for flight itself, why did it evolve? Previously mentioned aspects of metabolic physiology seem to rule out early selection for increased thermoregulatory capacity, and, in any case, Late Mesozoic Era climates were so mild and equable that even the earliest Jurassic birds were probably fully capable of relatively precise behavioural thermoregulation, similar to many extant reptiles. I have suggested previously that rather than evolving in association with flight itself, perhaps endothermy in birds appeared in response to selection for an increased capacity for long distance flight, an option unavailable to an ectothermic flyer. In this context, there is emerging evidence for widespread seasonal and/or chronic aridity around much of the Cretaceous world in which avian endothermy evolved (A. Boucot, personal communication). Many extant birds undertake longitudinal migrations to avoid regions experiencing extended dry periods. Perhaps modern avian metabolism and the capacity for long-distance flight evolved to facilitate similar migrations in the Cretaceous Era.

References

Bakker, R.T. (1971). Dinosaur physiology and the origin of mammals. *Evolution* **25**, 636–58.

Bakker, R.T. (1980). Dinosaur heresy-dinosaur renaissance. In *A Cold Look at the Warm Blooded Dinosaurs* (ed. R.D.K. Thomas & E.C. Olson), pp. 351–462. Boulder, CO: Westview Press.

Bakker, R.T. (1986). *Dinosaur Heresies*. New York: William Morrow.

Bang, B G. (1971). Functional anatomy of the olfactory system in 23 orders of birds. *Acta Anatomica* **79**, (suppl. 58), 1–76.

Barrick, R.E. & Showers, W.J. (1994). Thermophysiology of *Tyrannosaurus rex*: evidence from oxygen isotopes. *Science* **265**, 222–4.

Bennett, A.F. (1973). Ventilation in two species of lizards during rest and activity. *Comparative Biochemistry and Physiology* **46A**, 653–71.

Bennett, A.F. (1991). The evolution of activity capacity. *Journal of Experimental Biology* **160**, 1–23.

Bennett, A.F. & Dalzell, B. (1973). Dinosaur physiology: a critique. *Evolution* **27**, 170–4.

Bennett, A.F. & Ruben, J.A. (1979). Endothermy and activity in vertebrates. *Science* **206**, 649–54.

Bennett, A.F. & Ruben, J.A. (1986). The metabolic and thermoregulatory status of therapsids. In *The Ecology and Biology of Mammal-like Reptiles* (ed. N. Hotton, P.D. MacLean, J.J. Roth & E.C. Rothy), pp. 207–18. Washington: Smithsonian Institution Press.

Block, B.A. (1991). Endothermy in fish: thermogenesis, ecology and evolution. In *Biochemistry and Molecular Biology of Fishes* (ed. P.W. Hochachka & T. Mommsen), vol. 1, pp. 269–311. Amsterdam: Elsevier Science.

Bock, W.J. (1985). The arboreal theory for the origin of birds. In *The Beginnings of Birds* (ed. M.K. Hecht, J.H. Ostrom, G. Viohl & P. Wellnhofer), pp. 199–207. Eichstatt: Freunde des Jura-Museums.

Brand, D.B., Couture, P., Else, L., Withers, K.W. & Hulbert, A.J. (1991). Evolution of energy metabolism. *Biochemical Journal* **275**, 81–6.

Brink, A.S. (1956). Speculations on some advanced mammalian characteristics in the higher mammal-like reptiles. *Paleontologica Africaner* **4**, 77–95.

Buffrenil, V. de & Mazin, J.M. (1990). Bone histology of ichthyosaurs: comparative data and functional interpretation. *Paleobiology* **16**, 435–47.

Buhler, P. (1985). On the morphology of the skull in *Archaeopteryx*. In *The Beginnings of Birds* (ed. M.K. Hecht, J.H. Ostrom, G. Viohl & P. Wellnhofer), pp. 135–40. Eichstatt: Freunde des Jura-Museums.

Butler, P.J. (1991). Exercise in birds. *Journal of Experimental Biology* **163**, 233–62.

Camp, C.L. (1930). A study of the phytosaurs. *Memoirs of the University of California* **10**, 1–161.

Carrier, D.R. (1987). The evolution of locomotor stamina in tetrapods: circumventing a mechanical constraint. *Paleobiology* **13**, 326–41.

Chinsamy, A., Chiappe, L.M. & Dodson, P. (1994). Growth rings in Mesozoic birds. *Nature* **368**, 196–7.

Clausen, T., van Hardeveld, C. & Everts, M.S. (1991). Significance of cation transport in control of energy metabolism and thermogenesis. *Physiological reviews* **71**, 733–74.

Colbert, E.H. & Russell, D.A. (1969). The small Cretaceous dinosaur *Dromaeosaurus*. *American Museum Novitates* **2380**, 1–49.

Crompton, A.W., Taylor, C.R. & Jagger, J.A. (1978). Evolution of homeothermy in mammals. *Nature* **272**, 333–6.

Else, P.L. & Hulbert, A.J. (1987). Evolution of mammalian endothermic metabolism: 'leaky' membranes as a source of heat. *American Journal of Physiology* **253**, R1–R7.

Farlow, J.O. (1990). Dinosaur energetics and thermal biology. In *The Dinosauria* (ed. D. B. Weishampel, P. Dodson & O. Halszka), pp. 43–55. Berkeley: University of California Press.

Farlow, J.O., Dodson, P. & Chinsamy, A. (1995). Dinosaur Biology. *Annual Review of Ecology and Systematic* **26**, 445–71.

Feduccia, A. (1985). On why dinosaurs lacked feathers. In *The Beginnings of Birds* (ed. M.K. Hecht, J.H. Ostrom, G. Vohl & P. Welnhofer), pp. 75–9. Eichstatt: Freunde des Jura-Museum.

Feduccia, A. (1993). Bird like characters in the triassic archosaur *Megalancosaurus*. *Naturwissenschaften* **80**, 564–6.

Feduccia, A. (1995). Explosive evolution in Tertiary birds and mammals. *Science* **267**, 637–8.

Feduccia, A. (1996). *The Origin and Evolution of Birds*. New Haven: Yale University Press.

Garlick, P.J., Maltin, C.A., Bailee, A.G.S., Delday, M.I. & Grubb, D.A. (1989). Fiber-type composition of nine rat muscles. II. Relationship to protein turnover. *American Journal of Physiology* **257**, E828–E832.

Goldspink, F. (1972). Postembryonic growth and differentiation of striated muscle. In *Structure and Function of Muscle* (ed. G.H. Bourne), vol. 1, pp. 179–236. New York: Academic Press.

Goldspink, G., Marshall, P.A. & Watt, P.W. (1984). Protein synthesis in red and white skeletal muscle of carp (*Cyprinus carpio*) measured *in vivo* and the effect of temperature. *Journal of Physiology* **361**, 42.

Gudbarjarnason, S.B., Doell, S.B., Oskarsdottir, G. & Hallgrimsson, J. (1978). Modification of phospholipids and catecholamine stress tolerance. In *Tocopherol, Oxygen and Biomembranes* (ed. C. de Duve & O. Hayaishi), pp. 297–310. Amsterdam: Elsevier.

Hayes, J.P. & Garland, Jr., T. (1995). The evolution of endothermy: testing the aerobic capacity model. *Evolution* **49**, 836–47.

Hemmingsen, A.M. (1960). Energy metabolism as related to body size and respiratory surfaces, and its evolution. *Stereo Memorial Hospital Report* **9**, 1–110.

Hildebrand, M. (1995). *Analysis of Vertebrate Structure*, 4th edn. New York: Wiley.

Hillenius, W.J. (1992). The evolution of nasal turbinates and mammalian endothermy. *Paleobiology* **18**, 17–29.

Hillenius, W.J. (1994). Turbinates in therapsids: evidence for Late Permian origins of mammalian endothermy. *Evolution* **48**, 207–29.

Himms-Hagen, J. (1990). Brown adipose tissue thermogenesis: interdisciplinary studies. *FASEB Journal* **4**, 2890–8.

Horwitz, B.A. (1989). Biochemical mechanisms and control of cold-induced cellular thermogenesis in placental mammals. In *Advances in Comparative and Environmental Physiology* (ed. L.H. Wang), vol. 4, pp. 84–116. Berlin: Springer-Verlag.

Houlihan, D.F. (1991). Protein turnover in ectotherms and its relationships to energetics. In *Advances in Comparative and Environmental Physiology* (ed. R. Gilles), pp. 1–43. Berlin: Springer-Verlag.

Houlihan, D.F. & Laurent, P. (1987). Effects of exercise training on performance, growth, and protein turnover of rainbow trout (*Salmo gairdneri*). *Canadian Journal of Fisheries and Aquatic Science* **44**, 1614–21.

Hulbert, A.J. & Else, P.L. (1989). Evolution of mammalian endothermic metabolism: mitochondrial activity and cell composition. *American Journal of Physiology* **256**, R63–R69.

Hulbert, A.J. & Else, P.L. (1990). The cellular basis of endothermic metabolism: a role for 'leaky' membranes? *NIPS* **5**, 25–8.

Ismaili-Begi, F. (1988). Thyroid thermogenesis: regulation of (Na^{+}–K^{+})-adenosine triphosphate and active Na, K transport. *American Zoologist* **28**, 363–71.

James, R.S., Altringham, J.D. & Goldspink, D.F. (1995). The mechanical properties of fast and slow skeletal muscles of the mouse in relation to their locomotory function. *Journal of Experimental Biology* **198**, 491–502.

Josephson, R.K. (1993). Contraction dynamics and power output of skeletal muscle. *Annual Review of Physiology* **55**, 527–46.

Kemp, T.S. (1982). *Mammal-like Reptiles and the Origin of Reptiles*. London: Academic Press.

Kemp, T.S. (1988). Haemothermia or archosauria? The interrelationships of mammals, birds, and crocodiles. *Journal of the Linnaean Society, London* **92**, 67–104.

Lewis, E.M., Kelly, F.J. & Goldspink, D.F. 1984). Pre- and postnatal growth and protein turnover in smooth muscle heart and slow- and fast-twitch skeletal muscles of the rat. *Biochemical Journal* **217**, 517–26.

Lillegraven, J.A. & Krusat, G. (1991). Craniomandibular anatomy of *Haldanodon exspectatus* (Docodontia; Mammalia) from the Late Jurassic of Portugal and its implications to the evolution of mammalian characteristics. *Contributions to Geology, University of Wyoming* **28**, 39–138.

McNab, B.K. (1978). The evolution of homeothermy in the phylogeny of mammals. *American Naturalist* **112**, 1–21.

McNab, B.K. & Auffenberg, W.A. (1976). The effect of large body size on the temperature regulation of the Komodo dragon, *Varanus komodoensis*. *Comparative Bochemistry and Physiology* **55A**, 345–50.

Marden, J.H. (1994). From damselflies to pterosaurs: how burst and sustainable flight performance scale with size. *American Journal of Physiology* **266** (*Regulatory, Integrative and Comparative Physiology* **35**) R1077–R1084.

Munro, H.N. & Downie, E.D. (1964). Relationship of liver composition to intensity of protein metabolism in different mammals. *Nature* **203**, 603–4.

Nagy, K.A. (1987). Field metabolic rate and food requirements scaling in mammals and birds. *Ecological Monographs* **57**, 111–28.

Nagy, K.A. & Peterson, C.C. (1988). Scaling of water flux rate in animals. *University of California Publications in Zoology* **120**, 1–172.

Norberg, R.A. (1985). Function of vane asymmetry and shaft curvature in bird flight feathers; inferences in flight ability of *Archaeopteryx*. In *The Beginnings of Birds* (ed. M. Hecht, J.H. Ostrom, G. Viohl & P. Wellnhofer), pp. 303–18. Eichstatt: Freunde des Jura-Museums.

Norberg, R.A. (1995). Feather asymmetry in *Archaeopteryx. Nature* **374**, 221.

Ostrom, J.H. (1976). *Archaeopteryx* and the origin of birds. *Biological Journal of the Linnaean Society* **8**, 81–182.

Ostrom, J.H. (1990). Dromaeosauridae. In *The Dinosauria* (ed. D.B. Weishampel, P. Dodson & H. Omolska), pp. 269–79. Berkeley: University of California Press.

Ostrom, J.H. (1994). Wing biomechanics and the origin of flight. *Neues Jahrbuch Palaoritologie Abhandlungen* **19**, 253–66.

Padian, K. (1983). A functional analysis of flying and walking in pterosaurs. *Paleobiology* **9**, 218–39.

Parkes, K.C. (1966). Speculations on the origins of feathers. *Living Bird* **5**, 77–86.

Porter, R.K. & Brand, D.B. (1993). Body mass dependence of H^+ leak in mitochondria and its relevance to metabolic rate. *Nature* **362**, 628–9.

Putnam, R.W., Gleeson, T.T. & Bennett, A.F. (1980). Histological determination of the fiber composition of locomotory muscles in a lizard, *Dipsosaurus dorsalis. Journal of Experimental Zoology* **214**, 303–9.

Randolph, S.E. (1994). The relative timing of the origin of flight and endothermy: evidence from the comparative biology of birds and mammals. *Zoological Journal of the Linnaean Society* **112**, 389–97.

Regal, P.J. (1985). Common sense and reconstruction of the biology of fossils. In *The Beginnings of Birds* (ed. M.K. Hecht, J.H. Ostrom, G. Viohl & P. Wellnhofer), pp. 67–74. Eichstatt: Freunde des Jura-Museums.

Rowe, T. (1992). Phylogenetic systematics and the early history of mammals. In *Mammalian Phylogeny* (ed. F.S. Szalay & M.C. McKenna), pp. 129–45. Berlin: Springer-Verlag.

Ruben, J.A. (1991). Reptilian physiology and the flight capacity of *Archaeopteryx. Evolution* **45**, 1–17.

Ruben, J.A. (1993). Powered flight in *Archaeopteryx*: response to Speakman. *Evolution* **47**, 935–8.

Ruben, J.A. (1995). The evolution of endothermy: from physiology to fossils. *Annual Review of Physiology* **57**, 69–95.

Ruben, J.A., Bennett, A.F. & Hisaw, F.L. (1987). Selective factors in the origin of the mammalian diaphragm. *Paleobiology* **13**, 54–9.

Speakman, J.R. (1993). Flight capabilities in *Archaeopteryx. Evolution* **47**, 336–40.

Speakman, J.R. & S.C. Thompson. (1994). Flight capabilities of *Archaeopteryx. Nature* **370**, 514.

Spottila, J.R., O'Connor, M.P. & Paladino, F.V. (1991). Hot and cold running dinosaurs: size, metabolism and migration. *Modern Geology* **16**, 203–27.

Swope, S.J., Johnson, T.P., Josephson, R.K. & Bennett, A.F. (1993). Temperature, muscle power output and limitations on burst locomotor performance of the lizard *Dipsosaurus dorsalis. Journal of Experimental Biology* **174**, 185–97.

Van Mierop, L.H.S. & Barnard, S.M. (1978). Further observations on thermoregulation in the brooding female python *Python molurus bivattatus* (Serpentes: Boidae). *Copeia* **1978**, 615–21.

Vazquez, R.J. (1992). Functional osteology of the avian wrist and the evolution of flapping flight. *Journal of Morphology* **211**, 259–68.

Wittmer, L.M. (1992). *Ontogeny, phylogeny and airsacs: the importance of soft-tissue inference in the interpretation of facial evolution in* Archosauria. Ph.D. dissertation, Johns Hopkins University, Baltimore, MD.

A. CLARKE

The influence of climate change on the distribution and evolution of organisms

Introduction

The study of how organisms adapt to temperature has a long and distinguished history. During the past few decades much attention has been directed at understanding the mechanisms by which such adaptation is achieved. At the same time palaeobiologists have been attempting to unravel the influence of climate on the evolution and extinction of the Earth's biota, a concern heightened by the recent realisation that mankind's own activities may already have initiated a significant experiment in the impact of climate change.

In this chapter I discuss the impact of past climate change on organisms, using results from palaeobiology, environmental physiology and comparative ecology. A comprehensive review of these fields would require a whole book and so for this chapter I have necessarily been selective, concentrating on those areas that might help generate a coherent picture of how organisms respond to climate change. I have attempted to emphasise both areas of agreement and what appear, with current knowledge, to be significant mismatches between different fields of enquiry. In doing so I have tried to balance citation between key historical literature, comprehensive reviews and important recent work. Space constraints mean that usually only a single illustrative example can be used and I have quite deliberately chosen some of these from the older literature; it is not always the most recent work that is either the best or the most relevant.

Climate and temperature change

An increasing volume of evidence indicates that climate has varied significantly over historical time. The medieval warm period (roughly the 12th to 15th centuries) allowed the Vikings to colonise Greenland, and the Little Ice Age experienced in north-west Europe from the 15th to the late 18th centuries has been well represented in both literature and art. Such variability in climate is now becoming recognised wherever records go back far enough (for an example, see Hameed & Gong, 1994).

Studies of palaeoclimates have shown that temperature has never been stable; organisms have been faced with a climate that has changed continuously, albeit more slowly at some times than others. The resolving power of the different techniques for determining palaeotemperature varies and this scale-dependent resolution means that in general the further back one goes in time the smoother the variation in climate appears to be. Thus Milankovitch cyclicity is generally evident only in more recent deposits, although it is likely that the variations in the earth's orbit inducing this cyclicity have occurred throughout evolutionary time (Ruddiman *et al.*, 1989; Bennett, 1990).

Perhaps the most significant result of recent work has been the indication that atmospheric climate may shift very rapidly between periods of relative stability (Crowley & North, 1988). The first evidence for such rapid shifts in climate came from studies of subfossil Quaternary insects (Coope, 1979, 1995; Elias, 1994) but this important aspect of climate change only became fully established with the advent of more detailed information from analysis of polar ice-cores (Dansgaard *et al.*, 1989; Lorius *et al.*, 1990).

Until recently the only ice-core record of atmospheric climate reaching a significant period into the last glacial cycle was the 2083 m core drilled from 1972 to 1983 by the Soviet Union at Vostok, Antarctica, close to the Pole of Inaccessibility. In the summer of 1992, however, the European Greenland Ice-core Project (GRIP) reached the silty ice above bedrock below the summit of the Greenland ice sheet. A parallel core was drilled 30 km to the west by the United States Greenland Ice-Sheet Project (GISP2) and together these cores have changed our view of the Earth's climate over the past 35 000 years. Measurements of stable isotope ratio (O^{16}/O^{18} and H^2/H^1) of the ice, conductivity (a measure of dust content) and greenhouse gas concentrations in trapped air have shown that, in contrast to the relative stability of the Holocene, the previous interglacial (Emian) was characterised by a series of intermittent cold periods which ended abruptly (GRIP members, 1993). The end of the Younger Dryas glacial event in the Holocene also appears to have been sudden, with the time-scale of the switch to interglacial conditions being estimated variously at approximately 50 years (oxygen isotope data), approximately 20 years (dust records) or as little as 1–3 years from changes in accumulation rate (Alley *et al.*, 1993). Overall the emerging picture is one of a marked instability of climate, particularly during periods of change in mean climate, and quite likely of at least regional extent (Dansgaard *et al.*, 1993). This instability has been characterised as akin to a 'flickering switch' (Taylor *et al.*, 1993), and suggests the possibility that in terms of impact on flora and fauna, extreme events may be more important than changes in mean climate.

Although there are striking correlations between the Greenland and Vostok ice-core records, there are also strong contrasts. In particular the Vostok record exhibits long periods dominated by smooth changes (Jouzel *et al.*, 1987) which can be explained only partially by the coarser temporal resolution of the measurements in this earlier core.

The marine sediment core record also tends to show less variability than the Greenland ice-cores, although some evidence of sudden climatic events in the sea is now being detected (Kennett & Stott, 1991). These features are well illustrated by the example of the Cenozoic temperature record for the Southern Ocean which shows that seawater temperature has changed continuously through the past 65 million years (Fig. 1). In the Tertiary there have been five significant periods of warming, and at least one short-term temperature excursion, superimposed on an overall cooling since the early Eocene.

Overlayed on these longer term variations will be shorter term variations, including Milankovitch cyclicity, on time-scales not well resolved in Fig.1 but perhaps numbering up to 50 in the past 2.4 million years (Ruddiman *et al.*, 1989). At still shorter temporal scales

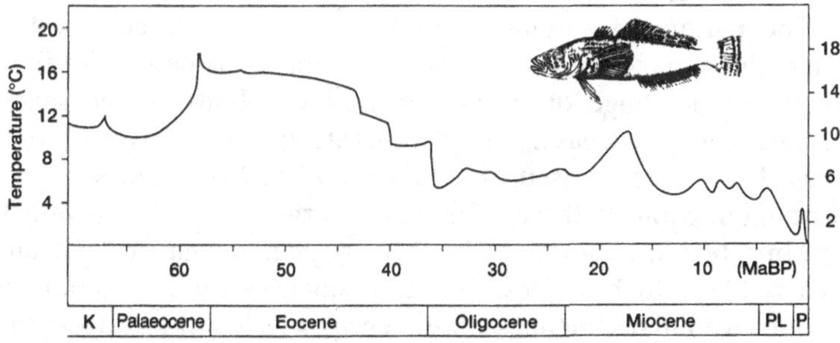

Fig. 1. Southern Ocean palaeotemperatures during the Cenozoic era. Temperatures were estimated from oxygen isotope fractionation in skeletal carbonate of foraminifera from sediment cores taken from a variety of locations in the Southern Ocean. The line shows a best fit for long-term temperature change. Superimposed on an overall cooling since the early Tertiary are several periods of both global warming and sharp cooling, and at least one short-term warming event. (After Crame (1993), where detailed citations to sources may be found.)

there are also sub-decadal interannual variations driven by atmospheric processes (and especially the El Niño Southern Oscillation; Murphy *et al.*, 1995), together with seasonal, diurnal and tidal changes.

This well-documented climate history demonstrates clearly that over evolutionary time the environment has posed a constantly shifting thermal challenge to marine organisms living around the fragment of Gondwana that now forms Antarctica. This is a general result, and it emphasises the critical point that coping with mean climate is more a question of tracking continual change than of adapting to a fixed temperature.

The combination of long-term records of climate from ice cores, sediment cores, peat cores and tree-ring analysis, together with the advent of high resolution temperature recorders has thus revealed that the thermal regimen experienced by terrestrial and marine organisms varies over almost all time-scales. These variations are not strictly fractal in nature for the extent of variability itself varies with temporal scale, and this points to a range of forcing variables. They also require a spectrum of responses by the organism.

There are, however, important differences between the thermal regimens of aquatic and terrestrial habitats. The lower viscosity and thermal mass of air compared with water means that the temperature changes experienced by terrestrial organisms are generally much faster and more extreme than for aquatic organisms. In some regions of the ocean (for example the polar oceans and the deep sea) the temperature may be very stable indeed, at least on the time scale of available data.

The pattern of variability itself also differs between land and sea. Thus the variance of the atmosphere appears to be effectively constant over a wide range of scales, whereas variability in the sea tends to decrease with increasing frequency (Steele, 1978). These two patterns have been characterised as 'white noise' and 'red noise', respectively. The interaction of these differing patterns of variability with the contrasting temporal scales of life-history phenomena on land and in the sea is likely to have powerful consequences for population processes in the two environments, consequences which will in turn influence the ways in which climate change will impact the organism living there, but which we are only beginning to glimpse.

Responses to temperature change

The only two options open to organisms in the face of temperature change are to move elsewhere, or to stay put and cope. Where temperature changes on a time-scale significantly shorter than the life-span of an individual (for example tidal, diurnal or seasonal temperature changes) the organism may respond by moving elsewhere, or by some form of phenotypic adjustment to maintain viability; for plants or sessile animals only the latter is an option. When rates of change are slower, significant shifts in mean temperature will clearly take place over time-scales far longer than the life-span of an individual organism. In this case the organism may shift its distribution, adjust through some form of evolutionary temperature compensation, or become extinct.

This distinction between phenotypic adjustment and genotypic (evolutionary) compensation is fundamental and yet is frequently ignored or misunderstood; for example when results of experiments involving acute temperature change to eurythermal organisms are used to predict the impact of long-term climate change or interpret past patterns of evolution. The way a eurythermal organism has evolved to deal with short-term temperature variation may give an insight into the mechanism of long-term evolutionary temperature adaptation, but equally it may not (Clarke, 1991). A phenotypic response (acclimatisation) to a short-term thermal challenge may involve metabolic costs that would render the organism uncompetitive in a less variable environment (Hoffmann & Parsons, 1991) or over evolutionary time.

This chapter is concerned primarily with long-term climate change and so acclimatisation will not be discussed further; thorough reviews of phenotypic acclimatisation are given by Hochachka & Somero (1984) and Cossins & Bowler (1987).

Changes in distribution

To understand how changes in distribution have resulted from past climate change it is necessary first to establish the role of temperature in regulating the distribution of organisms.

Does temperature influence the distribution of organisms at the global scale?

At the global scale perhaps the most striking pattern of the distribution of terrestrial organisms is the latitudinal cline in species diversity. There is no generally agreed explanation for this cline (for differing recent discussions, see Huston, 1994; Rosenzweig, 1995), although given the parallel cline in climate it is perhaps inevitable that some of the most frequent correlations examined are with some measure of temperature.

A number of features of the polar climate (of which temperature is only one) combine to create a genuinely harsh environment for high latitude terrestrial plants and animals, and some organisms are absent totally from these regions. If the ability of a species to expand its distribution in a poleward direction is determined by its ability to survive some critical temperature, then it might be predicted that taxonomic diversity would decline towards the poles. Woodward (1987) tested this hypothesis for plants by examining the latitudinal variation in the distribution of 313 families of angiosperms and arboreal gymnosperms. When data were pooled by bins of 15° of latitude there was a striking cline in diversity from tropics to poles (Fig. 2*a*). There was also a highly significant correlation between family level diversity and absolute minimum temperature recorded for that latitudinal bin, although the relationship differed between the northern and southern hemispheres (Woodward, 1987). Despite the intuitive reasonableness of a relationship between overall diversity and temperature, the difficulty with this argument was pointed out long ago by Hutchinson (1959): if one species can adapt successfully to polar regions, why cannot another?

Not all taxa exhibit quite as striking a symmetrical diversity cline as plants. In North America, birds, mammals and amphibians, for example, often show a peak in species richness at intermediate rather than tropical latitudes, and Stevens (1989) provides a list of taxa that show no obvious diversity cline. Although the relationship between diversity and latitude appears to be simple, its explanation may be complex (a not uncommon occurrence in ecology); starting with the list collated originally by Pianka (1966) at least 14 different explanations have been advanced to explain this deceptively simple pattern (Stevens, 1989; Pagel *et al.*, 1991; Huston, 1994). Despite the striking correlations which can be obtained between diversity and temperature, in the absence of a functional explanation these remain simply correlations between two variables that co-vary with latitude.

A recurring second theme in discussions of the latitudinal diversity cline on land has been the relationship between diversity and area. Indeed Rosenzweig & Abramsky (1993) argue that the explanation for global patterns of diversity is essentially the combined effect of the species/area relationship and allopatric speciation (Terborgh, 1973; Rosenzweig, 1975, 1977).

A simple model assuming a random association between latitude and the size and placement of species ranges can produce clear, bell-shaped diversity clines centred on the equator (Colwell & Hurtt, 1994). This result is a straightforward consequence of geometry, but does suggest that area may be a factor in large-scale patterns of diversity. Although simple (linear) correction for land area does produce a more or less

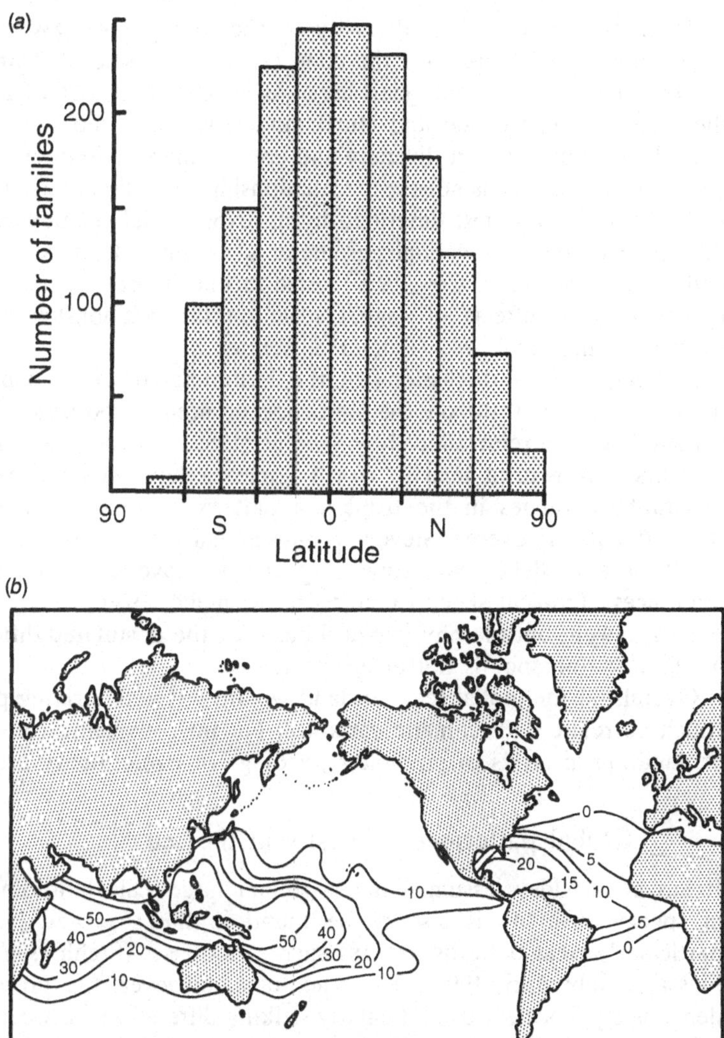

Fig. 2. (*a*) A clear biogeographical pattern in taxonomic richness that has often been explained in terms of direct control by climate: the latitudinal distribution of 313 families of angiosperm and arboreal gymnosperm plants, pooled by bins of 15° of latitude. (After Woodward, 1987.) There is no consensus on the explanation for this deceptively simple pattern. (*b*) Taxonomic richness (genera) contours for reef-building scleractinian corals, showing a concentration of richness in the Indo-West Pacific leading to a pronounced longitudinal (East–West) cline in diversity in addition to the well known latitudinal cline. (After Stehli & Wells, 1971.)

uniform pattern for plant diversity at the family level (Woodward, 1987), this does not remove the cline at the species level. Correcting diversity data for available land area (as against total surface areas of the globe in the Colwell and Hurtt model) is only valid, however, if the relationship between diversity and area is linear. Numerous studies have shown that this is not so; the relationship between species richness and area is usually best described by an exponential relationship with the value of the exponential parameter generally falling between 0.2 and 0.4 (Connor & McCoy, 1979). Also recent theoretical studies have revealed subtle effects of spatial scale on the relationship between species richness and area (Palmer & White, 1994).

Although recent discussions of the latitudinal diversity cline have tended to concentrate on the influence of area, a possible role for climate has been reintroduced by Stevens (1989). Stevens demonstrated that many terrestrial taxa in North America tend to have narrower geographical ranges in the tropics, a pattern he termed 'Rapoport's rule' after its discoverer. Stevens proposes that the pattern may be the result of a tendency for tropical species to have narrower climatic tolerances. Tropical species tend to have smaller overall geographical ranges as well, but the physiological basis for the postulated differences in climatic tolerance is not at all clear.

Overall, at the global scale on land we are left with striking patterns which correlate with climatic variables, but no convincing (or agreed) explanation in terms of physiology, ecology or evolutionary processes.

Global patterns of diversity in the sea

It has been widely assumed that a similar large-scale latitudinal cline in species diversity is also to be found in the sea, and there are particularly clear patterns for taxa such as molluscs (Stehli *et al.*, 1967; Flessa & Jablonski, 1995). For other taxa, however, the present evidence is equivocal, with particularly striking differences between different groups of organisms and also between the northern and southern hemispheres (Clarke, 1992; Clarke & Crame, 1996). These differences, and the pronounced longitudinal (east–west) cline in many marine taxa resulting from the exceptionally high species richness in the Indo-West Pacific (Fig. 2*b*), indicate that other factors are involved (Clarke & Crame, in press). The explanation for these distinctive patterns is still unclear, although the evidence suggests that we can rule out a simple regulation by temperature.

Stevens (1989) extended his examination of Rapoport's rule to the sea, but was unable to find many convincing examples. Although shallow water molluscs were suggested as a possible example, a recent detailed examination of the well-known eastern Pacific molluscan fauna found no evidence of any relationship between range size and latitude (Roy *et al.*, 1994). This emphasises the difficulties inherent in trying to map concepts developed in terrestrial ecology to the sea. Not only is the biogeographical database much less extensive, but we cannot be sure that the underlying explanation in terms of climatic tolerance would even predict the expected pattern: patterns of temperature variability in the shallow sea are very different from those on land.

Temperature and meso-scale distribution

The influence of temperature on the distribution of organisms appears to be much clearer at the meso-scale. Perhaps the most persuasive evidence for a role for temperature in the sea is that many marine species reach the edge of their distributions where temperature changes rapidly over a short distance; indeed in many cases these zones are used to define the boundaries to biogeographical provinces (Angel, 1994).

A particularly clear example is that of bivalve and gastropod molluscs on the Pacific seaboard of North America. Ignoring local anomalies, the coastal waters of the northeastern Pacific shelf range from cold temperate at the northern end to subtropical at the southern end. The species composition of the coastal molluscan fauna parallels this change with the assemblages containing predominantly cold water species to the north and predominantly warmer water species to the south. Detailed numerical analyses of the latitudinal distribution of these molluscs have shown that there is a narrow zone of rapid faunal change extending over about 5° of latitude and centred on Point Conception (Newell, 1948; Valentine, 1966). Point Conception thus marks a boundary zone where ecological conditions prevent the warmer water Californian Province species from living further north, and cooler water Oregon Province species from spreading further south. The generally accepted explanation for this pattern is the markedly steep temperature gradient in this region (Hedgpeth, 1957) and although the coincidence of rapid faunal change with a steep thermal cline does suggest a role for temperature, the impact of oceanographic factors on larval dispersal and recruitment will also likely be important.

A primary role for temperature is, however, suggested by changes to the distribution of this fauna both in recent historical times, and over geological time. Valentine (1961) has shown how the distribution of these molluscs shifted in response to climatic changes during the

Pleistocene, and Barry *et al.* (1995) have recently demonstrated shifts in distribution coincident with an increase in mean shoreline ocean temperature of 0.75 K over the last 60 years.

Although it is usually assumed that temperature influences the distribution of animals, there are relatively few examples where the distribution of a species has been shown to be related directly to some measure of thermal performance or tolerance. There are, however, a great number of examples where temperature physiology can be shown to correspond to local thermal conditions (for reviews, see Hochachka & Somero, 1984; Cossins & Bowler, 1987). In very few cases, however, has the underlying genetic mechanism been examined (for a review, see Gillespie, 1991), a notable exception being the study of the genetic basis of temperature tolerance in Australian species of *Drosophila* (Hoffmann & Watson, 1993; Watson & Hoffmann, 1995). Where such correspondence has been demonstrated, it is taken to reflect the expected adjustment of species to local thermal conditions. As such it is a trivial result, indicating no more than local adaptation. It is important, however, in demonstrating the outcome of evolutionary processes. Clines in genetic composition associated with temperature and believed to reflect local adaptation have also been demonstrated *within* species (see pp. 53–78; Powers *et al.*, 1986).

For plants, distribution is related to a whole suite of climatic factors in addition to temperature and sophisticated models have been developed to relate plant distribution to climate (Woodward, 1987; Gates, 1993 and references therein) and to climate change (Davis & Zabinski, 1992; Henderson-Sellers & McGuffrie, 1995). As with animals there is a general correspondence between physiology and local climate (for a review, see Crawford, 1989), although there are far fewer such studies than for animals.

Does temperature limit distribution?

Despite the existence of a general correspondence between thermal performance and geographical distribution, there are relatively few cases where temperature can be shown unequivocally to be the factor defining the limits to distribution. The problem is that we cannot yet define precisely what processes dictate the edge of a species' distribution; although temperature is often implicated we cannot be certain that it is not some other factor(s) that covaries with temperature. To take just one example, many of the most endangered species of terrestrial invertebrate in the UK are at the northern edge of their distribution. Many of these species are characteristic of the warmest habitats (often early seral stages and frequently man-made habitats). This suggests clearly an ultimate role for temperature in limiting distribution (Thomas & Morris, 1995), although the detailed distribution and the edges of the range in the UK are more likely to be set by proximate factors involving population dynamics.

May (1975) has suggested that geographical range sizes might best be described as logarithmically distributed; this would suggest that they are an example of the central limit theorem, the distribution arising from the interplay of many independent factors. It is now generally agreed that the abundance and range of an organism is dictated by a suite of factors of which climate is just one (albeit an important one in many cases), and it is likely that interaction with other species is also an important factor.

It is a general observation that an organism towards the edge of its range does not reproduce successfully and the population is maintained by immigration of propagules or juveniles from elsewhere. The edges of the distribution are thus occupied by *sink* populations maintained by more central *source* populations (Fig. 3*a*), which usually also show higher abundances (Lawton, 1993). The edge of the range appears to be set by competitive interactions with nearby species.

Even within the centre of the range, the distribution of organisms is often patchy, reflecting an underlying heterogeneity in the habitat (Fig. 3*b*). An excellent example of a natural population segregated into source and sink areas is the desert annual plant *Stipa capensis* (Kadmon & Shmida, 1990; Kadmon, 1993). This mosaic of metapopulations has profound consequences for population dynamics, and especially for the processes of dispersal, colonisation, evolution and extinction (Kareiva, 1990; Hanski, 1991; Doak *et al.* 1992).

Single species metapopulation models have largely confirmed the intuitive observation that an organism is better protected against stochastic events if it is present in a small number of large abundant populations than a large number of small and scattered populations. Analyses have, however, revealed powerful threshold effects for extinction (for a recent review, see Kareiva & Wennergren, 1995), and the difficult problem of multispecies metapopulation dynamics is now being tackled. Almost all work in this area is being undertaken in the more tractable terrestrial environment. We are still largely ignorant of the role of metapopulation dynamics in the sea, where scales of heterogeneity and the dynamics of dispersal are very different (Steele, 1991). It is likely, however, that an understanding of marine metapopulation dynamics will be central to determining what causes extinction in the sea.

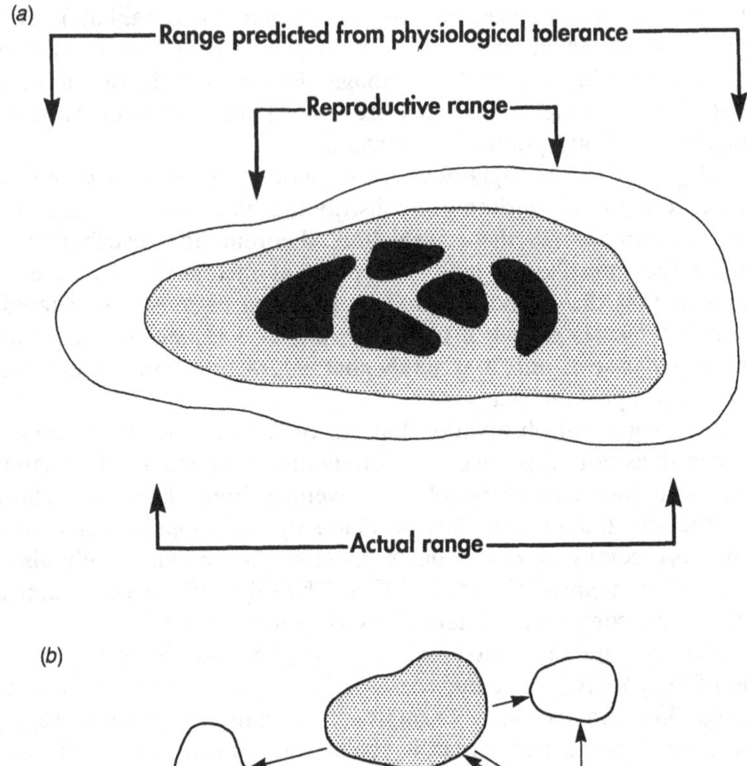

Fig. 3. (*a*) Conceptual model of the distribution of a species in relation to an environmental variable such as temperature. The range predicted from physiological tolerance (clear area) is significantly greater than that actually occupied. Furthermore, within the occupied range the outer *sink* areas (stipple) are maintained by immigration of individuals or propagules from the more central *source* populations (black). The latter themselves are patchily distributed in a mosaic of metapopulations. (*b*) Conceptual model of metapopulation structure, showing isolated populations occupying areas of habitat (stippled areas) and other areas of suitable habitat currently unoccupied; in between are areas of unsuitable habitat. The arrows show the processes of dispersal and colonisation. The dynamics of extinction, colonisation and dispersal between these metapopulations have profound consequences for population genetic structure, and overall population dynamics including the likelihood of overall extinction (Kareiva, 1990; Hanski, 1991).

Species borders

When a species' distribution changes, the borders shift. A number of studies have shown suboptimal performance in marginal populations (for a recent review, see Hoffmann & Blows, 1994), so the nature of species borders is of some evolutionary significance. It is relevant to ask for example, what prevents a marginal population from evolving and expanding its range, a question which echoes Hutchinson's (1959) query as to why there are not more polar plants. A distribution may well map onto mean environmental conditions (as in a cline of related taxa along an environmental gradient), but the edges may be defined by the frequency of extreme events or competition. Transplant and reciprocal transplant experiments have revealed differences in performance between marginal and core populations (Hoffman & Parsons, 1991) and marginal populations often display low fitness except during times of stress. It is possible that in periods of enhanced climate change it is the marginal populations that are most critical to a species' ability to cope (Blows & Hoffmann, 1993). As Hoffmann & Blows (1994) have emphasised, there is an urgent need for more work on the genetic basis of performance in marginal populations.

Has past climatic change induced changes in distribution?

Although the distribution of many species is relatively stable on the short temporal scale of human perception (and the even shorter time-scale of most ecological research: Pimm, 1994), the distribution of organisms is actually quite dynamic and ranges are changing continually in response to variation in many aspects of the environment. In recent years most of the more striking examples have related to changes induced by humans but it is the recent fossil record that provides the best evidence for climate-induced changes in distribution.

The vagaries of the fossil record mean that when a given taxon disappears from a particular rock or peat sequence it is always difficult (and often impossible) to distinguish between extinction and a shift in distribution. Throughout most of the fossil record of life, disappearance is taken to mean extinction. It is only where the record is exceptionally good that changes in distribution can be demonstrated, and classic examples are provided by the Pleistocene record of the molluscan fauna of the Pacific seaboard of North America, Coleoptera (beetles) from Europe, and vegetational changes in the northern hemisphere following the end of the last glacial period.

Valentine (1961) has documented in detail the changes in distribution of the marine molluscan fauna of California during the Pleistocene, and Barry *et al.* (1995) have shown that distributional changes are currently underway in the Californian intertidal. Both of these are associated with changes in seawater temperature, and temperature is also clearly implicated in the distributional changes associated with El Niño (Arntz & Fahrbach, 1991) and in changes of the rocky shore fauna of north-west Europe (Southward, 1991).

The Pleistocene was a time of rapid climatic change marked by the development of a glacial climate over most of the northern hemisphere. At the height of the glaciation a single ice-sheet covered much of northern Europe, including almost all of the British Isles. The development of the Pleistocene ice-sheet was episodic, and there have been a number of glacial advances and retreats throughout the past 1.5 million years. During the warmer interglacial periods the newly exposed ground was recolonised, and ecological communities shifted northwards as the climate warmed. These switches from glacial to interglacial may have been rapid at times; thus at about 13 000 years ago mean summer temperatures may have risen by at least 1 K per decade, with the switch from fully glacial to fully interglacial conditions possibly occurring in some areas within a single human lifetime (Coope, 1995).

The peats and muds deposited in northern Europe from Pleistocene bogs and lakes contain many subfossil beetles in an excellent state of preservation. All of the beetles examined so far are identical morphologically to living species, and closely similar to Miocene species. A detailed analysis of these remains has shown clearly that the response of the beetle fauna during the Pleistocene climatic cycles has been neither to become extinct nor to evolve (at least morphologically), but to shift distribution. As the climate warmed and cooled, the beetle assemblages simply shifted distribution. This does not mean that individual beetles moved great distances, for the rate of change was generally slow compared with the life-span of these beetles (which typically have an annual life-history), but that there was a slow shift in the whole assemblage of which these beetles formed an integral part.

Although the palaeontological record of plant distribution is extensive, the most complete picture of climate-related changes in plant distribution comes from the period since the last glacial maximum. The primary data come from analyses of pollen in lake sediments and bog peats, and these have provided a detailed picture of the changes in the distribution of plants in response to climatic variations in both North America and northern Europe. There is insufficient space to review this extensive literature here, but two generalisations are now firmly established.

First, the broad features of plant communities can be predicted in general terms from the current generation of vegetational distribution models (Prentice, 1990) but history has shown that communities *per se* do not migrate. It is individual species whose distribution changes, in relation both to climate and as modified by interactions with other species (Davis, 1989). The concept of plant communities as objective, repeatable entities is thus not supported by history, for some communities have had different compositions in the past (Gates, 1993). Although steady-state bioclimatic distributional models can reproduce the broad features of existing vegetational zones, we are unlikely to be able to model successfully either the impact of enhanced climatic warming, or to reproduce the detail of past distributional changes, without incorporating ecological interactions, dynamics and feedback effects into the models (Henderson-Sellers & McGuffrie, 1995).

A second generalisation is that changes in distribution may lag climate change, and colonisation of new areas by plants may be held back by factors such as geographical barriers to dispersal, or insufficiently developed habitat (Pennington, 1986; Mayle & Cwynar, 1995). Migration is, however, not an alternative to evolutionary change, for new areas may present new physiological or ecological challenges requiring adaptation (Geber & Dawson, 1993).

Fragmentation, refugia and speciation

There is thus ample evidence from the fossil record of marine invertebrates, terrestrial insects and plants of changes in distribution in relation to climate change. Such distributional changes are not always simple and straightforward, however, and physiographical factors may lead to range fragmentation.

Clearly assemblages of organisms can shift only if there is somewhere to move. Thus for tropical species a general cooling may mean that there are no suitable habitats available, and a simple shift in distribution is not an option. Historical biogeographers have long recognised the importance of refugia as areas where species can survive until conditions change once more (for a succinct review of this huge topic, see Lynch, 1988, and for a discussion of the role of refugia in relation to glaciation, see Pielou, 1991). Refugia are typically small in comparison with the original distribution, so confinement of a species to one or more such refugia involves a reduction in overall population size, and often fragmentation of that population. Both have important consequences for processes of evolution, speciation and extinction, particularly when a fauna is subject to repeated climatic cycles.

A good example of a fauna subject to such repeated climatic cycles is that of the Southern Ocean, which has probably been subject to glaciation for over 30 million years, a much longer period than the northern hemisphere. During this period the ice-sheet has expanded and contracted several times, occasionally extending to the edge of the narrow continental shelf around Antarctica. At other times the ice-sheet has contracted sufficiently to allow the development of shallow epicontinental seas between East and West Antarctica (reviewed by Clarke & Crame, 1989).

These climatic cycles have resulted in the repeated increase and destruction of habitat for continental shelf benthos and fish. At times of maximum ice extension many taxa will have been confined to small and scattered refugia; in some cases this will have involved a shift in distribution to deeper water (Brandt, 1991, 1992). This repeated fragmentation of populations followed by recolonisation is a process of major importance in allopatric speciation. Populations confined to scattered refugia may diverge genetically as a result of several factors including founder effects, drift and local adaptation. When these populations expand and once more come into contact they may have evolved sufficiently far for reproductive barriers to have formed, and speciation to occur (Fig. 4). Repeated cycles of expansion and contraction may thus lead to an enhanced rate of speciation, the so-called taxonomic diversity pump (Valentine, 1967, 1984; Crame, 1993). It is analogous to (but not identical with) the taxon cycle (Ricklefs & Cox, 1972). There may have been as many as 50 Milankovitch-driven climatic cycles in the past 1.5 million years, and there is good evidence that this has been an important process in the evolution of the Southern Ocean benthic marine fauna leading to a high diversity for some taxa (Crame, 1993; Clarke & Crame, 1992, 1996).

Although the operation of the diversity pump in the Southern Ocean may be related to the relatively high species richness of the shallow water benthic fauna found there, this is a process superimposed on the process of evolutionary adaptation to the overall decrease in water temperature which has characterised the Cenozoic history of the area.

Evolutionary adaptation

Experimental studies of eurythermal organisms can tell us about the mechanism of acclimation, but the best insights into mechanisms of

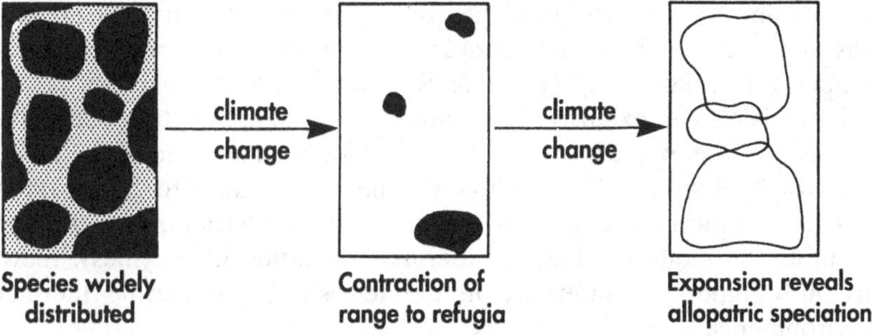

Fig. 4. Conceptual model of the impact of climatic cycles on population distribution. Initially a species is widely distributed over an extensive area (left), with the population exhibiting classic metapopulation structure comprising some areas of high abundance and successful reproduction (represented by the black areas) and sink areas of unsuccessful reproduction (stippled areas). Following climatic change the range contracts to a small number of refugia, such as areas protected from an expanded continental ice shelf (centre). Following further climate change, such as a contraction of the ice sheet, the refuge populations can once more expand, coming into contact (right). Under these circumstances, genetic differences accumulated during the allopatric contraction to refugia may be sufficient to prevent interbreeding and hence result in speciation. When repeated cyclically this process can lead to an increase in taxonomic richness, and has been termed the *climatic diversity pump* (Valentine, 1968; Crame, 1993).

evolutionary adjustment to temperature come from comparative studies of organisms from different thermal regimes. The present climate regimen is distinguished by a steep cline in temperature from tropics to poles both on land and in the sea; this cline reflects the Pleistocene glaciation and is not typical of evolutionary history as a whole. It does, however, provide an opportunity to examine the mechanisms of temperature adaptation, and the limits and constraints to such adaptation.

A classic example of thermal adaptation in physiology is the positive correlation between the optimal temperature for sprinting and the average field body temperature (itself a function of environmental temperature) of iguanid lizards (Huey & Kingsolver, 1989, 1993). This relationship holds over a range of optimal sprint temperatures of 15 K, and remains highly significant after control for the effects of phylogeny. Similar patterns are found in Australian lizards (Garland *et al.*, 1991). Interestingly, no correlation was found between critical thermal maxima and minima, suggesting that the process of evolutionary adjustment was more complex than a simple shift in the rate (performance)/temperature relationship (Huey & Kingsolver, 1993).

There is also an extensive literature on thermal acclimation in marine organsims (see pp. 53–78 and 127–52; Hochachka & Somero, 1984; Cossins & Bowler, 1987). Although there are likely to be features common to thermal adaptation in marine and terrestrial organisms (for example the molecular basis of thermal adaptation in enzymes), there are also important difference in the terrestrial and marine thermal environments.

The thermal challenges to physiology in terrestrial organisms differ significantly from those of aquatic organisms, largely as a result of the large thermal mass of water that constrains the rate of temperature change in the sea. Polar plants, for example, may experience daily variations in excess of 50 K (Lewis Smith, 1988), and seasonal changes may be in excess of 70 K (Miller & Werner, 1987), whereas marine organisms are subject to much slower changes and a narrower daily and seasonal range. The physiological challenges differ for organisms in terrestrial and marine environments, so we should not necessarily expect the evolutionary responses to be the same and we should extrapolate between the two environments with care.

Comparison of polar, temperate and tropical organisms shows clearly that extensive compensation for the effects of temperature on physiological processes has evolved. In some cases compensation appears to be very good, in others there is clear evidence of constraints which influence both the nature and extent of compensation (Clarke, 1991; Clarke & Johnston, 1996).

The extensive palaeotemperature data now available for the Southern Ocean (Fig. 1) provide the time-scale over which evolutionary adaptation to polar temperatures has occurred in the marine flora and fauna. During the past 60 million years the rate of temperature change has varied widely, and this prompts the question of the rate at which evolution can proceed in response to environmental change. This is a fundamental question (indeed Peters & Darling, 1985, have queried whether evolution will be able to match present climate change at all), and attempts to model evolutionary adaptation in terms of population genetics go back to Fisher (1930). Recent concerns over climate change have given a renewed impetus to these studies, and a sophisticated model of evolutionary change in response to climate change has been developed by Lynch & Lande (Lande, 1976; Lynch *et al.*, 1991; Lynch & Lande, 1993).

This model considered a single quantitative polygenic trait initially under stabilising selection, explicitly including both the genetic and environmental contributions to phenotypic variance. Utilising diffusion theory it was shown that an infinite sexual population under directional selection would develop a lag between the mean population performance and the optimum environmental value. The steady-state value of this lag was shown to be directly proportional to the rate of environmental change, and inversely proportional to the genetic variance in the trait and to the strength of stabilising selection (Lynch & Lande, 1993).

We have therefore the possibility of predicting rates of environmental change that will exceed the capacity of the population to respond and hence lead to extinction. Thus for a population with a maximum intrinsic rate of increase of 0.5 per generation and a realistic fitness function (a measure of how rapidly fitness declines with deviation of a phenotypic character from the optimum), then the model suggests that the population cannot sustain a prolonged change in the optimum value of the selected trait greater than about 0.1 phenotypic standard deviations per generation.

Huey & Kingsolver (1993) have used this model to consider the specific case of the evolution of thermal performance in the face of climate change. This analysis confirmed the intuitively reasonable expectation that the ability of a population to adapt to climate change will depend on, among other factors such as population size, the rate at which the environment changes; too rapid a change may result in a lag between performance and environmental optimum which reduces population fitness to zero and results in extinction. Examination of the effects of a trade-off between optimum performance and performance breadth (phenotypic variance) indicated a complex interaction between the rate of environmental change, performance breadth and genetic variance. The contribution of genetic variance to thermal performance breadth in particular appears to be critical in determining whether thermal specialists or thermal generalists will be favoured during periods of rapid climate change (Huey & Kingsolver, 1993).

Extinction: the failure of adaptation

The history of life shows a number of episodes where rapid climate change appears to have resulted in widespread extinction. The rate of extinction has varied widely through time, but statistical analysis demonstrates five distinct periods of enhanced extinction (Hubbard & Gilinski, 1992). These 'big five' mass extinction events have attracted a great deal of attention but there is still no general consensus as to whether mass extinctions are simply extremes of a continuum of extinction rates and hence are definable only in statistical terms (Raup, 1991), or whether they represent events different in kind from background extinction (Jablonski, 1986*a*, *b*).

Although the mass extinction at the end of the Cretaceous (the K/T event) appears to be coincident with the impact of a major bolide (Alvarez *et al.*, 1980), climate change remains a favoured hypothesis for many episodes of greater than average extinction. A recent compilation by Donovan (1989) lists climate change as a potential causative agent for five of the nine largest extinction events known.

The generally accepted model for extinction is that interaction with a competitively superior species reduces the population to a level where stochastic environmental or genetic processes result in the death of all individuals. Lawton (1995) has listed four proximate causes for the extinction of a small population, namely demographic and environmental stochasticity, genetic deterioration and social dysfunction. These processes, however, simply provide the *coup de grâce* to populations already at low abundance; they are quite distinct from those processes which made the species rare in the first place. Recent theoretical work on metapopulation dynamics has indicated that the population abundance threshold for such extinctions may be quite high (Kareiva & Wennergren, 1995). Human activities have resulted in the fragmentation of many populations and habitats, so the interaction of metapopulation processes with spatial heterogeneity is of great importance to conservation biology (Tracy & George, 1992; Pickett & Cadenasso, 1995).

Most current work on the process of extinction is based in the terrestrial environment where species' ranges and abundances are much better documented, and manipulative experiments (for an excellent example, see Ives *et al.*, 1993) are much easier than in the sea. Indeed, there are very few documented cases of extinction of a marine species in historical times (see the recent compilation by Pimm *et al.*, 1995). The marine environment does have, however, a rich record of historical extinctions over evolutionary time against which to test hypotheses. One striking result from this marine record is that in many of the cases where climate change has been implicated as a causative agent of extinction in the sea, it has been a cooling rather than warming of the climate, and tropical faunas appear to have been hit preferentially.

A well-documented example of a marine cooling event affecting primarily warm water species is that of the Pliocene molluscan fauna of the western Atlantic, which has been studied in detail by Stanley (1984, 1986). Stanley has shown that the molluscan fauna of the tropical western Atlantic suffered a severe extinction in the Pliocene (roughly 3 Ma BP). This extinction appears to have occurred in two waves, and eliminated over 65% of the bivalve fauna. Less severe extinctions also occurred around this time in the North Sea and the Mediterranean, whereas Pacific faunas in California and Japan suffered no more than background extinction (Stanley, 1984). Stanley has tied this regional mass extinction event to the onset of northern hemisphere glaciation and a cooling of tropical surface temperatures by at least 4 K. Furthermore, Stanley (1986) has shown that this extinction event acted as a thermal filter in that every one of the 57 early Pliocene bivalve species which survive today in Florida are eurythermal, whereas those species which became extinct were preferentially those of narrow geographical range (and by inference stenothermal tropical species).

Despite the intuitive reasonableness of explaining the preferential extinction of warm water species in terms of climatic cooling, the rates of temperature change involved are several orders of magnitude slower than those with which marine invertebrates living today are well able to cope (Clarke, 1993). The rates are also much slower than the critical rates of climate change suggested by models (Lynch & Lande, 1993), and this poses an intellectual challenge to evolutionary biologists. It suggests that either we are missing subtle effects of temperature that may take a very long time to take effect or which may have their effects on the generally more sensitive embryonic or juvenile stages, or the ultimate causes of extinction are ecological and mediated primarily through competition (Clarke, 1993; Lawton, 1995).

An increasing body of evidence is now pointing to extinction in the sea being the result of subtle ecological factors rather than a simple response to climatic change. Two examples are the relationship between the likelihood of extinction and geographical range (fig. 5: Jablonski, 1986*a*), and the relationship between survival of a mass extinction event and larval type (Jablonski, 1986*c*). In this context it is interesting that the eastern Pacific and tropical western Atlantic molluscan assemblages which differed in the extent to which they suffered extinction at the end of the Pliocene also differed in (inferred) larval development (Jackson, 1995), as did Caribbean reef corals in the early Miocene (Edinger & Risk, 1995).

Extinction is generally a slow process and so we have to turn to the fossil record for insight into what factors might be important. These studies are necessarily correlative, and also likely to be biased by selective preservation of numerous, widely distributed (and hence relatively extinction-resistant) taxa. They are also almost all marine. Nevertheless these studies reveal consistent trends for wide geographical

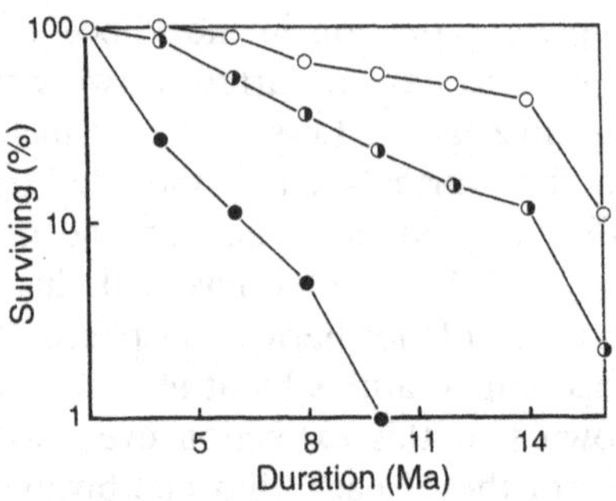

Fig. 5. Influence of geographical range on the survivorship of late Cretaceous bivalves and gastropods during periods of extinction. The time axis shows the elapsed time (millions of years: Ma) since the reference point, at which time the extant number of species is normalised to 100%. Thereafter the trajectories show the percentage of the original fauna remaining with time. The symbols represent species with differing estimated geographic ranges: ○ >2500 km, ◑ 1000–2500 km, ● <1000 km. (After Jablonski, 1986*a*.)

range, high abundance and a dispersing larval stage to confer resistance to extinction. These traits are, of course, often co-variates; they are also all factors of significance to metapopulation dynamics. The relationship between climate change and extinction is not simple, and is clearly mediated through population dynamics. It is likely that the susceptibility of tropical marine faunas to extinction at times of global cooling lies with their energetics and population dynamics, rather than simply that they live in warm water. Tropical species tend to have narrow geographical ranges (Stevens, 1989) and also low population abundance (because, in general, these two traits are correlated: Gaston & Lawton, 1988; Lawton, 1993). The preferential demise of tropical faunas during periods of more rapid climate change, if true, may thus follow from long-term proximate effects mediated through population dynamics with climate change only a distant, ultimate (although important) cause.

What can we learn from history?

It has long been recognised that organisms may respond to an environmental challenge in essentially one of four ways: they may move to somewhere else, they may stay and wait for better times, they may adapt to the changes, or they may go extinct. Although we can demonstrate historical examples of all these responses to climate change, we do not yet have any real understanding of the balance between them. The importance of phenotypic plasticity to short-term responses is clear; what we lack is any clear understanding of longer-term responses. Given a tropical shallow water molluscan assemblage, for example, and an estimated rate of climate change, we cannot predict which fractions of the taxa will migrate, adapt or go extinct. There are also difficulties in comparing, for example, the rates of evolution which can be obtained in the laboratory with those many times slower rates observed in the field (Haldane's paradox), or in comparing the rates of climate change which appear to have caused extinction in the past with the very much faster rates of temperature change which appear to cause little problem to species living now (Clarke, 1993). Part of the explanation for these mismatches may lie in the previous thermal history (Stanley, 1986), in subtle historical and ecological processes which take many generations to reveal their effect, and part in the impact of rare but extreme events.

Botanists are now developing sophisticated models which map the migration of vegetational assemblages in response to the changes in climate predicted by global circulation models (GCMs). These include both steady-state bioclimatic models and models based on population dynamics. The latest generation of GCMs now incorporate the effects of polar sea-ice and atmospheric sulphate aerosols, but have yet to tackle the difficult problem of feedback between vegetation metabolic activity, carbon dioxide concentration and climate change (Henderson-Sellers & McGuffrie, 1995).

In the interim it is clear that a full understanding of the impact of global climate change will only come from a combination of approaches. We need the detailed physiological comparisons of tropical, temperate and polar species to understand what evolution has achieved in the face of past climate change; we need to understand the genetic basis of physiological performance (both in embryological development and in adult forms) and individual variation in that performance; and we need to integrate these with the subtle effects of competition, and metapopulation dynamics and population genetic heterogeneity on the long-term viability of fragmented populations.

Acknowledgements

I thank the Company of Biologists for supporting my attendance at the Thermal Biology symposium SEB, St. Andrew's University, 7 April 1995. My research is supported by the British Antarctic Survey (Natural

Environment Research Council). I also thank my colleagues for many useful discussions on those topics over the past decade or so, and an anonymous reviewer for a penetrating review of a previous draft.

References

Alley, R.B., Meese, D.A., Shuman, C.A., Gow, A.J., Taylor, K.C., Grootes, P.M., White, J.W.C., Ram, M., Waddington, E.D., Mayewski, P.A. & Zielinski, G.A. (1993). Abrupt increase in Greenland snow accumulation at the end of the Younger Dryas event. *Nature* **362**: 527–9.

Alvarez, L.W., Alvarez, W., Asaro, F. & Michel, H.V. (1980). Extraterrestrial cause for the Cretaceous–Tertiary extinction. *Science* **208**: 1095–108.

Angel, M.V. (1994). Spatial distribution of marine organisms: patterns and processes. In *Large-scale Ecology and Conservation Biology* (ed. P.J. Edwards, R.M. May & N.R. Webb), pp. 59–109. Oxford: Blackwell.

Arntz, W.E. & Fahrbach, E. (1991). *El Niño: klimaexperiment der Natur*. Basel: Birkhauser Verlag.

Barry, J.P., Baxter, C.H., Sagarin, R.D. & Gilman, S.E. (1995). Climate-related, long-term faunal changes in a California rocky intertidal community. *Science* **267**: 672–5.

Bennett, K.D. (1990). Milankovitch cycles and their effects on species in ecological and evolutionary time. *Paleobiology* **16**: 11–21.

Blows, M.W. & Hoffmann, A.A. (1993). The genetics of central and marginal populations of *Drosophila serrata*. I. Genetic variation for stress resistance and species borders. *Evolution* **47**: 1255–70.

Brandt, A. (1991). Zur Besiedlungsgeschichte des antarktischen Schelfe am Beispiel der Isopoda (Crustacea, Malacostraca). *Berichte zur Polarforschung* **98**: 1–240 (in German).

Brandt, A. (1992). Origin of Antarctic Isopoda (Crustacea, Malacostraca). *Marine Biology* **113**: 415–23.

Clarke, A. (1991). What is cold adaptation and how should we measure it? *American Zoologist* **31**: 81–92.

Clarke, A. (1992). Is there a latitudinal diversity cline in the sea? *Trends in Ecology and Evolution* **7**: 286–7.

Clarke, A. (1993). Temperature and extinction in the sea: a physiologist's view. *Paleobiology* **19**: 499–518.

Clarke, A. & Crame, J.A. (1989). The origin of the Southern Ocean marine fauna. In *Origins and Evolution of the Antarctic Biota* (ed. J.A. Crame), pp. 253–68. London: Geological Society.

Clarke, A. & Crame, J.A. (1992). The Southern Ocean benthic fauna and climate change: a historical perspective. *Philosophical Transactions of the Royal Society, London, Series B* **338**: 299–309.

Clarke, A. & Crame, J.A. (1996). Diversity, latitude and time: patterns in the shallow sea. In *Marine Biodiversity: Patterns and Pro-*

cesses (ed. R.F.G. Ormond & J. Gage). Cambridge: Cambridge University Press (in press).

Clarke, A. & Johnston, I.A. (1996). Evolution and adaptive radiation of Antarctic fishes. *Trends in Ecology and Evolution* (in press).

Colwell, R.K. & Hurtt, G.C. (1994). Nonbiological gradients in species richness and a spurious Rapoport effect. *American Naturalist* **144**: 570–595.

Connor, E.F. & McCoy, E.D. (1979). The statistics and biology of the species-area relationship. *American Naturalist* **113**, 791–833.

Cossins, A.R. & Bowler, K. (1987). *Temperature biology of animals.* London: Chapman and Hall.

Coope, G.R. (1979). Late Cenozoic fossil Coleoptera: evolution, biogeography and ecology. *Annual Review of Ecology and Systematics* **10**: 247–67.

Coope, G.R. (1995). Insect faunas in ice age environments: why so little extinction? In *Extinction Rates* (ed. J.H. Lawton & R.M. May), pp. 55–74. Oxford: Oxford University Press.

Crame, J.A. (1993). Latitudinal range fluctuations in the marine realm through geological time. *Trends in Ecology and Evolution* **8**: 162–6.

Crame, J.A. & Clarke, A. (1996). The historical component of taxonomic diversity gradients. In *Marine Biodiversity: Patterns and Processes* (ed. R.F.G. Ormond & J. Gage). Cambridge: Cambridge University Press (in press).

Crawford, R.M.M. (1989). *Studies in Plant Survival.* Oxford: Blackwell Scientific Publications.

Crowley, T.J. & North, G.R. (1988). Abrupt climate change and extinction events in earth history. *Science* **240**, 996–1002.

Dansgaard, W., White, J.W.C. & Johnsen, S.J. (1989). The abrupt termination of the Younger Dryas event. *Nature* **339**: 532–3.

Dansgaard, W., Johnsen, S.J., Clausen, H.B., Dahl-Jensen, D., Gundestrup, N.S., Hammer, C.U., Hvidberg, C.S., Steffensen, J. P., Sveinbjørnsdottir, A.E., Jouzel, J. & Bond, G. (1993). Evidence for general instability of climate fauna 250-kyr ice-core record. *Nature* **364**: 218–20.

Davis, M.B. (1989). Insights from paleoecology on global climate change. *Bulletin of the Ecological Society of America*, 70: 222–8.

Davis, M.B. & Zabinski, C. (1992). Changes in geographical range resulting from greenhouse warming: effects on biodiversity in forests. In *Global Warming and Biological Diversity* (ed. R. Peters & T. Lovejoy), pp. 297–308. New Haven, CT: Yale University Press.

Doak, D.F., Marino, P.C. & Kareiva, P.M. (1992). Spatial scale mediates the influence of habitat fragmentation on dispersal success: implications for conservation. *Theoretical Population Biology*, 41: 315–36.

Donovan, S.K. (ed.) (1989). *Mass Extinctions: Processes and Evidence.* London: Belhaven Press.

Edinger, E.N. & Risk, M.J. (1995). Preferential survivorship of brooding corals in a regional extinction. *Paleobiology* **21**: 200–19.

Elias, S.A. (1994). *Quaternary Insects and Their Environments*. Washington: Smithsonian Institution Press.

Fisher, R.A. (1930). *The Genetical Theory of Natural Selection*. Oxford: Clarendon Press.

Flessa, K.W. & Jablonski, D. (1995). Biogeography of recent marine bivalve molluscs and its implications for paleobiogeography and geographical extinction: a progress report. *Historical Biology* **10**: 25–47.

Garland, T., Huey, R.B. & Bennett, A.F. (1991). Phylogeny and coadaptation of thermal physiology in lizards: a reanalysis. *Evaluation* **45**, 1969–75.

Gaston, K.J. & Lawton, J.H. (1988). Patterns in the distribution and abundance of insect populations. *Nature*, **331**: 709–12.

Gates, D.M. (1993). *Climate Change and its Biological Consequences*. Sunderland, MA: Sinauer Associates.

Geber, M.A. & Dawson, T.E. (1993). Evolutionary responses of plants to global change. In *Biotic Interactions and Global Change* (ed. P.M. Kaveiva, J.G. Kingsolver & R.B. Huey), pp. 179–97. Sunderland, MA: Sinauer Associates.

Gillespie, J.H. (1991). *The Causes of Molecular Evolution*. Oxford: Oxford University Press.

GRIP Members. (1993). Climate instability during the last interglacial period recorded in the GRIP ice core. *Nature* **364**: 203–7.

Hameed, S. & Gong, G. (1994). Variation of spring climate in Lower-Middle Yangtse River Valley and its relation with solar-cycle length. *Geophysical Research Letters* **21**, 2693–6.

Hanski, I. (1991). Single-species metapopulation dynamics: concepts, models and observations. *Biological Journal of the Linnaean Society* **42**: 17–38.

Hedgpeth, J.W. (1957). Marine biogeography. In *Treatise on Marine Ecology and Paleoecology*, (ed. J.W. Hedgpeth), Vol. 1, pp. 359–83. Boulder Co.: *Geological Society of America Memoir* **67**.

Henderson-Sellers, A. & McGuffrie, K. (1995). Global climate models and 'dynamic' vegetation change. *Global Change Biology* **1**, 63–75.

Hochachka, P.W. & Somero, G.N. (1984). *Biochemical Adaptation*. New York: Princeton University Press.

Hoffmann, A.A. & Blows, M.W. (1994). Species borders: ecological and evolutionary perspectives. *Trends in Ecology and Evolution* **9**: 223–7.

Hoffman, A.A. & Parsons, P.A. (1991). *Evolutionary Genetics and Environmental Stress*. Oxford: Oxford University Press.

Hoffmann, A.A. & Watson, M. (1993). Geographical variation in the acclimation responses of *Drosophila* to temperature extremes. *American Naturalist* **142**: S93–S113.

Hubbard, A.E. & Gilinsky, N.L. (1992). Mass extinctions as statistical phenomena: an examination of the evidence using χ^2 tests and bootstrapping. *Paleobiology* **18**: 148–60.

Huey, R.B. & Kingsolver, J.G. (1989). Evolution of thermal sensitivity of ectotherm performance. *Trends in Ecology and Evolution* **4**, 131–5.

Huey, R.B. & Kingsolver, J.G. (1993). Evolution of resistance to high temperature in ectotherms. *American Naturalist* **142** (suppl.), S21–S46.

Huston, M.A. (1994). *Biological Diversity*. Cambridge University Press, Cambridge.

Hutchinson, G.E. (1959). Homage to Santa Rosalia; or, why are there so many kinds of animals? *American Naturalist* **93**: 145–59.

Ives, A.R., Kareiva, P. & Perry, R. (1993). Response of a predator to variation in prey density at three hierarchical scales: lady beetles feeding on aphids. *Ecology* **74**: 1929–38.

Jablonski, D. (1986*a*). Background and mass extinctions: the alternation of macroevolutionary regimes. *Science* **231**: 129–33.

Jablonski, D. (1986*b*). Causes and consequences of mass extinction: a comparative approach. In *Dynamics of Extinction* (ed. D.K. Elliott), pp. 183–229. New York: Wiley.

Jablonski, D. (1986*c*). Larval ecology and macroevolution in marine invertebrates. *Bulletin of Marine Science* **39**: 565–87.

Jackson, J.B.C. (1995). Constancy and change of life in the sea. In *Extinction Rates* (ed. J.H. Lawton & R.M. May), pp. 45–54. Oxford: Oxford University Press.

Jouzel, J., Lorius, C. Petit, J.R., Genthon, C., Barkov, N.I., Kotlyakov, V.M. & Petrov, V.M. (1987). Vostok ice core: a continuous isotope temperature record over the last climatic cycle (160 000 years). *Nature* **329**: 403–8.

Kadmon, R. (1993). Population dynamic consequences of habitat heterogeneity: an experimental study. *Ecology* **74**: 816–25.

Kadmon, R. & Shmida, A. (1990). Quantifying spatiotemporal demographic processes: an approach and a case study. *American Naturalist*, **135**: 382–97.

Kareiva, P. (1990). Population dynamics in spatially complex environments: theory and data. *Philosophical Transactions of the Royal Society of London, Series B* **330**: 175–90.

Kareiva, P. & Wennergren, U. (1995). Connecting landscape patterns to ecosystem and population processes. *Nature* **373**: 299–302.

Kennett, J.P. & Stott, L.D. (1991). Abrupt deep-sea warming, palaeo-oceanographic changes and benthic extinctions at the end of the Palaeocene. *Nature* **353**: 225–9.

Lande, R. (1976). Natural selection and random genetic drift in phenotypic evolution. *Evolution* **30**: 314–34.

Lawton, J.H. (1993). Range, population abundance and conservation. *Trends in Ecology and Evolution* **8**: 409–13.

Lawton, J.H. (1995). Population dynamic principles. In *Extinction Rates* (ed. J.H. Lawton & R.M. May), pp. 147–63. Oxford: Oxford University Press.

Lehman, S.J. & Keigwin, L.D. (1992). Sudden changes in North Atlantic circulation during the last deglaciation. *Nature* **356**: 757–62.

Lewis Smith, R.I. (1988). Recording bryophyte microclimate in remote and severe environments. In *Methods in Bryology* (ed. J.M. Glime), pp. 275–284, Hattori Botany Laboratory, Nichinan. Proceedings of the Bryological Methods Workshop.

Lorius, C.J., Jouzel, J., Raynaud, D., Hansen, J. & Le Treut, H. (1990). The ice-core record: climate sensitivity and future greenhouse warming. *Nature* **347**: 139–45.

Lynch, J.D. (1988). Refugia. In *Analytical Biogeography* (ed. A.A. Myers & P.S. Giller), pp. 311–42. London: Chapman and Hall.

Lynch, M., Gabriel, W. & Wood, A.M. (1991). Adaptive and demographic responses of plankton populations to environmental change. *Limnology and Oceanography* **36**: 1301–12.

Lynch, M. & Lande, R. (1993). Evolution and extinction in response to environmental change. In *Biotic Interactions and Global Change* (ed. P.M. Kareiva, J.G. Kingsolver & R.B. Huey), pp. 234–50. Sunderland, MA: Sinaeur Associates.

May, R.M. (1975). Patterns of species abundance and diversity. In *Ecology and Evolution of Communities* (ed. M.L. Cody and J.M. Diamond), pp. 81–120. , Cambridge, MA: Harvard University Press.

Mayle, F.E. & Cwynar, L.C. (1995). Impact of the Younger Dryas cooling event upon lowland vegetation of maritime Canada. *Ecological Monographs* **65**: 129–54.

Miller, L.K. & Werner, R. (1987). Extreme supercooling as an overwintering strategy in three species of willow gall insects from interior Alaska. *Oikos* **49**: 253–60.

Murphy, E.J., Clarke, A., Symon, C. & Priddle, J.H. (1995). Temporal variation in Antarctic sea-ice: analysis of a long term fast-ice record from the South Orkney Islands. *Deep Sea Research* **42**: 1045–62.

Newell, I.M. (1948). Marine molluscan provinces of western North America: a critique and a new analysis. *Proceedings of the American Philosophical Society* **92**: 155–66.

Pagel, M.D., May, R.M. & Collie, A.R. (1991). Ecological aspects of the geographical distribution and diversity of mammalian species. *American Naturalist* **137**: 791–815.

Palmer, M.W. & White, P.S. (1994). Scale dependence and the species-area relationship. *American Naturalist* **144**: 717–40.

Pennington, W. (1986). Lags in adjustment of vegetation to climate caused by the pace of soil development. *Vegetation* **67**: 105–18.

Peters, R.L. & Darling, J.D.S. (1985). The greenhouse effect and nature reserves. *BioScience* **35**: 707–17.

Pianka, E.R. (1966). Latitudinal gradients in species diversity: a review of concepts. *American Naturalist* **100**: 33–46.

Pickett, S.T.A. & Cadenasso, M.L. (1995). Landscape ecology: spatial heterogeneity in ecological systems. *Science* **269**: 331–4.

Pielou, E.C. (1991). *After the Ice Age: the Return of Life to Glaciated North America* Chicago: University of Chicago Press.

Pimm, S.L. (1994). The importance of watching birds from airplanes. *Trends in Ecology and Evolution* **9**: 41–3.

Pimm, S.L., Russell, G.J., Gittleman, J.L. & Brooks, T.M. (1995). The future of biodiversity. *Science* **269**: 347–50.

Powers, D.A., Ropson, I., Brown, D.C., van Beneden, R., Cashon, R., Gonzalez-Villasenor, L.I. & Dimechele, L. (1986). Genetic variation in *Fundulus heteroclitus*: geographical distribution. *American Zoologist* **26**, 131–44.

Prentice, K.C. (1990). Bioclimatic distribution of vegetation for general circulation model studies. *Journal of Geophysical Research* **95**: 11 811–30.

Raup, D.M. (1991). A kill curve for Phanerozoic marine species. *Paleobiology* **17**: 37–48.

Ricklefs, R.E. & Cox, G.W. (1972). Taxon cycles in the West Indian avifauna. *American Naturalist* **106**: 195–219.

Rosenzweig, M.L. (1975). On continental steady states of species diversity. In *Ecology and Evolution of Communities* (ed. M.L. Cody & J.M Diamond), pp. 121–40. Cambridge, MA: Belknap Press of Harvard University Press.

Rosenzweig, M.L. (1977). Geographical speciation: on range size and the probability of isolate formation. In *Proceedings of the Washington State University Conference on Biomathematics and Biostatistics* (ed. D. Wollkind), pp. 172–94. Washington: Pullman.

Rosenzweig, M.L. (1995). *Species Diversity in Space and Time*. Cambridge: Cambridge University Press.

Rosenzweig, M.L. & Abramsky, Z. (1993). How are diversity and productivity related? In *Species Diversity in Ecological Communities* (ed. R.E. Ricklefs & D. Schluter), pp. 52–65. Chicago: University of Chicago Press.

Roy, K., Jablonski, D. & Valentine, J.W. (1994). Eastern Pacific molluscan provinces and latitudinal diversity gradient: no evidence for 'Rapoport's Rule'. *Proceedings of the National Academy of Sciences of the USA* **91**, 8871–4.

Ruddiman, W.F., Raymo, M.E., Martinson, D.G., Clement, B.M. & Backman, J. (1989). Pleistocene evolution of northern hemisphere climate. *Paleoceanography* **4**: 353–412.

Somero, G.N. (1996). Temperature and proteins: contrasting and complementary adaptations over evolutionary, acclimatory and acute timescales. In *Phenotypic and Evolutionary Adaptation to Temperature* (ed. I.A. Johnston & A.L. Bennett). Cambridge: Cambridge University Press (in press).

Southward, A.J. (1991). Forty years of changes in species composition and population density of barnacles on a rocky shore near Plymouth. *Journal of the Marine Biological Association of the United Kingdom*, 71, 495–513.

Stanley, S.M. (1984). Marine mass extinctions: a dominant role for temperature. In *Extinctions* (ed. M.H. Nitecki), pp. 69–117. Chicago: The University of Chicago Press.

Stanley, S.M. (1986). Anatomy of a recent regional mass extinction: Plio-Pleistocene decimation of the western Atlantic bivalve fauna. *Palaios* **1**, 17–36.

Steele, J.H. (1978). Some comments on plankton patches. In *Spatial Pattern in Plankton Communities* (ed. J.H. Steele), pp. 1–20. New York: Plenum Press.

Steele, J.H. (1991). Can ecological theory cross the land-sea boundary? *Journal of Theoretical Biology* **153**: 425–36.

Stehli, F.G., McAlester, A.L. & Helsley, C.E. (1967). Taxonomic diversity of recent bivalves and some implications for geology. *Geological Society of America Bulletin* **78**, 455–66.

Stehli, F.G. & Wells, J.W. (1971). Diversity and age patterns in hermatypic corals. *Systematic Zoology* **20**: 115–26.

Stevens, G.C. (1989). The latitudinal gradient in geographical range: how so many species coexist in the tropics. *American Naturalist* **133**: 240–56.

Taylor, K.C., Lamorey, G.W., Doyle, G.A., Alley, R.B., Grootes, P.M., Mayewski, P.A., White, J.W.C. & Barlow, L.K. (1993). The 'flickering switch' of late Pleistocene climate change. *Nature* **361**: 432–6.

Terborgh, J. (1973). On the notion of favourableness in plant ecology. *American Naturalist*, **107**: 481–501.

Thomas, J.A. & Morris, M.G. (1995). Rates and patterns of extinction among British invertebrates. In *Extinction Rates* (ed. J. H. Lawton & R.M. May), pp. 111–30. Oxford: Oxford University Press.

Tracy, C.R. & George. T.L. (1992). On the determinants of extinction. *American Naturalist* **139**: 102–22.

Valentine, J.W. (1961). Paleoecological molluscan biogeography of the Californian Pleistocene. *University of California Publications in Geological Science* **34**: 309–442.

Valentine, J.W. (1966). Numerical analysis of marine molluscan ranges on the extratropical northeastern Pacific shelf. *Limnology and Oceanography* **11**: 198–211.

Valentine, J.W. (1967). The influence of climatic fluctuations on species diversity within the Tethyan provincial system. In *Aspects of*

Tethyan Biogeography (ed. C.G. Adams & D.V. Ager), pp. 153–66. Systematics Association, Publication Number 7, pp. 153–66. London: Systematics Association.

Valentine, J.W. (1968). Climatic regulation of species diversification and extinction. *Geological Society of America Bulletin* **79**, 273–6.

Valentine, J.W. (1984). Neogene marine climate trends: implications for biogeography and evolution of the shallow-sea biota. *Geology* **12**: 647–50.

Watson, M.J.O. & Hoffmann, A.A. (1995). Cross-generation effects for cold resistance in tropical populations of *Drosophila melanogaster* and *Drosophila simulans*. *Australian Journal of Zoology* **43**: 51–8.

Woodward, F.I. (1987). *Climate and Plant Distribution*. Cambridge: Cambridge University Press.

Index